Stephen Powers

The American Merino

For wool and for mutton a practical treatise on the selection, care, breeding, and diseases of the merino sheep in all sections of United States

Stephen Powers

The American Merino
For wool and for mutton a practical treatise on the selection, care, breeding, and diseases of the merino sheep in all sections of United States

ISBN/EAN: 9783337256883

Printed in Europe, USA, Canada, Australia, Japan

Cover: Foto ©berggeist007 / pixelio.de

More available books at **www.hansebooks.com**

THE AMERICAN MERINO:

FOR WOOL AND FOR MUTTON.

A Practical Treatise on the Selection, Care, Breeding and Diseases of the Merino Sheep

IN

ALL SECTIONS OF THE UNITED STATES.

BY
STEPHEN POWERS.

ILLUSTRATED.

NEW YORK:
O. JUDD CO., DAVID W. JUDD, Pres't,
751 BROADWAY.
1887.

tered, according to Act of Congress, in the year 1886, by t
O. JUDD CO.,
In the Office of the Librarian of Congress, at Washington.

CONTENTS.

Letter of Request... 7
Letter of Presentation...................................... 8

Chapter I.
From Spanish to American.................................. 11

Chapter II.
Form.. 22

Chapter III.
Fleece.. 27

Chapter IV.
Blood... 43

Chapter V.
Breeding.. 47

Chapter VI.
Feed.. 59

Chapter VII.
Pasture in the West....................................... 67

Chapter VIII.
A Mutton Merino... 72

Chapter IX.
Lambing... 85

Chapter X.
Care of Ewes and Lambs.................................... 95

Chapter XI.
Tagging, Washing, etc.....................................104

Chapter XII.
Shearing and Doing Up Wool................................115

Chapter XIII.
Summer Management...125

Chapter XIV.
From Grass to Hay...138

Chapter XV.
Selection and Care of Rams................................142

Chapter XVI.
The Breeding Flock..155

Chapter XVII.
Sheep Houses and Their Appurtenances......................165

Chapter XVIII.
Winter Management...177

CONTENTS.

CHAPTER XIX.
Feeding for Mutton...................189

CHAPTER XX.
From Hay to Grass...................200

CHAPTER XXI.
Fodder for Sheep...................203

CHAPTER XXII.
Systems of Sheep Husbandry...................209

CHAPTER XXIII.
Systems of Sheep Husbandry (Continued)...................222

CHAPTER XXIV.
Systems of Sheep Husbandry (Continued)...................234

CHAPTER XXV.
Systems of Sheep Husbandry (Continued)...................251

CHAPTER XXVI.
Systems of Sheep Husbandry (Continued)...................264

CHAPTER XXVII.
Diseases of the Merino—"Paperskin"...................277

CHAPTER XXVIII.
Parasitic Diseases (Continued)...................287

CHAPTER XXIX.
External Parasites...................301

CHAPTER XXX.
Diseases of the Feet...................316

CHAPTER XXXI.
Diseases of the Respiratory Organs...................324

CHAPTER XXXII.
Diseases of the Alimentary System...................328

CHAPTER XXXIII.
Blood Diseases...................338

CHAPTER XXXIV.
Diseases of the Nervous System...................345

CHAPTER XXXV.
Diseases of the Urinary and Reproductive Organs...................348

CHAPTER XXXVI.
Miscellaneous...................353

CORRESPONDENCE.

Mr. G. B. Quinn, the President, and Mr. J. G. Blue, the Secretary, of the Ohio Spanish Sheep Breeders' Association, addressed a communication to Mr. Stephen Powers, in which they said:

OFFICE OF THE
OHIO SPANISH MERINO SHEEP
BREEDERS' ASSOCIATION.

Cardington, Ohio.

The breeders and owners of Merino Sheep find they are called upon to master new and, in many cases, fatal diseases not spoken of by the celebrated writers, Randall and Youatt. Among the writers on the Merino of to-day, we think some one should present to the public a practical treatise, which shall discuss the present management, diseases and breeding of Merinos and sheep of different bloods, comparing their merits in our States and Territories. We think the present magnitude of this industry demands * * * * the proper education of our shepherds and flock-masters in all the new diseases of Merinos which have been developed during the last decade, and in the older ones which yet, in some instances, infest our flocks.

The Ohio Spanish Merino Sheep Breeders' Association, by their President and Secretary, would respectfully request you, at your earliest convenience, to condense your ideas on this subject into a suitable volume, to be printed and presented to the public for their enlightenment. You have perhaps observed the need and demand for such a volume, properly written and illustrated, to be placed upon the market for the thousands of flock-masters of to-day.

Should you comply with this request, and should it be possible for you to give your time continuously to the volume until completed, we think the sheep fraternity of our country, and all who are interested, will freely patronize your work and appreciate your labors.

Mr. Powers replied as follows :

MESSRS. GEO. B. QUINN AND J. G. BLUE.

GENTLEMEN:— Together with your kind letter, inviting me to prepare a book on our National breed of sheep, I received a copy of the Register of your Association, containing a record of several hundred pure-blood flocks owned mostly in Ohio—a work carefully edited and printed, and substantially bound. Nothing could afford more convincing proof than this elegant volume, of the solidity and the prosperity of your ancient calling in our State.

I have undertaken to do what you ask, and offer you herewith a work on "The American Merino." I tender it modestly and without comment, except the simple remark that my task has been conscientiously performed, and that it is based on years of personal experience in sheep husbandry.

While it would be presumptuous in me to say that the volume herewith tendered to yourselves and the public, fully meets the requirements of modern shepherding in the United States, it is not too much to aver that our great industry has outgrown the manuals heretofore published.

Since the learned work of Dr. Randall was given to the world, the American Merino has not only crossed the Missouri and ascended the slopes of the Rocky Mountains, but has followed the dusty wagon-trail of the emigrant to California, where it attained a larger and hardier form and a new acclimation ; and, starting out thence afresh, north, south and east, it overspread the whole mid-continent. With a scholarly pen this distinguished author traced the development of the American Merino to the banks of the Mississippi ; but the traits and needs of that great new branch or type of the race, which may be called the California Merino—the "rustler," as it is termed in the expressive vernacular—were little understood by him.

The work of Mr. Henry Stewart is invaluable to the American breeder of the English races, with their long category of special wants and ailments ; but it would hardly be claimed, even by the candid and painstaking author himself, that it is fully abreast of the advance of the Merino in the Far West.

The present may seem to be a dark day for the breeder of Merinos, but the American future of this great race, potent from "long descent," is as well assured as that of the continent itself. In 1865, the Boston price of fine wool was one dollar and two cents per pound; of coarse, ninety-six cents. In 1885 the number of Merinos in the world is at least one hundred per cent. greater than then, while the number of coarse-wools (owing to the actual decrease in England) has increased very little, if at all. Yet to-day, the Boston price of Merino wool is thirty-four cents, and of coarse, it is thirty cents.

In spite of the enormous increase of Merinos, their wool is proportionately higher than it was then.

Even in 1866, before the tariff was increased, the actual annual revenue from the Merino sheep of the United States was two dollars and sixteen cents; from the mutton-sheep of England, one dollar and seventeen cents.

The breeder of the American Merino should not for one moment allow himself to be discouraged, *if he is a good shepherd*. He can abate much, and yet make more money than the flock-master of other lands.

Vermont, the mother of the American Merino, gave to Ohio and the West, a sheep incomparable in the whole world as a producer of wool; and which has well fulfilled its destiny in our younger civilization. Let it now be the work of Ohio, of your Association, and kindred societies in other States, to give to America what the disciples of Daubenton created at Rambouillet: the farmer's sheep, a "mutton Merino," presenting in itself the best attainable combination of flesh and pelage, which, as a writer in the *Breeders' Gazette* happily says, "stands ready for a partnership arrangement with any domestic animal or any sort of crop the farmer may choose to cultivate."

Against a National race of such a type, the American Government can never afford to enact hostile legislation.

While it is yours, gentlemen, to labor for the accomplishment of this highly desirable result, and to preserve in your several Registers that pedigree so highly valued by the breeders, let it be mine to give in the following pages, as well as I may, the present condition and directions for the rearing of the Merino.

STEPHEN POWERS.

THE AMERICAN MERINO.

CHAPTER I.

FROM "SPANISH" TO "AMERICAN."

There are two etymologies given for the word "Merino." One, put forth in the biography of Consul Wm. Jarvis, and adopted by the *Ohio Register*, traces it to two Spanish words meaning "from over the sea." The other, upheld by E. Ollendorf, a writer in the *Breeders' Gazette*, and some others, would derive it from *Merino*, the designation of a certain royal officer of Spain, years ago; one of whose functions was the assignment of their respective pasture grounds to the mountain sheep (*Serranos*), and the migratory sheep (*Trans-humantes*).

Mr. Seth Adams imported the first pair of Spanish Merinos to the United States for breeding purposes, in 1801, bringing them from France to Dorchester, Mass. In 1807 he became a citizen of Wacatomica (now Dresden), in Ohio, and brought with him twenty-five or thirty sheep, the descendants of this pair. He continued to breed them for several years, under the very discouraging circumstances which attended pioneer life in those days, but finally sold out the flock and moved to Zanesville. Though this importation was of great benefit to Ohio and also to Kentucky (the first pair Mr. Adams sold in Ohio was to Judge Todd, of Kentucky, for fifteen hundred dollars), yet the stress of pioneer life was too severe, and there are not now any descendants of it positively recognizable.

The credit of the first traceable importation, therefore, belongs to Col. David Humphreys, who brought from Spain to Derby, Conn., in 1802, twenty-one rams and seventy ewes. But this now celebrated flock would have been lost to recorded history, too, though not to the blood and stock of the country, had it not been preserved by the one ewe bought by Stephen

Atwood. A Humphreys' ewe and a Heaton ram, in the hands of this noted and careful breeder, alone preserved for modern registers the blood of this large and choice flock.

Still, for a time, Merino sheep were wonderfully popular. It is recorded that President Madison wore, at his inauguration in 1809, a coat made from wool grown on sheep from Col. Humphreys' flock, and a waistcoat and small clothes made from the Livingston French flock, of Clermont, N. H. Four lambs were sold in 1810 from the Livingston flock, at one thousand dollars each, and Col. Humphreys is said to have sold two pairs of Merinos at three thousand dollars a pair. (It should be borne in mind that one dollar then, represents at least two now). Col. Humphreys sold his half-blood Merino wool at seventy-five cents a pound; three-quarter-blood at one dollar and twenty-five cents; and his full-blood at two dollars a pound.

Accordingly, very large importations of Merinos began to arrive. Mr. Albert Chapman states that, in the years 1810 and 1811, one hundred and six vessels arrived at various ports of the United States, bringing in all, fifteen thousand seven hundred and sixty-seven sheep! Of these, the vast majority were Merinos from Spain; and of the latter, it is considered probable that the greater number were purchased by that indefatigable patriot, Consul Jarvis.

It is not certainly known from what *cabañas*, or flocks, in Spain, Col. Humphreys selected his purchase, nor does it appear that he considered it a matter of importance. Mr. Atwood said, in 1864, "The original Humphreys' sheep were, in color, lighter than my present flock," while those imported by Mr. A. Heaton, "were short-legged, dark and heavy-wooled."

The principal flocks of Spain from which Merinos were brought to America, were Infantados, Paulars, Escurials, Negrettis, Montarcos, Guadaloupes and Aguirres. It has generally been believed that Col. Humphreys selected his sheep from the Infantados, while Consul Jarvis bought from all the other flocks above named, except Infantados.

Col. Humphreys mentions that a ram bred on his farm cut seven pounds and five ounces of washed wool. Mr. Jarvis says: "From 1811 to 1826 * * * * * * my average weight of wool was three pounds and fourteen ounces, to four pounds and two ounces — varying according to keep. The weight of the wool of the bucks was from five and a quarter pounds to six and a half pounds, in good stock case, all washed on the sheeps backs."

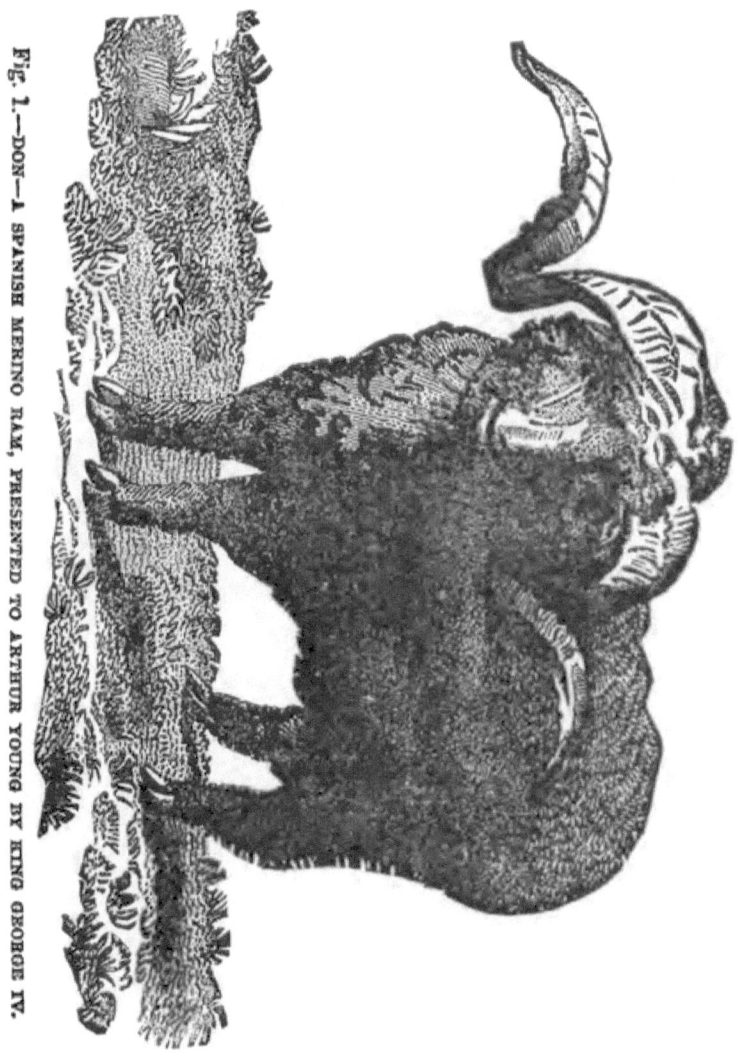

Fig. 1.—DON—A SPANISH MERINO RAM, PRESENTED TO ARTHUR YOUNG BY KING GEORGE IV.

Many acrimonious controversies have been waged by the partisans of the different flocks in the United States, as to their respective merits and their purity. It is now acknowledged by the authors of the *Ohio Register*, that we probably have no pure sheep of any one of the above named celebrated *cabañas* in America; they have all been more or less mingled. But we have, perhaps, a million pure American Merinos of undoubted Spanish descent; and this one fact, which alone is of practical importance, should satisfy every breeder of this great and ancient race. This, I take it, is the true purport of the following sentence (p. 28) in the *American Register*, of Wisconsin: "The imperfect records of the Spanish Merino sheep, from their early importations until 1860, have been such that an absolute certainty is an impossibility, but the march of progress has been so grand, and the improvements so great, that any imperfections that may have stained the blood of those early breeders, does not and cannot stain the blood of to-day."

The Paulars were undoubtedly one of the handsomest flocks in Spain; the wool was compact, soft and silky, and the surface not so much covered with gum. The Aguirres had more wool about their faces and legs than either of the other flocks. The wool was more crimped than the Paular, and less so than the Negretti, and it was thick and soft. They were short-legged, round and broad-bodied, with loose skins. The Negrettis were the tallest sheep in Spain, but were not handsomely formed; the wool was somewhat shorter than the Paular, the skin more loose and inclined to double; many of them were wooled well on the face, and on their legs down to their hoofs. All the loose-skinned sheep had large dewlaps. The Guadaloupes were rather large-boned, but not handsome; wool thick and crimped; skins loose and doubling; generally more gummy than any of the other flocks. The Escurials were about as tall as the Paulars, but slighter, and their wool not so thick; they were plainer than the Negrettis and Aguirres, and not so well wooled on the faces and legs. The Infantados were the largest and most popular flock; their lambs, like the Paular, often have a hairy coat when born; a mark of a good shearer. The Paular lambs often have butter-nut-tipped ears at birth. A black lamb is oftener yeaned from the Paular strain of blood, than any other; but the best-informed shepherds nowadays, do not consider a black lamb any evidence of impurity of blood, though the color itself is objectionable.

When Col. Humphreys first began to sell pure Merinos, the

price did not generally exceed one hundred dollars per head; but, as we have seen above, they afterward commanded enormous prices. This was in consequence of the embargo and the war of 1812, during which, full-blooded wool at one time, brought two dollars and fifty cents a pound, (two dollars even as far west as Marietta, Ohio). But after this war closed there was a disastrous collapse; many pure Merinos were sold for one dollar a head; and many of the best flocks of the country were sold and dispersed. The extensive importations of Consul Jarvis also contributed to this cheapening.

This country, therefore, owes an inextinguishable debt of gratitude to that plain, simple man, Stephen Atwood, who, with an abiding faith in the future of this breed of sheep, in 1813 paid one hundred and twenty dollars for a full-blood Humphreys ewe, and in 1819, bought five more of the same descent; and with this little band as a foundation, breeding to Humphreys rams until 1838 (after which he could find no more that were pure, and was obliged to depend on his own), for more than half a century, whether wool was up or down, tariff or no tariff, he kept his small flock together on his small farm, and bred it so pure that, in this day of many Registers, and of much "crookedness," the very highest warrant that can be given any sheep is, to pronounce it a "straight Atwood."

His first fleece from this noted ewe, shorn in June, 1814, was three pounds and nine ounces. That he was a progressive breeder appears from the fact that, in June, 1857, he cut from a ram of the same blood, nineteen pounds and eleven ounces, though the same animal, next year, with another owner, yielded thirty-two pounds. Recent investigations by the *Ohio Register* leave it doubtful whether this ewe of Atwood's was a Paular or an Infantado. They also show that Atwood was less careful in his records than in his breeding, and that the present blood of the American Merino is much purer than its recorded pedigree.

While Mr. William Jarvis deserves the highest praise for the indomitable energy, perseverance and sagacity, which led him, as Consul to Lisbon, amid the conflicts of the Napoleonic wars, to gather up the wrecks of the ancient flocks of Spain, and dispatch ship-load after ship-load to America; yet he ranks below Mr. Atwood in the singleness and steadfastness with which the latter held to his purpose and practice of breeding for fifty years. Mr. Jarvis put on his farm at Weathersfield, Vt., three hundred sheep of the Paular, Aguirre, Escurial, Montarco and Negretti flocks. According to the Spanish custom, he bred each of these separately until 1816 or 1817, when he mixed them together. In 1826 he committed the mistake of crossing with the Saxony Merinos, a mania for which was at that time over-sweeping the country. But this country is indebted to Mr. Jarvis for most of the admirable Paular blood it has received; and there were men who bought of him pure Spanish Merinos, and who were not swept away by the Saxony mania which passed over the country. To him, ultimately, we

Fig. 2.—MERINO EWE.

are indebted for the fine flock of the Messrs. Rich, of Vermont, which has been a prolific mother of Western studs. Mr. Chapman says, with the fervor of a strong partisan: "Let us all especially revere the memory of Thurman and Charles Rich, whose firmness and judgment were not shaken, and who have left unto their heirs and the land, the goodly heritage of the Rich flock, without even a smell or rumor of Saxony upon its outermost skirts."

At this point I will present a sketch of a Paular Merino ewe, figured in the *Albany Cultivator*, December, 1840, of which the owner says, "Her form at any rate is genuinely *Merino*," though he complains further on: "Still it must be acknowledged that the Merino, compared with the *improved* breeds of sheep, is an ill-formed animal."

By way of contrast, I give next a group of two ewes of the American Merino, owned by G. B. Quinn, Esq., Brown's Mills, Ohio. (*See Frontispiece.*)

The greatest breeder America has yet produced, Edwin Hammond, of Vermont, now appeared upon the scene, to give that improvement to the Merino form, which the contributor to the *Cultivator* had sighed for. Before Hammond, there was only the Spanish Merino; after Hammond, there was a truly American Merino.

We may believe that this great specialist began with about such material as that figured above; for "Old Black," which he bought of Atwood, in 1849, is thus described by Mr. Randall: "He was long, tall, flat-ribbed, rather long in the neck and head, strong-boned, a little roach-backed, deep-chested, moderately wrinkled; his wool was about an inch and a half long, of medium thickness, extremely yolky, and dark-colored externally; face a little bare, and not much wool on shanks. He did not possess a very strong constitution." He weighed about one hundred and thirty-five pounds, and cut about fourteen pounds of wool, unwashed. This was certainly not very promising material.

I would not, if I could, trace Mr. Hammond's wonderful career through all the intricacies and niceties of his art. He developed ultimately three lines, or sub-families, in his flock—"the dark or Queen line," "the light-colored line," and the "intermediate." The best sheep of his flock were, almost invariably, produced by crossing between these lines.

But we may profitably trace a few of his foot-steps, as they are imprinted on the records. "Old Black" lived nineteen years, attesting the vigor created by Atwood's open-air shepherding; but Hammond soon found (or created) better material. His own ram, "Wooster," bred in 1849, weighed only one hundred and five pounds, but sheared nineteen and one quarter pounds, unwashed. He served three hundred ewes when he was a year old! He was compact and short-legged; head short and thick; very wrinkly; wool about two inches long. "Old Greasy," bred in 1850, weighed one hundred and ten pounds and cut twenty-two pounds. "Old Wrinkly," bred in 1853, weighed one hundred and thirty pounds and sheared twenty-three pounds. In breeding, next, from "Little Wrinkly," Mr. Hammond suffered a backset in weight of fleece, though his wool was very fine and even. But "Sweepstakes" (1856) went up to one hundred and forty pounds in weight of carcass, and twenty-seven pounds in fleece. In this noble animal, perhaps the art of the master reached its culmination; he united in himself the blood of the three lines, and is believed to have produced more scoured wool in one fleece, than any other ram which Hammond ever owned.

"Young Matchless" was a model of compactness, strength, and symmetry; had immense constitution, and did more than any other ram to impart the short, thick, round carcass so conspicuous in the American Merino; while "Long Wool" improved the fleece above any other, perhaps, especially in length. But "Sweepstakes" combined both or all these excellences, and transmitted them to his progeny.

Mr. Hammond died in 1870, but the year 1856, which marks the birth of "Sweepstakes," may be assumed as the starting-point of the American Merino. In 1861, Mr. Randall instituted certain measurements of carcass on a ram and three ewes of his flock (which was of the Hammond blood); and a few of these, with the Austrian figures reduced to English, will be of interest here as showing the points in which the American Merino is an improvement over the Spanish.

	Weight with wool on.	From horns to shoulders.		Total length		Girth.		Distance of hip bone apart.
	lbs.	ft.	in.	ft.	in.	ft.	in.	ft. in.
Infantado.								
Ram....	104	1	10½	5	8	5	2¼	7¼
Ewe.....................	73	1	10	5	4½	4	11½	7
Negretti.								
Ram........	100½	1	11½	5	7¾	5	1¾	7½
Ewe..	70	1	9½	5	4	5	2	6
American.								
Ram....	122		10	3	11	4	4¼	9
Ewe...	114		10	3	11½	4	4½	8
Ewe....................	123		10	4	0	4	3	8
Ewe....................	100		11	3	11	4	0¾	8

From these figures we learn the almost incredible fact that, while the Spanish Merinos were nearly two feet longer in all, and a foot longer in the neck, they weighed from twenty to twenty-five pounds less, and were not so broad across the hips by about two inches! Their fore-legs were also six or seven inches longer than those of the American Merino!

Livingston gives the weight of the unwashed Spanish fleeces at eight and one half pounds for the ram, and five pounds for the ewe. Even in Randall's day, the American Merino unwashed fleeces were nearly double these weights.

In Spain, the best rams yielded only about six or eight per cent. of their weight in wool; in America, about 1844, it had increased to fifteen per cent.; and in 1861, Hammond's celebrated ram, "Twenty-one Per Cent." had increased the proportion to the figures which gave him his name. There were forwarded to the Paris Exposition, American Merino fleeces (twenty-one rams, forty-six ewes), of which the per cent. of wool, to live weight for the whole, was 22; of the best thirty, 25.2; of the best six, 30.1; of the best one, 36.6.

With this notable improvement in compactness of form, and wool-bearing capacity, there has been no deterioration in the fineness of the fiber, but, perhaps, the reverse. Youatt gave as the average diameter of the Merino wool-fiber in his day, one-seven hundred and fiftieth of an inch; of Saxony wool, one-eight hundred and fortieth of an inch. In 1878, measurements of wool made in Vermont, as stated by Hon. Henry Lane, of that State, showed rams' fleeces with a fiber of the diameter of one-nine hundred and thirty-fourth of an inch; and ewes' of one-one thousand and fifth of an inch.

But it may well be questioned whether sheep yielding such a fine fiber as this, and such an enormous percentage of wool to live weight, are desirable; they are generally lacking in vigor. The best shepherds are beginning to acknowledge that the hot-house forcing of the American Merino's wool-bearing aptitude, has been, in many instances, carried too far. Thus, in reporting the annual "State Shearing" of Vermont for 1885, Mr. Albert Chapman says: "It will be remarked that there is a falling off in the weights attained by rams and ewes one year old, a very good indication that our breeders are becoming convinced that the forcing system to attain large size and heavy fleeces the first year, is neither desirable or profitable, and the gains in the mature sheep show that slower development tends to much better and larger improvements in the end."

In the percentage of scoured wool, per fleece, there has been, perhaps, a slight improvement over the Spanish, in the great mass of American Merinos and high-grades throughout the country; but the enormous development of yolk, under the housing and other artificial treatment of the stud flock, has tended to prejudice the breed in the minds of many careful conservative wool-growers. I have the authority of Messrs. Coates Brothers, of Philadelphia, for saying that fleeces have been shorn in this country which yielded only twelve and one-half per cent of pure wool. In 1876, "Patrick Henry," bred by L. P. Clark, of Vermont, yielded a fleece of thirty-seven pounds, which turned out nine pounds and ten ounces of clean wool, or twenty-six per cent. "Bascom," owned by Capt. J. G. Blue, of Cardington, Ohio, once gave a fleece of twenty-nine and one-quarter pounds, which scoured nine and one-quarter pounds, or thirty-one and six-tenths per cent.

The heaviest known fleece yet cut from an American Merino, was one of forty-four pounds and four ounces, which was yielded by "Buckeye," a ram owned partly in Huron County, Ohio, partly in Michigan, at the "State Shearing" of the latter State in 1884.

For detailed histories of noted breeders and their flocks, the reader must consult the voluminous registers of the various National and State Associations. But there are a few items, which may be given here as landmarks in the progress of the American Merino. While the written or printed histories of the Adams, Humphreys, Heaton and Jarvis importations are practically lost, owing to numerous transfers, the flock of C. S. Ramsey, Castleton, Vt., has an unbroken traditional record from the Humphreys' importation to the present time. In 1809, Israel Putnam, of Marietta, Ohio, bought of Seth Adams some full-blood Merinos, and founded a flock, which was continued by his son, L. J. P. Putnam, substantially to the present time, but without registration. June 13, 1811, Dr. Increase Mathews, of Putnam, Ohio, bought an Infantado ram and two ewes, just imported into Alexandria, Va., and had them brought in a wagon to his farm in Ohio, where he kept up a pure flock until about 1850. In 1811, Col. Humphreys sold a ram for sixteen hundred acres of Ohio land to Paul Fearing and B. I. Gilman, of Marietta, Ohio, and this ram was brought on and laid the foundation for a flock which was kept up many years. In 1826, Col. John Stone and George Dana, of Belpre, Ohio, bought a number of pure Merinos from the celebrated Wells flock, of Steubenville, Ohio; and Col. Stone kept up a flock over half a century. The Wells flock, just mentioned, was founded in 1815, and continued to 1829, when it was a grand flock of three thousand head, shearing about five pounds of washed wool per fleece. It was then sold and scattered.

For some reason a cloud has always rested over the importations made subsequently to 1812; hence the fine flocks of "Black-top" or "Delaine Merinos"—locally known by way of emphasis as "the big Merinos"— found in Washington County, Pa., tracing to the Meade importation, and with some admixture of Saxony in several cases, founded about 1826–30, have been regarded as Pariahs and outcasts, whose abode was without the camp. But, in view of the fact that these same flocks have contributed, perhaps more than any others within its borders, to set Washington County at the very forefront of the United States in the production of sheep and wool, they can rest tranquilly under this bar sinister on their escutcheon. Presenting themselves with a modest register, in which no special effort is made to conceal the stain in their blood (if it be one) the "Victor-Beall Delaine Merinos" ought to be recognized as an excellent variety of the American.

The flock of Daniel Kelly, Wheaton, Ill., has a record dating from 1829. That belonging to Alex. Fraser, East Troy, Wis., originating from Atwood Merinos, has a record reaching back to 1846.

The spread of the Merinos over the Far West is traced to some extent in subsequent chapters.

CHAPTER II.

FORM.

CORRELATION OF CARCASS AND FLEECE.—In the appendix to *The Practical Shepherd*, Mr. Randall gives some valuable tables, which go to show that small sheep produce proportionately more wool than medium or large ones. I shall abridge these somewhat, and give, first, a table which is based on six hundred and fifty-five sheep, divided into lots according to age and sex. These tables represent the results of three years' observations.

Age.	Sex.	Average Weight of Body.	Average Weight of Fleece.	Pounds of Body to One of Wool.	Per Cent of Wool to Weight of Body.
1	E	55.74	5.07	11.01	8.10
2	E	67.08	4.94	13.54	6.90
3	E	75.99	5.18	14.58	6.41
4	E	89.49	5.06	16.33	5.88
5	E	74.67	4.75	15.68	6.00
6	E	79.00	4.78	16.49	5.70
1	W	64.23	5.16	12.43	7.50
2	W	84.23	5.09	14.77	6.49
3	W	88.86	6.45	14.57	6.58
4	W	103.94	7.04	14.04	6.65
5	W	97.72	7.12	13.71	7.00

From this table, it appears that ewes shear their heaviest fleece at three years old, but gain in weight until they reach the age of four. The percentage of wool to live weight decreases every year (with the exception of one) until they are six years old. It shows also that, for the first two years, ewes are more profitable as shearers than wethers; but after they begin to bear lambs, of course, they fall a little behind in their percentage of wool to carcass. The second table is based on the same number of sheep, classified by **weight**, for the same number of years.

No. in Lot.	Weight of Lots.	Average Weight.	Average Weight of Fleece.	Pounds of Body to One of Wool.	Per Cent of Wool to Live Weight.
52	34 to 51	44.63	4.08	11.36	8.16
89	50 to 61	55.78	4.71	11.90	7.80
129	60 to 71	66.03	5.09	12.96	7.13
160	70 to 81	75.52	5.31	14.21	6.53
92	80 to 91	85.25	5.78	14.77	6.33
75	90 to 101	95.90	6.10	15.44	5.85
58	100 to 140	111.31	7.17	15.56	6.04

It will be observed from this table, that the percentage of wool to live weight, decreases steadily with the increase in the size of the sheep, until the last lot is reached, where there is an increase of the fifth of one per cent. But there were only seven sheep in this heavy lot, and if there had been a large number to average from, the result might have been different. At any rate, the conclusion is irresistible, that young sheep are the most profitable as wool-producers; also, the further conclusion, that a wether at four years of age will yield more mutton, on an average, than he ever will afterward. Hence, that flock will pay best which has every year the highest percentage of lambs, notwithstanding the fact that lambs are subject to more accidents and fatalities than older sheep. Furthermore, since a ewe is more profitable as a wool-bearer than a wether, up to the time when she bears a lamb, and is more profitable afterward, by reason of her lamb, ewes are a better paying class of sheep than wethers. This would indicate the policy of selling off wethers closely, and buying ewes for breeders.

I may add that M. Bernardin, the superintendent of the Rambouillet flock of France, in a letter to Mr. W. G. Markham, states that: "Dividing a flock according to weight into four sections, we find the smallest sheep will yield twelve and thirty-eight hundredths per cent of their live weight in wool; the next largest, eleven and forty-one hundredths; the next, ten and thirty-eight hundredths; and the heaviest, nine and fifty-one hundredths."

A small Merino is hardier and more prolific than a large one. One hundred and twenty sheep, weighing ten thousand pounds, will not consume any more feed than one hundred weighing a like amount. On the score of mutton, the medium sheep is not objectionable, because the butcher considers size as secondary, and seeks for the carcass which is thoroughly well fattened. In proof of this, I give a list of the sales of mutton sheep made on two consecutive days in the last week of March, 1885, at the Union Stock Yards, Chicago, the second column showing the average live weight, and the third column the price per hundred pounds. (Note how quality rules instead of weight) :—

No.	Av.	Pr.	No.	Av.	Pr.
30 inferior	66	2 30	81 good	106	4 00
10 inferior	80	2 75	52 good	95	4 00
126 inferior	60	2 95	125 good	90	4 15
67 common	81	3 35	506 Western	124	4 20
60 common	67	3 50	177 Western	100	4 25
89 common	73	3 50	169 good	118	4 25
287 common	70	3 55	58 good	103	4 40
171 fair	102	3 60	98 choice	95	4 40
40 fair	87	3 75	171 choice	123	4 50
103 Western	71	3 75	10 choice	100	4 50
96 Western	87	3 75	280 choice	139	4 75
19 medium	83	3 75	75 good	109	4 75
179 medium	95	3 90	18 choice	112	4 75
110 medium	90	3 90	139 extra	117	5 00
62 common	87	3 60	170 medium	94	3 90
80 common	88	3 25	137 medium	88	3 90
102 common	83	3 37½	94 medium	87	3 60
100 fair	82	3 40	74 medium	91	4 12½
100 fair	82	3 50	63 choice	114	4 50
73 fair	91	3 50	95 good	88	4 25
30 fair	75	3 50	70 good	108	4 25
55 fair	78	3 60	64 good	121	4 25
74 fair	77	3 65	83 good	113	4 25
27 fair	71	3 70	10 good	156	4 40
136 fair	90	3 75	59 choice	105	4 50
110 fair	73	3 75	66 choice	135	4 75
32 fair	90	3 75	47 lambs	93	5 50

RACE TYPE. — A perfect animal should be symmetrical and well-rounded, without angularity; the top and bottom lines straight, and nearly parallel to the root of the scrag or neck. Back straight; ribs well sprung out, giving a round barrel, thick through the heart; shoulders deep, chest broad, breast bone or brisket extending well in front and down; hips long, straight and broad; thighs well let down, and heavy; neck short and powerful, without droop on top; head broad, nose short and wrinkly, nostrils not flat, but round and open; legs stout, bony, standing wide apart at knee and hock.

Experience has demonstrated, that great weight of fleece (if not the greatest), can be combined with constitutional vigor. The greatest amount of yolk compatible with perfect physical development, is admissible in a ram ; so long as the skin remains a bright rosy pink color, and the yolk colorless, or nearly so, it is difficult to develop too much of the latter in the fleece. A fleece opening buff or orange, is the choice of many breeders, but a yolk tinted lemon, or nankeen, is objectionable, and still more so, one of a greenish tinge ; they evince a morbid habit of body which is associated with clot or induration of the fleece.

Wrinkles are not a distinctive race characteristic of the American Merino ; for full-blooded and very fine specimens can be found which are perfectly plain. They are an individual characteristic, and are generally (not always) associated with the highest development of the wool-bearing aptitude, Nature, uncontrolled in her breeding operations, seeks to perpetuate race characteristics alone, so that the labor and skill of man must continually intervene to preserve certain desirable features in the individual. Hence a somewhat greater degree of wrinkliness is permissible in the ram than is desirable in the progeny, as a counter-check to this tendency toward reversion. But, whatever the keeper of the stud-flock may choose, the judicious wool-grower, knowing that a nearly plain sheep is best fitted to cope with wind, and rain, and snow, and is easiest to shear, will look well to it that his rams shall not have the skin too heavily folded.

The breeders and wool-growers of Vermont, Western New York, Northern Ohio, and Michigan, carried the wrinkly habit of the Merino to a higher pitch than did those of Ohio, Pennsylvania and West Virginia ; and coupled with this was a shorter and more yolky staple. These facts have established for the clips of the first-named States, a lower price, by two or three cents, in the Boston and Philadelphia markets, than is paid for the latter. To this, however, there is one exception, namely : that the wools of Northern, or rather Northern Central Ohio, sell from one to three cents higher than those of Southern Ohio, which is due to their greater uniformity of breeding, and more thorough preparation for market.

DELAINE MERINO. — The longer stapled and plainer sheep of the three States mentioned above, find their culmination in Washington County, Pa., in the "Victor-Beall Delaine Merino," which is a cross between the old Pennsylvania "Black-top" and the "Spanish Merino." Their "scale of points" numbers one hundred, distributed as follows: Constitution, ten ; heavy round the heart, six ; short, heavy neck, six ; good dewlap, five ; broad back, eight ; well-sprung rib, five ; short legs, six ;

heavy bone, eight; small, sharp foot, ten; length of staple, one year's growth, three inches, eight; density of fleece, eight; darkish cast on top, five; opening up white, five; with good flow of white oil, five; good crimp in staple, five. Weight of rams at maturity not less than one hundred and fifty pounds, weight of ewes at maturity, not less than one hundred pounds. This family of sheep has been bred and kept in large flocks, without housing and without pampering. They have been bred also, to produce a short, sharp, and shapely hoof, in order to avoid one of the greatest curses of the Merino, a spongy, clubby hoof, and a consequent predisposition to foot-rot.

NATIONAL IMPROVED SAXONY.—This is the designation adopted by the present breeders of this fine class of sheep, whose seat is also in Washington County, Pa. They have a scale of points numbering one hundred, eighty of which admit to register, though no animal is eligible whose fleece grades in fineness below XXX (the two grades above being picklock and picknic). The points in the scale are otherwise about the same as those in the "Delaine" Register, though they tolerate no wrinkles, and only a slight dewlap. Constitution and evenness of fleece ("well covered on belly, face and legs"), are each fifteen points, which is well, in view of the ancient, hereditary defects of the Saxon. Mr. J. G. Clark, Secretary of their Register, writes me that their rams, when full-grown, weigh from one hundred and twenty-five to one hundred and fifty pounds, and some have gone over that; ewes, eighty to one hundred pounds and over.

BLACK-TOP MERINO.—Mr. W. G. Berry, Secretary of the Association, writes me that the "Black-top Merino" breeders have in press a Register of about seventy flocks, principally in Washington county, Pa. In default of more accurate information, I append the following extract from *The Shepherd's National Journal*. The editor, Mr. E. J. Hiatt, a veteran breeder of American Merinos, says: "The purity of the blood has not been questioned." He adds: "We have been acquainted with flocks for more than twenty-five years in which this blood predominated. The quality of the wool was good. Specimens were exhibited at the Pittsburgh and Wheeling Fairs last fall. In some respects they showed a marked improvement over their ancestors of twenty or twenty-five years since. The strongest improvement is noticeable in their increased size and their heavier fleeces. In size, they are possibly a little heavier than the American Merinos, and longer in the legs, neck and head. They also **shear lighter** fleeces. Much of it grades XX and XXX **delaine**. The head, legs and belly are not covered with as long or compact a fleece as would be desirable."

As this work goes to press, the expected **Vol. I.** of the *Blacktop Spanish Merino Register* **makes its appearance.** I quote two paragraphs, as to blood and scale of **points**:

"**Sheep must be purely** bred from the importation of Merino sheep from Spain in the year 1802, as bred by W. R. Dickinson.

"**Constitution, fifteen points**; size, twelve points; general appearance, three points; body, fifteen points; head, five points; neck, four points; legs and feet, ten points; covering, eight points; **quality, seven** points; density, seven points; length, eight points; oil, six points; total, **one** hundred points."

CHAPTER III.

FLEECE.

STRUCTURE OF THE WOOL-FIBER.—The **wool-fiber is made up** of a root, and a stem **or** shaft, continuous **with, and growing out of the root.** The root exhibits a flask-shaped enlargement, **which fits down** somewhat socket-like upon a very small papilla or bulb in the bottom of the fiber-sack; and this little bulb is the feeder of the fiber, the germ of it, which is able to produce a new one if the old one is plucked out. The shaft of the fiber is composed of an outer cortex, **an inner medulla, or** marrow (though in **a** majority of wool-fibers **this marrow is** hollow nearly to the tip), and, thirdly, an intermediate fibrous portion, constituting two-thirds of the substance of the fiber.

The **cortex** is formed by the growth of cells; these cells, lengthening out and becoming flat, assume the form of scales, these being produced one after another, just as a roof is made by the laying of one course of shingles after another, overlapping each other. (Vegetable fibers grow **at the top, but** hair and wool fibers grow **at** the root, the new portion constantly pushing out the old). The scales, overlapping each other, with free edges, constitute the felting property of **wool,** which hair, being **smooth, possesses to a** very limited degree, or not at all.

This lapped arrangement of the scales of the cortex, can be detected by the touch, but not by the eye; let a fiber be drawn between the thumb and finger from root to tip, and it will pass smoothly and sweetly through, but if it is drawn the other way it will go roughly. A hair will go about as smoothly one way as the other. When a quantity of wool is pressed, rubbed or beaten, the free edges of the scales interlock in an infinite number of places, and the whole is bound together in a close, dense mass; it is felted.

ROUND AND FLAT FIBERS.—Although wool, as well as hair, is of a tubular construction, yet the cylindrical form varies with the climate. A cross-section of a fiber of wool, if strictly circular, denotes that it has been grown in a cold northern climate, and is lank, long, and soft; but if the cross-section shows a flat-sided or oval hair in the extreme, then the wool or hair is of tropical growth, and is crisp and frizzled. There is a change in these animal downs as we ascend from the equator to the higher latitudes; hence our better class wool can only be grown in temperate climates. Too hot a climate yields a wool too crisp and too frizzly; while, on the other hand, too cool a climate, though yielding a wool that is soft to an extreme degree, gives too little of the curl or frizzle for many manufacturing purposes. This curl or waviness varies in different kinds of wool. The long Leicester wool has about eight or nine of these waves or curls per inch, but in a fine Ohio wool there are as many as thirty to thirty-three waves or curls per inch.

THE CRIMP.—This is one of the nice points of a first-rate Merino fleece. While the hairs of the horse or the ox are straight, the wool-fibers of the Merino are beautifully wavy or crimped, and in the best-bred fleeces this crimp is perceptible by the naked eye to the very tip of the fiber, not being lost in a dark clot or induration. This crimp is caused by frequent, but somewhat irregular, well-marked, and more or less spirally arranged thickenings of the cortex of the fiber. These thickenings of the cortical layer occur first on one side of the fiber, then on the other, which gives it its wavy and sinuous character.

LENGTH AND DIAMETER.—The difference in the length of staple or fiber of the different breeds of sheep is very remarkable, extending from the longest combing wool to the shortest clothing staple. There is a gradation of seventeen and a half inches; the longest staple being eighteen inches, and the shortest half an inch long, and the different breeds and crosses fill up a graduated scale between these extreme points. While the finest Silesian will yield a fiber one-fifteen hundredth of an inch in diameter, a Cotswold fiber will be double the size.

This idea of measuring the size of fibers of wool with a microscope, is not a new one; it was done thirty years or more ago, but was of no practical benefit. A wool sorter, who has worked at the business from his youth up, without intermission, and whose eyes have failed him so that he cannot read a daily paper without glasses, will tell without their aid the relative size of the fibers of wool so that the different qualities of cloth will be uniform, far more so than the Commissioner of Agriculture could select them with his microscope. But the microscope has shown us one very interesting fact, which the finest touch of the expert would hardly have detected, namely: that while the hair from the ox or horse, which falls out yearly, tapers its whole length, the Merino wool fiber tapers only for a short distance at the top; and when this hoggetty point has been shorn off with the first or lamb's fleece, the fibers ever afterward remain of the same diameter throughout their whole length.

How the Wool Fiber is Planted.—We have considered the fiber itself, somewhat; now let us turn to the follicle, or sack, out of which it grows. This is formed of the epidermis and the dermis of the sheep's skin, turned inward and prolonged in a very minute cylinder, which sometimes penetrates the tissues of the body one-twelfth of an inch. The bloodvessels are distributed in minute branches in the walls of this follicle, thickest at the bottom of it; and they supply nourishment to the germ at the bottom of the sack, which molds it into the substance of the fiber. Besides the wool-follicle, or fiber sack, there are two other kinds: the sweat follicle, and the yolk follicle, both of which are only about half as deep as the wool follicle. The sweat follicle has its mouth directly on the surface of the skin, this mouth being a pore: but the yolk follicle empties into the wool follicle near the mouth of the latter. The shaft of the fiber does not fit perfectly tight in its sheath or sack; this leaves space for the yolk to surround the fiber down to its very root. In this space, also, parasites sometimes harbor, such as the scab insect. The yolk is for the lubrication of the fiber, to prevent it from felting with its neighbors, while on the sheep's back. The free edges of the scales on the fiber, like little barbs pointing toward the tip, continually work the yolk outward toward the tip and at the same time expel dirt from the fleece. Thus we see how it is that the Merino, which has the finest and best felting wool (in others words, fibers with the greatest number of scales to the linear inch), needs also the greatest amount of yolk.

SORTS IN THE FLEECE.—The keen-eyed professional sorter tears a fleece into several sorts. "He will rapidly break off the coarse skirts for one sort; then the head, and, perhaps, if the fleece is a cross between a native and some fine wooled sheep, he will discover a coarse streak running down the nape of the neck nearly to the shoulders. This also he breaks out, and places with the sort to which it belongs. The shoulders yield the finest sort of the fleece, and the sides a sort below. In this way each fleece is broken into at least three and often five different sorts. A large factory has generally as many as eight clothing sorts, to which, where worsteds are made, are added as many long sorts—combing and delaine. On the sorting board the fleece loses its former identity. It is no longer known as fine, XX, half-blood, or by any other familiar name. Each fleece is resolved into first, second, third, etc., down to the bottom. The shoulder of the quarter blood rests in the bin with the skirt of the full-blooded fleece, and the skirt of the half blood may mingle in yarn with the shoulder of a common fleece. So unerring is the discernment of the best sorters that under microscopic tests it has been demonstrated that they can assort fibers as to fineness within a ten-thousandth of an inch."
—*American Sheep Breeder*.

In the Merino fleece, the wool on 1 and 2 in the diagram (Fig. 3), is finest, longest, and strongest; 3 and 12, short, but close; 4, rather longer, a shade lower than 3; 5 and 6, slightly coarser, not so close, and apt to be weak in fiber; 8, lower still, and termed the britch or breech; 7, good length, but slightly lower in quality than 1 and 2; 9, shorter, and loses vitality as compared with better parts; 10, short and generally frowsy; 11, shorter than 12; 13, the cap; dry and harsh; 14, fribby, and and of little value; 9, 10, 11, 13 and 14 constitute the skirt.

To economize the number of sorts, is very injudicious; as good, even sorts cannot be made without strict adherence to the division of the fleece into its separate parts, so as never to allow the ridge to adhere to the shoulder, or similar errors. Exception may be made in cases of high-bred wool, up to seven-eighths blood and above, as here the distinction is so trifling between the shoulder and other parts that a mere skirting is needed.

GRADES OF WOOL.—There are two great divisions into which wool is graded—carding, or clothing wool, and combing wool. The clothing staple may be very much crimped, and very short, since the fibers are mingled in every way by the cards, leaving the ends to project in a nap which conceals the warp and woof; but combing wool should be long and straight, since the fibers are to be laid side by side, end to end, and spun into yarn for

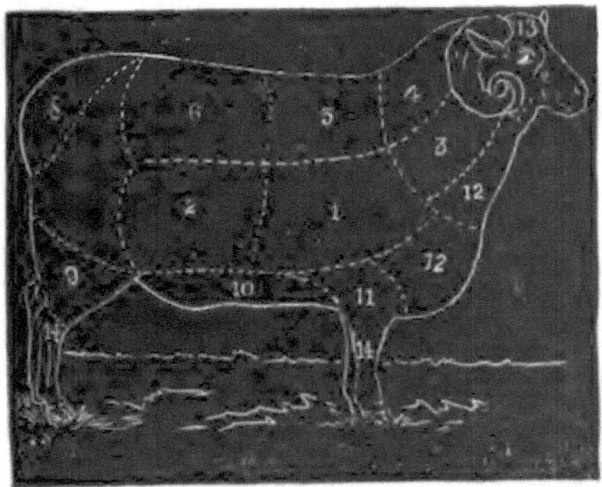

Fig. 3.—DIAGRAM OF FLEECE.

worsted goods. The finer the fleece the shorter the staple which can be used for combing. A coarse combing wool ought to be six inches long, while a fine XX staple only two inches long could be combed, though it should be two and one-half. Coarse fleeces are not graded very closely, while fine fleeces are subjected to closer grading. A staple an inch and a half long could be worked upon French combs; but for English machinery, the length of the staple must be determined by the machines following the combs. The ability of a Noble comb to handle short wool is not the guide for the buyer of wool for the English system. The good qualities of pure breed, soundness and evenness of wool, from well-fed and carefully handled young sheep, allow a margin in selection in favor of the minimum length of staple. A diagram will help to an understanding of this subject:

Superfine...............................	{	Picklock. XXX.	} Saxony.
Fine.... { 10 per cent. combing (delaine) 25 " "	{	XX. X.	} Merino.
Medium { 50 per cent. combing .... 75 " "	{	No. 1. No. 2.	} Quarter - blood and Downs.
		Coarse- Combing. Carpet.	} Cotswold, Leicester, Lincoln. } Chourro.

WHENCE THESE GRADES COME.—Picklock and XXX are now confined almost exclusively to Washington county, Pennsylvania, and the "Pan-handle," where a few flocks are still carefully bred, yielding the Electoral or "noble" wool. In the last three decades, the amount of Saxony has much decreased, while the proportion of fine and medium to the whole product of the nation, has vastly increased, owing to the spread of the American Merino. Occasionally, a Merino fleece grades XXX ("XX and above" is intermediate), but here a touch of Saxon blood may be suspected.

The greater part of American Merino wool grades XX, and of this the best samples come from Ohio, Pennsylvania, and West Virginia. The grade known as X is generally obtained from the full-blood Merino of Wisconsin and Michigan, and also from the finer crosses (half-blood and above), though the latter are often graded into "combing," and designated as delaine. Central Wisconsin has many choice delaine flocks; those of Messrs. Rich & McConnell, of Ripon, have yielded staples four and seven-eighths and four and five-eighths inches in length, respectively. Michigan wools long suffered from the same faults as those of Vermont and Western New York, short and gummy; but lately much improvement is manifest, and they often rank with the best Ohio.

Missouri now furnishes a considerable quantity of combing wool, the Merinos of that State having been crossed on the large native stock. The greater part of No. 1 and No. 2 now come from this cross. Kentucky yields a large percentage of combing wool. Kansas, Nebraska, and Colorado wools have greatly improved in the last decade; they now fall only five to ten cents the scoured pound, behind the best Ohio. California and Texas wools rarely ever yield any combing wool, but would do so, were it not for the semi-annual clippings. At the close of the war, Texas wool was fit only for carpet filling or, at best, for coarse blankets; now there are large flocks that grade up to X, and above. Twenty years ago, Santa Fe wool was fit only for carpets or horse blankets; but the introduction of Merino blood has raised the grade up to No. 1 and low X. Montana fleeces are the best of the whole Territorial clip.

A considerable combing and delaine wool comes from Oregon and Washington Territory. Vermont always furnishes a class of wool which will yield a large percentage of fine combing, together with a good deal of unwashed and un-merchantable.

EFFECT OF CLIMATE ON THE FIBER.—It was the opinion of Dr. Randall, that a hot climate, with its consequent rankness of vegetation, would coarsen the fiber of sheep taken to it from a colder region. This opinion does not appear to be sustained by modern investigators. Mr. G. W. Bond, an eminent expert of Boston, exhibited to a scientific society some skins of Arabian sheep; some of them covered with hair alone, and others with similar hair, but having a thick undergrowth of wool, which proved to be as fine as the finest Saxon. Mr. Mark R. Cockerill imported some Saxons in 1824 or 1826, and kept them in Mississippi (Madison County), a quarter of a century. At the World's Fair, in London, 1851, samples of their fleeces were brought into competition with German wools. The latter were recognized by the jury as the finest and longest on exhibition, but those of Mr. Cockerill received two prize medals, of the same grade as German, and were reported by the expert employed by the jury "as most approximating to the character of German wools." Mr. Graham, the author of a popular hand-book on Australian sheep, states it was the general belief that the climate of Darling Downs, a region within the tropics, was too hot for the growth of good wool; but that the superintendent of the Clyde Company, by a "careful and judicious system of selection," succeeded in producing "as good wool as any grown in Australia, although it still bore the name of *hot-country wool*."

EFFECT OF HIGH FEEDING.—Prof. Sanson, an eminent zoötomist of France, in a report to the French Academy of Sciences, which is perhaps the highest scientific body in the world, gives the following summary of the results of his investigations on this subject:

"1. The precocious development of Merino sheep, having the effect to carry their aptitude to produce flesh to the highest degree that sheep can attain, exercises no influence on the fineness of their wool. This preserves the diameter which it would have had it developed in normal conditions, for the reason that this diameter depends upon the individual and hereditary aptitudes.

2. The influence exercised by the precocious development upon the hair of the wool exhibits itself by an augmentation of the length of the same hair; its growth, resulting from the formation of epidermic cellules in the hair-bulb, being more active. There is, therefore, more woolly substance produced in the same time.

3. The precocious development does not vary the number of hair or wool bulbs existing for a determinate extent of the surface of the skin. It produces, therefore, no change in what is vulgarly called the *tasse* (density of staple). The modifications which the staple of wool presents in this respect are only apparent. By increasing the length of the hairs, the precocity necessarily increases that of the locks of wool which they form, which makes the fleece appear less dense."

To sum up all in a word, high feeding increases the length of the staple and the secretion of yolk, but not the diameter of the fiber, or the number of fibers to the square inch of skin. Very high feeding, or pampering, increases the yolk in a geometrical ratio to the fleece. This ought to operate as a safeguard for the protection of the sheep from this pernicious practice; but it does not wholly, for, unfortunately, the "big fleeces" of the fairs and public shearings are always weighed "in the grease," instead of scoured. The chief defense instituted by Nature against this evil of pampering is, that sheep so treated often suddenly and mysteriously die.

LENGTH AND DENSITY.—A fiber two and a half inches long which is perfectly sound and true, is better every way than one which is three inches long, but has a "joint" or weak spot caused by poverty or sickness in the animal, which will cause one-half inch to break off in the combs. It is better, because the existence and nourishment of the sheep during the growth of that half inch cost something, while that half inch is practically a total loss to the manufacturer, and tends to discredit both the fleece and the grower.

No one will dispute the proposition that it should be the cardinal object of the wool-grower to produce a sheep having the greatest possible number of fibers to the square inch of skin, and those fibers of the greatest possible length. The striving for the attainment of either of these objects has more or less tendency to defeat or repress the other; yet not so much as a certain school of breeders would have us believe. Density, though the expression of an exceedingly valuable constituent of the best fleece, is an illusive term, or perhaps more accurately, a term liable to be misapplied. When a fleece on the back of a sheep is grasped in the hand and offers a firm resistance to compression, presents a good handful, it is called dense by superficial men, without further examination. But it may not be true density at all; it may be yolkiness carried to the exaggerated state. A fiber is entitled to so much yolk as will thoroughly lubricate it from end to end and make it glisten; but this substance should not collect in lumps. A fleece with a fiber three inches in length, may carry as much weight of yolk as one of only two inches, yet not feel or be pronounced by the awarding committee as "dense" as the latter.

STRENGTH OF DRY AND YOLKY WOOLS.—Some years ago I addressed inquiries to a number of experts, as to the comparative strength of dry and yolky wools; and, contrary to my previous belief, they all replied that yolky wools are the stronger of the two. The explanation of this seeming paradox is this: Good feeding makes good wool, and it also makes yolk. Where you find a yolky flock of sheep, you are almost certain to find a liberal feeder and a pains-taking shepherd. The burden of yolk, especially if it is collected into pasty lumps, is distasteful to the manufacturer, and causes a higher percentage of loss in the scouring-tub; but at the end of the cleansing process, we are tolerably certain to find a staple true and sound.

The wools on the Vermont side of the Connecticut river, are much more yolky than those on the New Hampshire side; but the New England manufacturers are well aware that they are stronger. Vermont wools lose most in the scouring-tub; New Hampshire wools most in the cards or combs. A certain amount of yolk conduces to soundness and strength by protecting the fiber from dust, alkali and rain.

COTTING.—We shall perhaps best arrive at a definition of first-rate Merino wool by a statement of the faults which are liable to occur in a fleece. Cotting generally develops itself in the winter, as a result of a diseased condition of the sheep, and when shearing-time comes the animal is pronounced "fleece-grown." Some part or all of the fleece is completely fulled or matted together, so that it can be thrown about or held up by one lock as if it were a pelt. But, as the sheep has by this time generally recovered from its illness, most of the fibers will be found to have parted from the body, and the fleece will be clinging to the skin by a very few fibers, so that the shearer can strip it off rapidly. Such a fleece has a very low value, and it is a clear fraud to throw it into the pile and attempt to sell it for sound wool.

The nature of the felting quality has already been explained, and this unnatural felting is caused by the drying-up of the yolk-glands from disease. When we consider the innumerable particles of dirt, chaff, seeds, etc., which fall upon the fleece in a year, it is wonderful what a clean, bright interior is exhibited when we open up the unwashed fleece on the sheep's back in the spring. The useful purpose of the minute barbs, or free edges of the cortical scales is manifest; without them the fleece would become a mass of filth.

The most frequent cause of cotting, the ammoniacal exhalations of an uncleaned stable, will be further discussed in a subsequent chapter.

BLACK-TOP AND CLOTS.—A perfect Merino fleece will show the crimp to the extreme outer end of the fiber; it is needless to remark that there are not many perfect fleeces. There are two kinds of indurations: one is called the "Black-top," and the other may be designated as the "Gray Shoulder-Clot." The Black-top, extends the whole length of the fleece, being densest along the back-bone, and extending down the sides more or less, to the belly. In some sheep, the Black-top renders the fleece almost water-proof; during the summer no amount of rain will dissolve the gummy, pasty, tar-black sheeting of the fleece, or wash this down into it, as common yolk is washed. In the summer, this gummy top is soft to the touch, but in winter, especially if the animal is confined, and allowed little exercise to warm its blood, it becomes separated into horny lumps, each one tipping a lock of wool, and as hard as a board.

Though bad enough, this is not so objectionable as the gray Shoulder-Clot. This is more pronounced on the shoulders or withers, but frequently extends half-way down the shoulders, and more or less along the backbone. Sometimes this seems to be a constitutional defect, but generally it is caused by poverty; the circulation of the blood is so feeble, where the shoulders are sharpened thin, that there is not enough animal heat to keep the yolk liquid. The rain disorganizes and de-vitalizes the yolk, washes away the softer parts of it, and the residuum coagulates, and gums the locks together. In a short-fibered fleece (and they generally occur in this sort), the locks will be glued together for the outer half of their length, and so hard as to require a hammer to break them down. A sheep with these clots in its fleece ought to be rejected from the flock; they are an abomination to the manufacturer.

"JAR."—This is properly *gare*, but "Jar" seems to be permanently incorporated into the shepherd's vocabulary. It is simply hair, but it is hair of the worst kind; wild, coarse, and frizzly, utterly refractory in the cards, combs, and dye-tub. It is oftenest seen on rams, on the outer surface of neck-folds, and less frequently on the side-folds and hips. In Cotswold sheep and Angora goats, it occurs on the hips, and is called "Kemp." Jar is not found on a feeble sheep, it is an excrescence of a vigorous animal. Though highly objectionable in itself, it marks a very desirable quality; a good constitution. Unless it is excessive (short, curly jar is the worst), and the ram is imperfect otherwise, the breeder need not trouble himself much about it, as it seldom occurs on ewes and wethers. He could afford to throw away part of the ram's fleece for the sake of his constitution.

"JOINTED" WOOL.—Whatever keep the shepherd gives his flock, good, bad or indifferent, it ought to be regular. Amid all the conflict of opinion among wool-growers, as to the effect of different feeds on the staple, one thing is certain: regular feeding makes a true (even) fiber. To employ a somewhat fanciful illustration, the fiber is a delicate rod on which every attack of disease, every protracted spell of hunger cuts a notch, and thus weakens it. The reason is obvious. Like every other part of the body, the wool is nourished by the blood. If there is a lack of feed for a day or two, or an attack of disease, the blood becomes impoverished to that extent, and secretes in the wool-follicles less matter for the building-up of the fiber. It is an erroneous belief of some flock-masters, that a period of starving or disease weakens the staple along the whole length. This can not be. As previously described in this chapter, the fiber grows by the continual formation of new cells at the bottom. Every day's growth is a complete and finished product; nothing that can happen afterward can change its diameter or structure. When the sheep dies, the fiber stops growing, but so long as life lasts, it must keep growing. only in a case where the blood is impoverished, it can not furnish the usual quantity of matter, and there is a weak place formed.

In the Far West, a heavy snow covering the grass for several days, sometimes weakens the fibers so that most of them break near the skin, from their own weight, and the fleece falls off.

CLOUDED FLEECES.—The first choice of fleece rolls off from the shears elastic, voluminous, and white as snow; the second choice is a rich buff-yellow, or golden tint. Fleeces from the prairies or the adobe flats of California are stained dark by the soil. In East Tennessee they are reddened with clay. But sometimes there are fleeces shorn from sheep that are kept on the cleanest soils, and with the greatest care, that are disfigured by large saffron-colored, or lemon-colored patches along the back and down the sides.

These may be produced by rain-water trickling down through straw-roofed sheds, which, for this reason, are a nuisance. But generally, there is no other assignable reason than a disordered circulation of the blood, consequent upon a lack of exercise and an irregular system of housing. If an attempt is made to house a flock through the winter, it ought to be carried out. Systematic exposure is better than a housing from one storm, and a wetting in the next. These cloudy places do not necessarily injure the staple, but they detract from the beauty and salableness of the fleece.

MOLD.—Energetic flock-masters, especially in the Far West where they have vast flocks to handle, are sometimes tempted to push on the shearing, when the sheep are wet with rain or dew. This is a grave error. The dry parts of the fleece will not absorb the moisture sufficiently to prevent mold, and this is justly offensive to the manufacturer. The wishes of the manufacturer are generally based on sound business considerations, not emanating from caprice, and the farmer is bound by his own interest to give them reasonable attention.

STUFFING, STRINGS, ETC.—American flock-masters are too prone to stuff their fleeces with unwashed, with tags, dead wool, parts of rams' fleeces, here a little, there a little. Our national record in this regard, falls below the Australian, even below that of the Argentine Republic. The buyer or commission agent in the country, may pass this matter over lightly, fearing to estrange his men; but he knows that here is a sure menace of claims and discounts, and so quietly operates to cover, if he can. The buyer for the mill feels that here is a point where his vigilance, though sleepless, may be entirely inadequate. The following figures relate to wools carefully selected to *avoid burrs;* the "Fribs" include all locks too short for combing, taken off before skirting, but not the skirts themselves. The yield of two million pounds of American washed combing fleeces, mainly from Ohio (one million pounds being Ohio fine delaine), sorted in one year in a worsted mill, was over one hundred and fifty thousand pounds of fribs, twenty thousand pounds of burry clips, and twelve thousand pounds of strings; all of this was paid for, of course, as combing wool. The same treatment, the next year, of three hundred and thirty-four thousand pounds (one-sixth the quantity) of Australian unwashed Merino and Cross Bred wool, yielded four thousand pounds of fribs, six hundred of burry clips, and less than one hundred pounds of strings. If done up as the American wool was, this Australian lot would have contained twenty-five thousand pounds of fribs and two thousand pounds of strings.

BURRS, THISTLES, ETC.—Burrs, thistles and tar-marks, are more objectionable than the same weight of natural yolk. Scouring may take away all the yolk, while some remains of the former may obstinately resist all machines and all treatment, and appear as an incurable defect in a high-priced fabric. Besides this, the labor of the sorter (who generally works by the pound), is greatly increased by burrs, thistles, etc., and they may thus prove a tax that is a serious inroad upon his wages.

UNEVENNESS —One of the greatest errors the farmer can commit is, to grow mixed sheep. If it were profitable to grow carpet wools at all, it would be better for him to have a flock of carpet-wool sheep, rather than one containing some coarse, some medium, some fine, some superfine; because the clip would all practically grade and be sold as coarse, while the maintenance of the fine-wooled sheep, would be more expensive than that of the coarse-fleeced. Still worse than this *mixed flock*, is a *mixed sheep;* that is, one, which from ill-judged attempts at crossing, has a coarse streak running down the nape of the neck, or one, which from mismanagement in breeding, has white and long wool on the shoulder, but short, yellow and frowzy wool on the belly. It is the crowning excellence of the pure-blood American Merino, that it has wool very nearly of the same length all over the body.

SECTIONAL PRICES OF WOOL.—The following table shows the prices of wool at the date given, in different parts of the United States, with some foreign kinds:

CHICAGO, Oct. 20, 1884.

WASHED FLEECES.

OHIO, PENNSYLVANIA, AND WEST VIRGINIA.

XX and above	34 @	36
X	32 @	33
No. 1	32 @	34
No. 2	29 @	31
Common	24 @	26

NEW YORK, MICHIGAN, INDIANA AND WISCONSIN.

X and above	28 @	30
No. 1	30 @	32
No. 2 and Common	23 @	28

COMBING AND DELAINE.

Washed Fine Delaine	33 @	38
Washed Medium	34 @	37
Washed Coarse	26 @	30
Unwashed Medium	24 @	26
Unwashed Coarse	20 @	24

PULLED WOOLS.

New York City extra	25 @	30
New York City super	28 @	32
New York City Lambs	25 @	30
Eastern and Country extra	30 @	32
Eastern and Country super	32 @	35
Western extra and super	23 @	27

UNWASHED.

INDIANA, MISSOURI AND KENTUCKY.

	Bright.	Ordinary.
Fine	20 @ 22	17 @ 19
Medium	23 @ 26	19 @ 20
Coarse	19 @ 21	17 @ 19

KANSAS, NEBRASKA AND TERRITORY.

	Choice.	Average.
Kansas and Nebraska Fine	17 @ 18	15 @ 16
Kansas and Nebraska Medium	18 @ 20	16 @ 18
Utah and Wyoming Fine	18 @ 20	15 @ 17
Utah and Wyoming Medium	20 @ 21	17 @ 20
Montana Fine	20 @ 22	18 @ 20
Montana Medium	22 @ 24	20 @ 21
Nevada	17 @ 20	13 @ 15
Colorado and New Mexico Fine	17 @ 18	15 @ 17
Colorado and New Mexico Medium	18 @ 19	16 @ 18
Coarse and Carpet	14 @ 16	13 @ 14
Black	14 @ 16	12 @ 13

TEXAS.

	Choice.	Average.
Fine Eastern	20 @ 22	17 @ 18
Medium Eastern	21 @ 22	18 @ 20
Fine Western	16 @ 17	14 @ 15
Medium Western	16 @ 18	15 @ 16
Improved Mexican	15 @ 16	13 @ 14

(SPRING.)

CALIFORNIA AND OREGON.

Spring Clip, Northern	20 @ 24
Spring Clip, Southern	15 @ 18
Spring Clip, low grades and burry	10 @ 15
Fall Clip, A 1	10 @ 13
Fall Clip, low grades and burry	9 @ 10
Valley Oregon, A 1	21 @ 23
Valley Oregon, A 2	21 @ 22
Eastern Oregon, A 1	16 @ 19
Eastern Oregon, No. 2	16 @ 19

GEORGIA, LAKE, ETC.

Georgia	22 @ 23
Lake	21 @ 22
Virginia Medium	26 @ 28
Virginia Coarse	20 @ 23

FOREIGN WOOLS.

Cape of Good Hope	27 @ 28
Montevideo	29 @ 31
Australian	34 @ 38

WOOL PRODUCTION.—The following table, prepared from estimates of Mr. James Lynch, of New York, shows the recent enormous development of sheep husbandry beyond the Mississippi:

Year.	Washed.	Rocky Mountains.*	Texas.	Southern.	Aggregate.
1867	140,000,000	11,000,000	7,000,000	2,000,000	160,000,000
1868	150,000,000	16,000,000	8,000,000	3,000,000	177,000,000
1869	134,000,000	17,250,000	7,000,000	3,000,000	162,250,000
1870	130,000,000	23,000,000	7,000,000	3,030,000	163,000,000
1871	110,000,000	25,000,000	8,000,000	3,000,000	146,000,000
1872	120,000,000	27,000,000	9,000,000	4,000,000	160,000,000
1873	125,000,000	37,200,000	9,000,000	3,500,000	174,700,000
1874	120,000,000	44,500,000	10,000,000	3,500,000	178,000,000
1875	125,000,000	52,000,000	12,000,000	4,000,000	193,000,000
1876	110,000,000	70,250,000	13,000,000	5,000,000	198,250,000
1877	117,000,000	70,250,000	14,000,000	7,000,000	208,250,000

*Including Pacific Slope.

The following record of the quarterly average prices of Ohio clothing wool (the best average product of American Merino grades), as sold in the Boston market during the last seventeen years, is furnished by Mr. George William Bond, of Boston:

Year.	January.			April.			July.			October.		
1860	$0.60	$0.50	$0.40	$0.52	$0.45	$0.40	$0.55	$0.50	$0.40	$0.50	$0.45	$0.40
1861	45	40	37	45	37	32	40	35	32	47	47	52
1862	....	....	....	....	....	*50	....	....	*47	....	....	*58
1863	....	....	*62	....	....	*76	....	....	*73½	....	....	*70
1864	....	....	*74¼	....	....	*79	....	....	*83½	....	....	*1.03¼
1865	1.02	1.00	96	80	80	75	75	73	65	75	75	65
1866	70	65	50	65	60	48	70	67	60	63	60	56
1867	68	53	50	60	55	50	55	49	45	48	46	40
1868	48	43	38	50	48	45	46	45	43	48	48	45
1869	50	50	48	50	50	48	48	48	47	48	48	46
1870	48	46	44	48	47	46	46	45	43	48	48	45
1871	47	46	43	50	52	47	62	60	55	63	62	58
1872	70	67	66	80	80	76	72	70	65	66	60	57
1873	70	68	65	56	53	48	50	48	44	54	53	47
1874	58	54	47	56	56	47	53	53	46	54	54	47
1875	55	56	47	54	52	46	52	49	46	48	50	42
1876	48	52	42	46	49	40	38	35	31	45	40	38

*Average price.

The Boston record of Ohio wool prices, from the same source, is, from 1840 to 1861, as follows:

Years.	Fine.	Middle.	Long.	Years.	Fine.	Middle.	Long.
1840	$0.45	$0.36	$0.31	1851	$0.41	$0.38	$0.32
1841	50	45	40	1852	49	45	40
1842*	....	....	....	1853	55	50	43
1843	41	35	30	1854	41	36	32¼
1844	42	37	32¼	1855	50	42	34
1845	36½	30	26	1856	55	47	37
1846	34	30	26¼	1857	56	47	41
1847	47	40	30	1858	53	46	36
1848	32	28	24	1859	58	47	35
1849	41	37	32	1860	54	47	37
1850	47	42	36	1861	45	45	50

*Price all round, 33½ to 35 cents.

These tables show that long wool (from the mutton breeds), has shared in the fluctuations of fine; has risen and fallen with considerable uniformity, when the latter has done so. The amount of fine or Merino wool produced in the world, since the settlement of Australia and the American Territories, has increased enormously out of proportion to coarse wool; yet the price of the former has nearly held its old percentage of superiority in the general decline. In other words, the addition of two hundred million Merinos to the world's flocks, has depressed the price of their wool very little more proportionately, than the addition of twenty-five million mutton sheep to the world's supply has reduced the price of long wool. If Merino wool can endure this vast expansion, and hold its own under it so well, what may we not expect of it in the future?

I will add a brief table, giving a comparative view of wool and cotton, showing that wool has declined little more in three-quarters of a century, than the great staple of the South:

Year.	Wool Price in Boston.	Cotton Price in Boston.
1801—'5...	....$ .37 @ $ .45....	....19 @ 23
1806—'10...	.... 1.00 @ 2.00....	....14 @ 22
1811—'15...	.... 2.00............	....10.6 @ 16.5
1816—'20...	No record.	....17.4 @ 33.8
1821—'25...	.... 60	....11.8 @ 20.9
1829—'30...	.... 38 @ 70......	....10.4
1834—'35...	.... 60 @ 70......	....17.45
1840—'41...	.... 46 @ 52......	.... 9.50
1845—'46...	.... 36 @ 45......	.... 7.87
1850—'51...	.... 41 @ 47......	....12.14
1855—'56...	.... 40 @ 60......	....10.30
1860—'61...	.... 45 @ 60......	....13.01
1865—'66...	.... 70 @ 1.02.....	....43.20
1870—'71...	.... 47 @ 48......	....16.95
1875—'76...	.... 48 @ 55......	......
1885......	.... 33 @ 34½.....	....11 9/16

CHAPTER IV.

BLOOD.

Blood, breeding, and feed, are the three great factors with which the wool-grower, by judicious combination, can work out success. Money will buy blood, but breeding and feeding require art, or at least skill. The superficial thinker, might therefore conclude at once, that blood is of less importance than either of the other elements. This view is erroneous. Blood is the outcome of the breeding and feeding of a hundred years—in the Merino, of a thousand, or for aught we know, of two thousand years. Hence, with money we can buy the labor and skill of thirty generations of men. Certainly it would not be the part of wisdom to neglect to do so.

In a race of high antiquity, whose characteristics have long been established, blood is of higher proportionate value than in a breed of more recent origin. In the latter, blood is of less value, except as it is of individual excellence.

FULL-BLOOD AND THOROUGHBRED.—In popular language these terms are synonymous. When used in reference to horses, there is a well-defined difference between them, which it would argue ignorance to neglect. Some writers seek to establish a difference also, when they are used in relation to sheep and in this way: A full-blood is one in whose veins there is no admixture or stain of any other blood but the Spanish, while a thoroughbred, is all that and something more. A sheep may be a full-blood (pure-blood would be a better term), and yet be so deficient in form or fleece, as to be unfit for a breeder. But a thoroughbred, is the outcome of a long line of ancestors, which, beginning with pure blood, have been so consummately molded by man to a special purpose, that this last and finished product is, so to speak, incapable of begetting or bearing a progeny different from itself. While these ought to be, and with accurate men are the definitions of the two terms, in popular usage they are not.

All lions, all tigers, all animals in a state of nature are full-bloods, pure-bloods, average types of their respective races; but not all of them are thoroughbreds; that is, not all of them are so even in all their qualities, and so sound in their constitutions, as to be able to produce progeny up to the level of the race-standard. They are weeded out by natural selection; they are ill-formed, or weak, or lacking in cunning, and they perish in the struggle of life, leaving the best individuals behind to perpetuate the race. Under a state of domestication in which man seeks to preserve all the individuals, good and poor, he must himself conduct this selection of his breeders.

Pedigree may have a very high value, or it may have none at all. If a sheep with an unbroken ancestry of a thousand years, or two thousand years, has a very poor constitution, or a bald head, it is more likely to impart those faults to its offspring, than if it belonged to a breed of more recent origin. It may, for this reason, be even less valuable in every respect, than a high-grade. Every official Register is seriously at fault which does not require individual merit, a "scale of points," as well as unquestioned purity of blood, as qualifications for registry.

Pedigree, is like a long train of cars; it runs with strong momentum, and it runs straight. An animal without pedigree, originating yesterday, is like a single car; it rocks to and fro, it is liable to swing off the track.

Breeders like to claim for their favorite stock, something akin to the Papal infallibility; they say, in effect: Given a thousand years' pedigree in your breeding flock, and you can not get an inferior animal. But this logic can not stand. Twin rams, twin bulls, own brothers in a family, disprove it every day.

Yet I would not be thought to detract anything from the transcendent value of pure blood. Often a grade of three-fourths or seven-eighths blood, sired by a strong-blooded ram, will to all appearance possess all the desirable qualities of a thoroughbred, and reproduce himself in his progeny; but the next generation, or the next, or at the first ill usage, his descendants will "breed back" to his low original. The thoroughbred Merino produces a fleece of very nearly the same length all over the body, while a grade only approximates this, and that, when young and full-fed. The thoroughbred fleece is almost uniform in strength and fineness all over the body, so that a great part of it can enter into the fabric of the same garment.

In 1882, Dr. A. H. Cutting, of the Vermont Board of Agriculture, made a microscopic examination of the number of fibers of wool on a square inch of the undried pelt of a sterile full-blood ram, slaughtered for the purpose. He reported officially as follows: "The mean result of all my experiments is, that there are two hundred and seventy-six thousand, four hundred and eighty pores to the square inch, from which wool may grow, but they do not all contain a wool fiber, as the fibers per square inch are two hundred and twenty-two thousand and three hundred. Of course, either of these is liable to a small error, but I compared this with the ordinary open-wool sheep, and find that there are about thirty on this pelt, to one on the common sheep; and yet I examined what would be called a good-wooled sheep."

Such facts as are above recited, explain and make reasonable the enormous price paid for very choice thoroughbred rams and ewes. They are based on great individual prepotency, coupled to a long pedigree. No ram, however faultless in form and fleece, and illustrious in descent, could justly be valued at one thousand or two thousand dollars, until he had given proof of his own powers of transmission by actual service as a stock-getter.

THE RAM IS MORE THAN HALF THE FLOCK.—It is customary to emphasize the necessity of a careful selection and purchase of rams, by the statement that the ram is half the breeding-flock. He is more than half, as may be shown.

It is a great general law of biology, that animals in a state of domestication, and more especially when ill-fed and cared for, have a constant tendency to revert to their original condition. The frequent reappearance of "jar" is one, among several proofs of this. Under this tendency, we can breed down from a thoroughbred to a "scrub," sooner than we can the reverse. A diagram will illustrate.

    Thoroughbred ⎫   Half-blood ⎫
     Scrub.   ⎬   Scrub. ⎬   Scrub.

That is, as the result of the second cross, we have, for all practical purposes, in a majority of cases, a scrub, where by the rules of arithmetic, we should have a quarter-blood. Hence, either the ram or the ewe, or both, must constantly be so selected as to *breed up*, else the progeny will steadily *go down*. Two and two do not make four—in breeding. Either the ram or the ewe must represent three, if we wish to secure a steady uniform result of four. It is more convenient, and generally less expensive, to get a ram of very high standard than it is to get a flock of ewes of the same standard.

Many farmers have an unjust prejudice against thoroughbred Merinos. The scarcity, and consequent high price of these animals, for many years, led to the perpetration of gross frauds. In many cases, sheep of low degree, by unscrupulous pampering and artificial preparation, were so embellished, as to be palmed off upon the unsuspecting wool-grower, as full-bloods. When they, or their progeny, were compelled to "rough it" a little (for the average wool-grower of the United States is not yet prepared to house and blanket his sheep all summer), they speedily collapsed, and revealed the cheat.

The full-blood Spanish Merino was exposed freely on its native mountains for a thousand years. The thoroughbred American Merino would do well on the deserts of the Far West, if only a plain, hardy type was selected, and judiciously acclimated. I formerly shared the general belief as to the constitutional delicacy of the thoroughbred; but experience has taught me that, if equally well-fed with the grade, it is equally tolerant of the severest weather.

The Texas or California flock-master, generally holds that he must stop with a three-fourths or seven-eighths Merino, for the hard life of the plains. Let him fight clear of wrinkles, and he will be perfectly safe with a pure-blood.

CHAPTER V.

BREEDING.

AT WHAT AGE TO BREED EWES.—There has been much heated controversy on this point—between those who believe that a ewe should bear her first lamb at the age of two years, and those who advocate three years as the proper time; since no breeder of Merinos would bring a ewe into service at the immature age of one year.

There are several points to be considered:—

First: It was shown in a preceding chapter, that it is of importance to the wool-grower, to have the largest practicable proportion of young sheep in his flock, because of their greater profitableness. For this reason, it is desirable to bring into service all the ewes suited for it, as young as possible, thereby to enlarge the crop of lambs, and enable the owner to constantly weed out all the sheep that have passed the meridian of profit.

Second: One of the greatest defects of the Merino ewe is her lack of fertility and prolificacy. It is a constitutional defect, to begin with, and it is augmented by very high and artificial keeping. There is no doubt that a ewe which passes her heat a considerable number of times without conceiving, is rendered thereby more uncertain as a breeder; she is less likely to become impregnated when at length brought to the ram. Now, a healthy, thriving ewe, will frequently come in heat before she is a year old; indeed, this event sometimes occurs at the age of six months. If, then, a ewe goes a whole year after her heats have begun, without conceiving, she is more likely to "miss" at coupling, than one which is brought to the ram younger.

Third: The ewe's fleece is affected unfavorably, both in bulk and in strength of fiber, by lamb-bearing; but this deterioration occurs equally, whether she bears her first lamb at two, or at three years of age. At any rate, the loss is greater in yolk than in wool.

Fourth: The fact is unquestionable, that the greater part of the lambs born from two-year-old ewes are smaller, weaker and harder to winter, than those born from ewes three-year-old. This is my own experience, and I think it will be corroborated by every observing flock-master.

Fifth: The fact that wild animals begin to reproduce their kind before they are mature, and yet the race does not degenerate, is no criterion for the conduct of sheep husbandry; for the weaker animals in a state of nature, are relentlessly weeded out by natural selection, by the struggle for life, and the superior ones are left to perpetuate the race.

Sixth: A ewe bearing her first lamb at two years old, will subsequently generally become a better milker and nurse, than she would be if she had been withheld from service a year longer.

From all these facts, the following rules may fairly be deduced: In the case of a large flock, especially if the supply of feed is somewhat scanty, the traveling required to collect it is considerable, and the development of the sheep tardy, it would probably not be advisable to breed from two-year-olds, unless it might be from a very few exceptional animals, attaining at two years the size and maturity commonly reached only at three. But in a small and well-kept flock, it would generally be good policy to breed a majority of the ewes at two years old (leaving a few of slower growth, a year longer), since, in this case, they would probably be as large and strong as the three-year-olds in a great flock. Still, the two-year-old ewes ought either to receive richer and more succulent feed than the older ones, or their coupling ought to be so timed as to bring their lambing season on grass.

CONSTITUTION.—At the best, the sheep is a weak and frail animal. As the French shepherds graphically say: "the wool eats it." When we consider the enormous product of fleece (in the best shearers running up to the wonderful figure of thirty-six per cent. of the live weight!), what wonder is it that such a draft on its system, weakens it? It is stated by Chauveau that the weight of the secretions and exhalations from the yolk-glands and sweat-glands, in the skin of a healthy sheep, exceeds all the evacuations from bladder and bowels together! Not even the hog, with his two hundred per cent. increase in fat, is so heavily taxed every day, as the well-fleeced sheep of the Merino breed. It is all the while literally sweating itself to death. It may almost be said of the Merino, as of the silk-worm, that the web it spins is its death.

How important, then, to choose for breeders only those sheep that have robust constitutions. Without constitution, the finest-fleeced sheep ever bred, is of no value as a lamb-getter, or lamb-bearer.

CHOOSE A SHEEP WITH THE FEWEST DEFECTS.—The art of selecting a sheep for a breeder is vouchsafed to very few men. As Darwin remarks: "not one man in a thousand, has accuracy of eye and judgment sufficient to become an eminent breeder." The vast majority of wool-growers, lacking the special gifts of Bakewell or Hammond, must be content to be able to choose fairly good, money-making animals. The great Vermont specialist might not find more than one ram in his State that would suit him; another man (and perhaps too, a man capable of making more money), would find a thousand.

The average wool-grower can not expect to ride a hobby, to "breed to points," as does the keeper of the stud-flock, the man with special gifts for the occupation. His great safety lies in selecting the animals that have the fewest defects, that are well and symmetrically developed. In a "Lecture on Breeding Merino Sheep," kindly sent me by the author, Henry Lane, Esq., of Cornwall, Vt., I find the following:

"The twin, three-year-old rams, exhibited and shorn by B. B. Tottingham & Son, Shoreham, at the public sheep shearing at Middlebury last spring, were in size and general appearance as near alike as twins generally are. The fiber of wool on one was four and one-half inches in length, on the other, three and three-fourths inches, a difference of three-fourths of an inch. The longest staple ram weighed, after shearing, one hundred and fourteen pounds, the shortest staple ram, one hundred and thirty-five pounds, a difference of twenty-one pounds. The longest staple, sheared twenty-three pounds and twelve ounces, the shortest staple, thirty-three pounds and ten ounces, a difference of ten pounds, less two ounces. This extra ten pounds came mostly from a much denser fleece. Now, you might possibly find one breeder in twenty, whose hobby was long staple, that would select the longest fleeced ram, but the other nineteen would select the one having so many good points, to be found in the shortest staple ram, and that were lacking in the other."

"FANCY."—"Breeders' fancy" is not wholly to be ignored. Wool on the leg is of no value, any more than cows' hair; but it is a point of breeders' fancy—and it is something more. It is a mark of blood, and therefore it is of high value. The wool on the upper eyelid (or, rather, on a fold of skin which doubles down over the eyelid, which fold is lacking in a plain sheep), is not only of no value, but it is a positive defect; but it is "fancy," it denotes blood, and therefore it is highly esteemed. The soft, silky face, without spot or blemish (the smallest black spot on the lip, face or ear, being objectionable), covered with wool to just such a point, making a cap rounding down with just such a a curve; the ears woolled out just so far, with white, silky hair the rest of the length (for a woolly ear is a bad mark); the precise number of wrinkles across the nose—all these are fancy points, oftimes sought after, to the neglect of substantial merit; but they are, nevertheless, matters of importance, because they are typical. *They ought to be there.* They show blood, culture, a "long descent;" they are like the almost invisible water-marks which the Government incorporates in the paper upon which bank-bills are printed, to prevent counterfeiting.

INFLUENCE OF THE SEX.—The question whether the ram or the ewe exercises the greater influence over the progeny, also whether one determines a different set of qualities from the other, is not of the slightest practical consequence to any wool-grower, except as considered under the following heading.

PREPOTENCY.—An animal of great force will impress itself on the offspring more strongly than one which is weaker. This power in an animal, whether male or female—for either may possess it—by which it marks its progeny conspicuously in its own likeness, is called *prepotency*. It is a curious fact that a fault, as, for instance, a deficient cap, or bare legs, will reappear in the lambs more persistently than a merit. It is customary to say that the ram is more prepotent than the ewe, but there are many exceptions to this rule. The ram's traits are the more generally remarked in a year's get of lambs, because he is one out of a hundred, chosen with the greatest care; but if a hundred lambs, equally divided as to sex, were suffered to grow to maturity and then used as breeders, it would be found that there were, out of fifty, as many prepotent ewes as rams.

But the important point is this: The ram costs less money than the flock of ewes, is oftener changed, and is frequently about the only item of expense which the farmer is willing to incur for the purpose of bettering his stock. More than that, if the ram is inferior, his faults will be reproduced many times, while in a ewe, they will be reproduced only once. Therefore, it is more important to make a careful selection of a ram than of a ewe, unless—which ought to be the case—the farmer is willing to exercise the same diligence in selecting all the breeding stock.

There is no criterion of prepotency except use. Pedigree and constitution, even form, may be present without prepotency; but the latter can hardly exist without the first two, though it may without form, since it sometimes happens that a thin-shouldered or steep-rumped or flat-ribbed ram is powerfully prepotent. Mr. George B. Quinn's "Red Legs," was an instance. The only safe rule, therefore, is to select a ram-lamb combining the three excellences, and then test him. A good way would be to buy no ram without some age, and a history behind him ; but to do this might be more expensive than to breed and test a number of ram-lambs for one's-self.

A notable instance of a potent ram perpetuating his high type and line of excellence for generations, is seen in Hammond's "Old Black," followed by "Wooster," "Old Greasy," "Old Wrinkly," "Little Wrinkly," "Sweepstakes," "California," "Gold Drop," "Green Mountain." Sanford's "Eureka" and "Comet," and R. J. Jones' "All Right," have each produced a valuable line of prepotent rams.

VARIATION.—There is a law of biology, that animals under domestication exhibit more variability, a greater tendency to "sport," in their offspring, than those in a state of nature. In the latter condition, they are subject to the same influences of climate, soil, feed, and the same habits of life, from year to year, and the law, " Like begets like," goes on without interruption. But when they pass under the dominion of mind, of intellect, the caprice of man, changes of ownership, changes of habits, of feed, of climate, interfere with the sway of heredity ; and even the most ancient race will now and then suddenly throw out a scion which is a remarkable departure from the type.

Thus "Sweepstakes," at a bound, surpassed his sire eight and one-half pounds in fleece, and ten pounds in carcass, and surpassed all his ancestors at least five pounds in fleece. "All Right" went beyond all his ancestors ten pounds in fleece, and at least twenty-five pounds in carcass—a remarkable variation.

Now, when such a variation in a desirable direction occurs, it ought to be carefully examined before we attach too much value to it. Does the great gain in fleece consist in yolk or in wool? If it is principally in yolk, the variation may possess little or no value. If it is a gain in pure wool, the animal is a great acquisition. Variations in a useful direction ought to be carefully followed up, for thereby comes improvement. Still, it is not well to expect too much, for an animal departing so far from the standard may not be able to carry his stock with him to his high pitch of excellence. He may be what the breeders term an accidental sheep. The keeper of the stud-flock would by all means retain this accidental sheep, in the hope that he might produce something equal to himself, and thus take one step toward making this variation permanent. But the ordinary wool-grower might make a mistake, if he bought him at a price greatly above the average. It is quite possible that one hundred lambs gotten by a ram shearing thirty pounds (analogous instances have not seldom occured), might not shear as high an average as one hundred lambs from a ram yielding twenty pounds.

It is the mission of the stud-keeper to experiment with these exceptional sheep; he may develop thereby a great public benefit. We look to him to hold up the Merino standard, and to advance it constantly higher and higher. But to the average flock-master, I would repeat and emphasize the advice, before given, never to place his main dependence on a ram which has not been tested.

CROSSING AND CROSS-BREEDING.—Prof. W. H. Brewer has so correctly given the general results which come from crossing, that I append two paragraphs from his writings: "I know of no case where a new breed has been made of *two* well-defined breeds, the new breed having the excellences of the others, or even the excellences of the first cross. It is a common experience, not only as you have shown with sheep, but with cattle, with horses, with *everything* so far as I know, that while the first or earlier crosses are reasonably uniform, successive crosses vary greatly. Numerous new breeds have been formed by the crossing of *several* older ones. Noel's experiments on the old French breeds of coarse-wooled sheep are interesting. The formation of the English thoroughbred horse from three or possibly more distinct branches of the Oriental horse; that is, the Arabian, the Turk, and the Barb. The Poland China swine, so called, from several earlier and perhaps ill-defined breeds, and so on. Numerous examples can be given of new breeds being formed from the crosses of several, and then by long-continued selection of animals having the desired qualities, from three several breeds; but I know of no example where this has been done with only two breeds in the original stock.

"Again, it is a common experience, particularly in breeding for flesh (but it is true of all characters), that in cross-bred animals for one or two generations, the cross breeds may be better as animals of use than either of the present stocks. But this excellence cannot be maintained with sufficient uniformity to insure profit. In truth, the whole and sole reason of the enormous prices which thoroughbred animals of various kinds bring of a long-proved pedigree, is not because of the superior excellence of those animals themselves as animals of use, but simply because their characters are transmitted, and that of equally good mongrels are not. The crossing of different breeds of sheep for mutton or for particular grades of wool, will long be continued, and is very profitable in many directions; but it is only profitable, so far as I have been able to hear, where these rules are obeyed, and we frequently go back to the pure breed on one side or on the other, or on both, for keeping up the excellence."

The celebrated Improved Kentucky breed was formed by the union of the native sheep with the Merino, the Leicester, the Southdown, the Cotswold and the Oxford-down. This is considered one of the established and permanent breeds of the United States, capable of propagating itself with the certainty of the average thoroughbred.

The even more celebrated Cross-bred sheep of New Zealand, are the result of a combination of the Merino, the Leicester, the Lincoln and the Cotswold. A recent writer in *Agriculture*, a London newspaper, after describing the breed, adds:

"What to the practical breeder is still more interesting, is the fact that efforts have been so far successful as to establish large flocks in which uniformity is as prominent a characteristic as in the average of either of the breeds from which they spring. At the time of my visit (December and January) these had all been sheared, so that I was unable to see the fleeces outside of the storage lofts, but from these I was enabled to procure samples seven inches in length and exceeding in fineness anything I had deemed possible from such a cross, while much of the lustre so esteemed in combing wools was preserved. This cross-bred sheep is now so 'fixed' in type that breeding animals are bought and sold with the same faith that they will reproduce with the uniformity attaching to the recognized breeds."

Crosses between the various branches of the Merino race have been, in many cases, eminently successful. In fact, the American Merino of to-day, in its incomparable excellence, is the result of a fusion of the Paular and Infantado flocks, together with others, which have become so blended as to be practically one race. The "Victor-Beall Delaine," the "Black-top," and the "Improved Saxony"—all of them worthy sub-families of the American, whose standing and excellence are established beyond the reach of cavil—are the result of crosses between the Spanish and Saxon. In the Eastern States, the attempted crosses between the French and American Merinos have generally resulted disastrously; but in Southern California, it has been and still is, widely popular. The French gives a rangy carcass, and long, but rather coarse fiber, while the American braces up this somewhat shambling anatomy, with a heavy bone, gives density and fineness to the staple, and a hardy, self-supporting habit, enabling the sheep, as the Western men say, to "rustle" successfully for its living. In Oregon and California the American and Australian have been bred together with highly satisfactory results. In the *American Sheep Breeder*, Hon. John Minto thus speaks of this cross:

"There was but little difference in the size of the two strains, and I think there would have been no perceptible difference in the yield of scoured wool of first quality in proportion to live weight. The Vermont sheep covered the shanks and head more and had much more oil in the fleece of a coarser quality. The Australians were more apt to give twins, and the lambs were more robust when dropped. After a few crosses none but the most expert could tell it was cross-bred sheep, the signs being preserved longest in the superior quality of the wool. One of our breeders who started with a few sheep of such a first cross, bred continuously toward the Vermont Spanish, and soon had a very uniform and very superior flock; but classed it as *pure* Spanish. He sent a card of beautiful specimens to the 'Centennial,' 1876, which the judges said was 'of very superior quality, much resembling Australian wool.'"

As to crosses between the Merino and the English breeds, it ought to be borne in mind, that there is a radical difference between crossing and cross-breeding, or amalgamation (*metissage*, as the French call it). The first is often profitable in special cases; the second is found nearly always to be a mistake. The mutton breeds are not adapted to the free and wide ranging of the West; they travel and scatter too much, and an admixture of their blood with the Merino, reduces the self-supporting power of the latter. Merinos are gregarious, while the English sheep desire to spread out widely. Another objection is, when this process of amalgamation is continued for any length of time, it destroys the uniformity of the fleece, the evenness of grade and density, which it is one of the foremost objects of the intelligent breeder to produce. Then, too, the English breeds cannot withstand the summer heat and the winter rains so well as the Merino; their fleeces are too open.

Where there is special demand for cross-bred wool or mutton, the unilateral cross (that is, with a pure-blooded race on each side) will always give more satisfactory results than the cross between mongrels. On the great plains of the West, where the grower of the Merino flocks desires to meet a special demand, and where the soil and climate are suitable, he may advantageously make one cross of pure Cotswolds or Downs on his Merinos (preferably a Merino ram with Cotswold ewes); but he should be careful to keep a reserve of pure stock constantly on hand, from which to make the cross afresh each year, and sell off all the cross-bred animals as fast as they reach maturity, without breeding from them.

The testimony of the most eminent wool-growers of the West, among whom may be mentioned as having spoken or written on the subject, Mr. Benj. Flint and Mr. M. D. W. Ap Jones, of California; Hon. John Minto, of Oregon; Mr. F. W. Schaeffer, of Texas, and Mr. D. H. McKellar, of Australia, is strongly against the amalgamation in this line. But in Nebraska, Minnesota, Dakota, etc., on account of their nearness to the great mutton market of Chicago, it is regarded with more favor, that is the unilateral cross.

The cross between a Merino ram and a Southdown ewe, or *vice versa* (though the latter is not so favorably regarded by practical men, since the Southdown ewe is a better milker than the Merino), produces the best mutton known, with a single exception perhaps, of the little Welsh mountain sheep. It is conceded to be superior to the pure Southdown mutton. The Merino, also, crosses kindly with the Shropshire and the other middle-wools. The Merino-Cotswold, if bred beyond the first year, is apt to have an uneven, streaky fleece; neither are the mutton qualities so good as those of the Merino-Southdown. A Cotswold ram bred upon a Merino ewe produces the most objectionable cross. The ewe has not milk enough for the large lamb, even if she is able to give birth to it; and it grows up leggy, light in the flank, gaunt and bony.

A Merino ewe once crossed upon a British ram is not likely to "breed true" thereafter from a ram of her own race; her progeny are apt to be marked with coarse-wool traits.

IN-BREEDING.—This, too, is a subject which concerns the practical flock-master very little, especially since the families and strains of the Merino have become so numerous, and afford such a wide range to choose from. In-breeding tends to refine the bone, impair the constitution, and induce sterility. Even the advocates of it virtually admit this, for they say cautiously, that breeding "too close" must not be continued too long. An eminent breeder once said: "In-breeding was sure to produce an uncommonly good, or an uncommonly poor animal." However necessary in-breeding may be in the hands of a great specialist, to establish and perpetuate a certain desirable trait, the average flock-master had better avoid it altogether. Still, there is very little doubt, that in-breeding between the closest relatives, which are widely dissimilar (as a very yolky and a very dry-topped one), might be less injurious to the progeny than the crossing of two sheep exactly alike, but not related. Hammond recognized and acted on this principle, in keeping separate and distinct his "dark or Queen line," and his "light-colored line," between which to take out crosses. In the old Saxon flocks of southern Ohio, in-breeding used to produce the the much-dreaded "kinky shoulder" wool.

WRINKLES.—No candid breeder will deny that wrinkles are a great nuisance to the shearer. Why else do the fancy breeders, the stud-keepers of Ohio and Vermont, have to pay from twenty-five cents to one dollar for the shearing of a single ram? I have known a good average shearer to spend nearly half a day in getting the fleece off of a very wrinkly ram. Of course, this is an entirely exceptional case; I cite it merely to show how obstructive wrinkles are to the shearer.

But there is another respect in which they are even more mischievous in a flock which is not housed. The wool between the wrinkles, and even the skin, in hot, wet weather, becomes parboiled; a quantity of rancid yolk accumulates there, and becomes a home for flies and maggots. A very wrinkly sheep is not fit to run in the rain.

A high-bred lamb, when born, has a fine, soft, spider-web crinkle in the skin, running all over the body, which disappears in full fleece; that is, it creates no ripple in the exterior of the fleece. It is still there, however, but gradually disappears with advancing years. But this sort of wrinkle, so far from being objectionable, is a point of merit.

A very wrinkly sheep is generally slower of development and maturity than a plain one. Thus, in a party of seven two-year-old ewes, Vermont-registered, owned by Mr. L. W. Skipton, of Washington County, Ohio, No. 159, a very heavily marked but

FOR WOOL AND MUTTON. 57

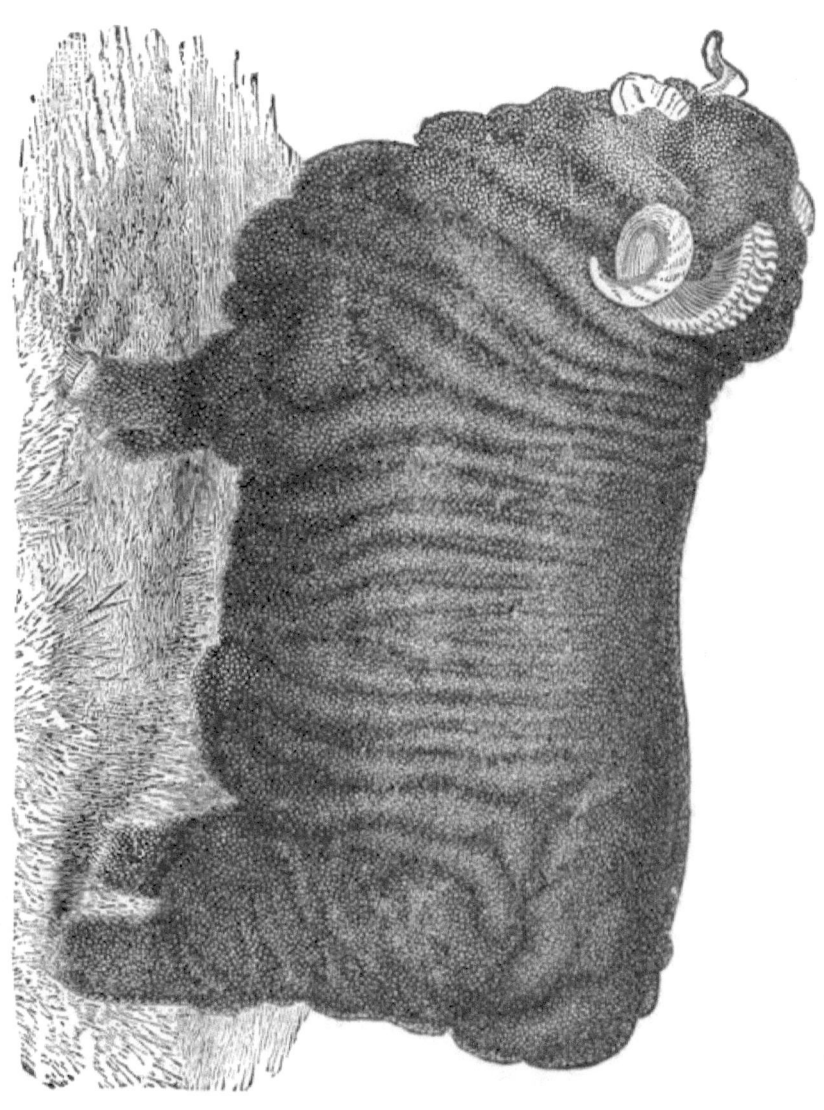

FIG. 5.—RIP VAN WINKLE.—A FAMOUS RAM, BUT OBJECTIONABLE ON ACCOUNT OF HIS WRINKLES.

undersized ewe, cut twenty pounds and two ounces; while No. 163, a plain ewe, but at least ten pounds heavier than the other, yielded twenty pounds and fourteen ounces. But the proprietor insisted that she had beaten the wrinkly ewe for the last time.

A very wrinkly sheep, seldom, if ever, yields a fiber long enough to be classed as delaine. This fact is illustrated by the high percentage of delaine wool in the clips of Messrs. R. and A. F. Breckenridge, of Brown's Mills, Ohio, who breed pure-blood Merinos notable for their plainness. On the other hand, Mr. C. C. Smith, of Waterford, is equally well-known for the heavily folded type of his pure-bloods; but, for his out-door wool-bearing flocks, he uses rams of medium wrinkliness, and he shears also a high percentage of delaine wool.

This fact, however, would not absolutely turn the scale in favor of a plain sheep over a wrinkly one, for delaine wool commands very little, if any, higher price than a clip of ordinary length, other qualities being equal, unless the prevailing type of wool throughout a large section of country is delaine. If there is a delaine "fashion," there will be a delaine price, but not otherwise, at least not until a time arrives when every clip is much more carefully graded and individualized than it is at present.

But there is a fact which works powerfully against wrinkly Merinos; and that is, that buyers and feeders discriminate against them, on account of the "sheepy" mutton which they yield. I am credibly informed that the butchers of Baltimore stigmatize them as "leather hides," and that prominent dealers, like S. Frankenstein and W. Finn, of that city, demand a concession of one-half to one cent a pound on very wrinkly sheep.

In sheep highly developed, the skin accumulates on the body to such a degree that a fold sometimes forms over the eye, and causes the eyelid to turn under and irritate the ball. This has been held to be a great objection to wrinkly sheep, but it can be remedied by a simple operation. With a pair of sheep-shears a piece of skin half or three-fourths of an inch long and a quarter of an inch wide (or narrower, if the case is not bad), is snipped out of the eyelid, lengthwise of the eye, and just far enough back from the eyelashes not to interfere with their roots. The cut being shallow, very little, if any, blood will flow from it. In healing the skin shrinks, and the eyelid is turned right side out again.

CHAPTER VI.

FEED.

Of the three departments of sheep husbandry, the most important is feeding. Blood and breeding may be compared to the field and line officers of the army, but feed is the common soldier. And, as in all well-regulated armies the officers come up out of the ranks by promotion, so, in respect to sheep; the blood of a thousand years, the longest and most celebrated pedigree, is after all, nothing but the outcome of good feeding. For under the term "feed" should properly be included care, management, choice of soils, etc.

The skillful breeder can select an animal with consummate insight into its good points (blood); but he cannot change its former fleece one iota except by combination with another sheep (breeding), and especially by care and management (feeding). In fact, Nature makes all the changes herself; man only supplies the conditions. Good feeding simply gives Nature or heredity a chance to do her best.

The coarse grasses and the roving, careless, "rustling" life of the Far West develop muscle, but weaken the fiber, and heredity tends to perpetuate these; but the rich pastures, the fragrant hay-racks and the faithful daily care of Ohio, produce a strong, uniform and yet fine staple.

A PERFECT FEED.—Nature has given us the formula for a perfect feed for the domestic animals. This is furnished in grass of different kinds, as seen by the following table, which gives the constituents of the principal pasture grasses and clover:

100 parts.	Water.	Albuminoids.	Fat.	Carbo-hydrates.	Woody fiber.	Ash.
Timothy..........	57.21	4.86	1.50	22.85	11.32	2.26
Orchard-grass....	70.00	4.06	.94	13.30	10.11	1.59
Barley-grass......	58.85	4.59	.94	20.05	13.03	2.54
Red clover........	76.00	2.00	...	7.00	13.9	1.00

The red clover and the orchard grass are rather watery, as every farmer knows; the best formula perhaps is that of the timothy. Counting the albuminoids as flesh-formers, and the carbo-hydrates (sugar, gum, starch) as fat-formers, and including with the latter half the woody fiber and all the fat (one pound of which is equivalent to 2.44 of starch or sugar), we have a total of 4.86 flesh-formers and 31.88 fat-formers. This gives an albuminoid or flesh-forming ratio of about 1:6.5.

Now, if the ingenuity of man could devise a feed as good as grass, it would leave nothing to be desired. The dried grass or hay is not such a feed; because, although we have abstracted from it nothing but the water, its magical quality is gone; it no longer possesses either the same nourishing or the same fattening properties.

Most of the nitrogen of feed is in the albuminoids, and both science and experience have taught that animals must have a certain ratio of nitrogenous matter to do well. But these nitrogenous matters are the most expensive element of feed (in most cases, though the commercial value is not always correctly adjusted); hence it behooves the farmer to know about what proportions of grain and hay it is most profitable to give to his sheep. If either the flesh-formers or the fat-formers are in excess of the formula above indicated by grass, the excess is practically wasted. It requires fixed amounts of oil and lye to make soap—to use a homely comparison—so it takes approximately measured proportions of the above named elements to feed stock advantageously.

THE BASIS OF FEED.—First, let us consider the basis or groundwork of feed, which, of course, is always some one or more of the varieties of hay, straw, etc.:

In 100 parts.	Water.	Ash.	Total organic matter.	Flesh formers.	Fat formers.	Crude fiber.
Meadow hay	14.3	6.2	79.5	8.2	41.3	30.0
Red clover hay	16.7	6.2	77.1	13.4	29.9	35.8
Pea straw	14.3	4.0	81.7	6.5	35.2	40.0
Bean straw	17.3	5.0	77.7	10.2	33.5	34.0
Wheat straw	14.3	5.5	80.0	2.0	30.2	48.0
Rye straw	14.2	3.2	82.5	1.5	27.0	54.0
Barley straw	14.3	7.0	78.7	3.0	32.7	43.
Oat straw	14.3	5.0	80.7	2.5	38.2	40.0
Corn fodder	14.0	4.0	82.0	3.0	39.0	40.0

Taking meadow hay first, we find that it has 8.2 of flesh-formers to 63.8 fat-formers, or an albuminoid ratio of 1:8, which falls little below the correct formula. And we know from experience that thoroughly good meadow hay will of itself many times support sheep in fair condition. But, taking wheat straw, we find it has an albuminoid ratio of about 1:29, which makes it a very poor article of feed.

GRAIN FEEDS.—To supplement this, we must have recourse to something richer in albuminoids. By far the most common grain-feed throughout the United States is Indian corn. But corn itself is not rich enough to make good the deficiencies of wheat straw. The composition of one hundred parts of corn is as follows: Flesh-formers, 10.0; total fat-formers (counting one part fat or oil equal to 2.44 of starch, sugar or gum), 75.8; woody fiber, 5.5; ash or mineral matters, 2.1; water, 14.4. In digestion the starch, sugar and gum are converted into fat or oil, and this, together with the vegetable oil existing in the feed, go to support respiration in the animal, and to the formation of fat. As to the woody fiber, it is known that a very considerable part of it is digested by an animal which performs mastication thoroughly, and has a vigorous stomach. The sheep will digest half or two-thirds of it.

What, then, do we find as the albuminoid ratio of corn? Flesh-formers, 10.0. Total fat-formers (sugar, starch, gum, oil=75.8; woody fiber, say 3.0; ash, 1.0), 79.8. This gives corn, therefore, an albuminoid ratio of 1:8, about. Now, let us add the wheat straw and corn together. We will suppose that one hundred pounds of corn are given to the sheep along with four hundred pounds of straw. Adding together the two terms of the proportion, we find the albuminoid ratio of the corn and straw together, to be 1:18.5. Manifestly this is too poor. The sheep must consume too large an amount of straw to obtain the necessary percentage of albuminoids, the excess of flesh-formers and woody fiber going to waste, that is, passing undigested.

The farmer in practice can partially remedy this by giving more corn and less straw, but he can do still better by using another kind of feed, for instance, Cotton-seed meal. The composition of one hundred parts of this is as follows: Water, 8.3; flesh-formers, 41.0; fat-formers, 33.4; woody fiber, 9.0; ash, 8.3. Of the fat formers, sixteen parts consist of oil, which is equivalent to 38.4 of sugar or starch; hence the total of fat-formers (adding half of the ash and woody fiber), is 64.5. This gives an albuminoid ratio of 1:1.57. Straw and cotton-seed meal together have an albuminoid ratio of 1:15, which is more nearly correct than that of the straw and corn. One hundred pounds of cotton-seed, therefore, has a higher value for feeding in connection with straw (or, for that matter, with any coarse feed), than one hundred pounds of corn. It will be for the farmer to determine whether it would not be good policy for him to sell his corn and buy cotton-seed.

VALUE OF ANALYTICAL TABLES.—All these calculations as to the value of different feeds and grains, however, are like the tables of values given for commercial fertilizers. The Government or State chemist works out the value in dollars and cents of so much nitrogen, so much phosphoric acid, so much lime, etc. But there is an infinite number of soils, and an infinite variety of circumstances under which these fertilizers are applied, all of which tend to bring different results and sometimes seem to belie absolutely the chemist's analysis. No judicious farmer will depend at all on these arithmetical values of hay and grain, except in this way: He will consult them to get a general idea of their comparative richness to start on, then he will adjust his feed-rations somewhat nearly as they indicate, and thoroughly scrutinize the outcome. No sea-captain will neglect his charts, but he will keep a most vigilant lookout for rocks and coasts and icebergs, nevertheless.

ROOTS.—Probably the most remarkable instance of the danger of depending on these tables without the "light of experience" to guide us, is afforded by the "nutritive values" of roots. The chemist gives a turnip such a high percentage of "water," and yet we know from practice that, when given in connection with dry feed, it has such a marked value, that it is a mere waste of space to print the "nutritive equivalents" of roots for the guidance of a practical feeder. When taken into the sheep's stomach, there is something in the water of a sugar-beet, as there is in the water of grass, which belies all chemistry.

The chief point of excellence claimed for roots is, that they supply the amount of water which all animals need when on dry feed, in a moderate and gradual way. If cut or pulped and mixed with bran, oats, or mill-feed, they furnish a soft, semi-liquid mass, which does not irritate the coats of the stomach, and does not overload it or dilute its solvent juices as a copious draught of cold water taken all at once would be apt to do. There is force in this argument. Sheep ought to be compelled, as much as possible, to eat their feed dry, as the saliva thereby secreted and mingled with it is of far more efficacy in assisting the stomach in digestion than any juice of roots, or any other moisture could be; still, it is undoubtedly injurious to the sheep to be obliged to drink at one time all the water it requires in twenty-four hours, especially if it is ice-cold.

Roots are not so necessary for Merinos as they are for the mutton-breeds; they are principally useful for ewes when giving milk, and for a short period before they begin. Sugar-beets, mangels, ruta-bagas, yellow turnips, white turnips, are valuable in the order in which they are here given.

MIXED FEEDS.—There are two cardinal principles in relation to mixed feed; first, that mixed feeds are better than plain; second, that all the elements of the mixture should be fed each day, instead of one element for one day or one week, and another for another day or week. Thus, for instance, the experiments at Rothamstead, England, showed that eight pounds of peas would make a pound of live weight, or six pounds of oil-cake meal; while, of peas and oil-cake meal mixed, four and one-half pounds would suffice. It is as an element of mixed feed that roots attain their greatest value. Thus, in a great majority of cases, it will be found that a sheep receiving three pounds of bright wheat straw, and six pounds of turnips per day, will increase as much in weight, or keep in as good condition, as another receiving three pounds of the best timothy hay; while the latter ration will be the less expensive of the two.

AMOUNT OF FEED PER SHEEP.—It has been ascertained that to keep a sheep in good thriving condition, fifteen pounds of perfectly dry feed (of average good quality), is required per week for each one hundred pounds of live weight. But since hay and grain, in their ordinary condition, contain about fourteen per cent. of water, from eighteen to twenty pounds per week will be necessary, or about three pounds per day. To facilitate digestion and prevent constipation, it would be well if an equivalent of this amount of nutriment could be expanded in bulk, so as to weigh seven or eight pounds.

PRACTICAL CORRELATION OF FEEDS.—In the following table I have given the albuminoid ratios of several varieties of feed, singly and combined, as a hint to the practical feeder:

	Ratio.		Ratio.	Together.
Corn	1:7.5	Meadow hay.	1:8	1:7.8
Oats	1:5.7	Clover hay.	1:4.2	1:4.5
Rye	1:5.2			
Barley	1:7.4			
Cotton-seed	1:1.6	Wheat straw	1:29	1:15.3
Linseed cake	1:2.3			
Wheat bran	1:4.2	Corn fodder	1:12.4	1:8.3
Shorts	1:5.8			

From this we see that oats and clover hay would be a combination too rich in albuminoids, some of which would be converted in the stomach into carbo-hydrates; consequently the farmer who should give these feeds together would not be pursuing an economical course, unless it was young lambs or ewes giving milk which he was feeding. For either of these classes of sheep it would be an admirable combination; but to mature stock-sheep, or even to fattening wethers, he would do better to give meadow hay, and with it corn, or cotton-seed, or linseed-cake meal. Another fact revealed by this table is, that wheat bran is better, weight for weight, for young lambs and suckling ewes, than corn meal—so much better that, in most cases, it would probably pay the farmer to exchange his meal for bran, even if he had to transport it some distance.

If the bran has been ground by the "new process," it is not so rich in starch as old-fashioned bran was, but richer in proteine. Linseed-cake meal, owing to improved machinery and the higher degree of pressure now employed in extracting the oil, has only six to nine per cent. of oil (sometimes only two and one-half), where it used to have ten to twelve. But this actually increases its albuminoid ratio (which gives it its great value); does not diminish the proportion of mucilage and digestible fiber; and the hard-pressed cakes keep better than those which were more loosely made under the old process.

EXPERIMENTS IN FEEDING.—But, after all that the most careful scientific investigators may ascertain for our guidance, there is nothing equal in value to actual experience, what might be called the testimony of the sheep. M. Moll, a noted French writer, thinks fine-wooled sheep reach their best estate in the region of the vine and the mulberry. In America, I would substitute for this the latitudes adapted to Indian corn. This is to the Western farmer what the turnip is to the English shepherd.

My experience for years in feeding sheep on fodder (which is better every way than fodder-corn, except for nursing ewes), has given me the highest opinion of its value for this purpose. The silk-worm-like closeness with which they pick every shred of the foliage from the canes obviates the necessity of cutting the stalks, which is an operation of dubious profitableness with the coarse Western corn. Besides that, fodder has a most admirable effect on the respiratory and circulatory systems. A horse may often be cured of a mild case of heaves by the substitution of fodder for hay in his manger. Thus, by quickening and stimulating the circulation, fodder is a better feed than hay for increasing the wool product. It is more laxative than timothy or any other hay, except clover. A perfect ration for sheep should include at least one daily feed of bright fodder; it is far preferable to rye and (bearded) wheat straw, the beards of which are liable to cause great irritation to the coats of the stomach.

One winter I fed a flock of two hundred and twenty-five young sheep, mostly yearlings, one and a half bushels of shelled corn and eighteen bundles of fodder per day. With a run of two or three hours a day on an old sod, they wintered remarkably well. The current local price for fodder is ten cents a shock of eighty hills. Planted in rows three feet ten inches apart, there would be thirty-three and four-fifth shocks to the acre. Four bundles make a shock. The fodder on an acre is worth $3.38. The flock consumes forty-five cents' worth per day. Of corn, at forty cents a bushel, they require sixty cents' worth per day. They are fed, say, four and a half months (this will allow for the diminished ration at the beginning and end of the season). This will make their winter's supply of fodder cost $60.75, and their corn $81; total, $141.75. This flock would have required three hundred pounds of hay per day, which, at $10 a ton, would be worth $1.50. Against this, the daily ration of fodder and corn cost $1.05.

The best way to employ wheat straw for sheep is in connection with fodder, details of which are given in the chapter on "Winter Management."

Mr. Arvine C. Wales, of Stark County, Ohio, who grew annually about seventy acres of fodder corn for sheep, gave in *The Shepherd's National Journal*, the following experience:

"I selected out three hundred ewes, and divided them into two lots as evenly as I could. One hundred and fifty were put into one shed, and a like number into a shed near by. Between the two sheds there was a set of stock scales. Each lot of sheep was carefully weighed at the beginning of the experiment, and were weighed again each week for eight weeks. During the continuance of the experiment I was asking questions for my own information, and had no interest except to get at the truth. One lot of sheep was fed with one feed per day of good clover and timothy hay, and one feed of sheaf oats. The other lot received two feeds of the fodder corn, which I cut up by horsepower, mixed with a bushel and a half of bran. I poured hot water over it and let it soak from morning until evening, or from evening until morning. I have not the minutes of these experiments before me, but I remember that the sheep fed on the fodder corn showed a marked gain over the other lot. The dung was about of the consistence of that of sheep on dryish pasture. They drank very little water, and I thought the growth of wool was healthier and stronger than that on the other lot."

In the Far West the question of a feed-supply is becoming yearly of more pressing importance. Those vast regions are fitted for pastoral pursuits; they will be the stronghold and refuge of the American Merino. But some artificial provision of feed must be made for the occasional heavy snow-storm, or else the wools of those regions will continue to be "jointed," untrue, unfit for combing purposes, and falling ten to twenty per cent. below the price they might by good management be made to command.

Prairie hay is generally excellent from most localities; not surpassed by that made from cultivated grasses. These natural meadows can be cheaply enclosed with wire and iron posts. Great hay-barracks, like those of Northern California, holding two hundred and fifty to three hundred tons each, could be made with iron roofing and siding. If not filled in one year, they could be in two or three; then they would be ready for an emergency.

Alfalfa is full of promise to the Western flock-master. It was the growing of Alfalfa in California which checked the flow of sheep from that State to Colorado. A hundred days' feeding on Alfalfa, with a half-pound of oats per head, daily, makes very fat sheep and exceptionally sweet, tender mutton. It will completely remove from the flesh the flavor of the Black Sage, and other offensive shrubs and plants of the West. It is sometimes slightly productive of scours and hoven, if allowed to grow too rank before the sheep are turned on it; but lumps of rock-salt kept constantly within reach of the flock, have been found in California to be a preventive of these troubles. Bermuda grass, so common and so dreaded by the cotton-planters of the South, has been found to succeed in alkali soil, even where the deposit was very strong—and this grass is admirable for sheep.

THE SHEEP NEEDS MINERAL MATTERS.—The art of feeding takes account of all that the sheep requires to promote its health and growth. Not only the feed, but the water must be considered. The sheep needs a large proportion of mineral matter, either in its feed or in the water. Five per cent of clean wool is sulphur; two per cent of the sheep's urine is mineral; thirteen and one-half per cent of the dung is mineral; the bones contain sixty to seventy per cent of phosphate and carbonate of lime; the yolk has a large proportion of potash, and the flesh and blood contain the following mineral substances: Phosphorus, sodium, potassium, chlorine, magnesia, iron and lime. The bones of the sheep contain eleven per cent more of the carbonate of lime than those of the ox, five per cent more of the phosphate of lime, and a fraction more of magnesia, lime, potash, etc. This shows the necessity of supplying the sheep with mineral matters. Soft water is not so healthful and nourishing to a flock as hard water. The healthiest flock I think I ever saw was one on the Nascimiento River, in California, which had no water to drink during the six months' drought of summer, except that from some strong sulphur springs in the bed of the river. Ashes, lime, sulphur, copperas, in small quantities, form an excellent addition to their salt. On the great alkali plains of the Far West sheep frequently have to be kept away from alkali ponds and deposits; they eat so much of it as to do them harm, though perhaps they would not, if allowed to visit them often and regularly.

CHAPTER VII.

PASTURE IN THE WEST.

In subsequent chapters detached notes are given on the grasses peculiar to the regions west of the Mississippi; but it may be well to present here a condensed view of the various grasses, plants and shrubs, with which the shepherd has to do throughout the United States.

The grasses of the East, both wild and cultivated, generally form an unbroken turf; but those of the remote West seem to protect themselves against the great aridity of the climate by growing in bunches or tufts, with the spaces between either naked, or overgrown with weeds. Hence the popular name, "Bunch-grass," which is extremely indefinite, describing nothing except the habit of growth, and applied to several different species. The grasses of the great arid region are scanty, but are mostly sweet and nutritious, both for summer and winter grazing. A considerable part of their value arises from their greater richness in seed than the Eastern grasses. The winter aridity is so great, generally, and the stems of the grasses so stiff and strong that, when touched by frost, they do not become broken down by the rains and snows to decay on the moist soil, but stand firmly on the ground all winter long and "cure," forming a sort of uncut hay.

To a considerable degree, the greater or less abundance of the grasses is dependent on latitude and altitude; the higher the latitude, the better are the grasses, and they improve as the altitude increases. In the mid-continental region, in low altitudes and latitudes, the grasses are so scanty as to be of little or no value; here the true deserts occur. Also, on the Atlantic slope of the Rocky Mountains (especially the eastern part of it) and on the Pacific cost, the grasses are coarser than those of the great central basin. In the prairie section, proper, the native grasses are coarse and rough; they have to be kept closely depastured, or they become unpalatable to sheep. They form a very fair article of hay, often scarcely inferior to Eastern hay; but the leaves are so harsh that they frequently give sore mouths to the sheep. These grasses are also rather susceptible to frost; in the latitude of Southern Kansas they are generally cut off by November 1, thereafter becoming "dry feed."

On the Pacific Coast, several of the more important grasses are noted for their rankness of growth and their prolificacy in seeds, these constituting a large share of their value. Furthermore, being annuals, they depend on their seeds for propagation, and the consumption of these by sheep curtails their volume from year to year.

The finer and sweeter perennial grasses which are more characteristic of the central continental region are better suited for sheep, though the available pasturage areas are much lessened by low alkali and sandy deserts, on the one hand, and by rocky, broken mountain chains, on the other.

East of the Mississippi the Blue-grass or June grass (*Poa pratensis*) is, of all the grasses, cultivated or self sown, the best for sheep, and it is likewise the most widely spread. It yields a scanty crop of hay, but it grows so early and so late, makes so tough a sod, is so rich and so eagerly sought after by sheep, that it is always good policy to allow it to take possession of a considerable part of the land devoted to pasture. *P. compressa* is the Wire-grass; in Ohio it is better for pasture than the June grass. *P. annua* has become naturalized in many parts of the South. In Northern Texas it becomes a bright green in the fall, withstands the cold better than other grasses; in February and March makes a strong growth and furnishes good grazing when no other grass does; but it does not resist drought well in the summer. It is highly relished by sheep, but is so small that it is chiefly valuable in the winter. *P. serotina*, False red-top, is found in Oregon and the Rocky Mountains; eastward to the Atlantic it is common where the soil is moist. The Kentucky Blue-grass is becoming common in California and the Rocky Mountains. *P. alpina* is found in the Sierra Nevada and the Rocky Mountains, nearly up to the line of perpetual snow. *P. tenuifolia* is one of the valuable bunch-grasses of the West.

Timothy becomes naturalized wherever the soil and climate permit; it is already sown and is common through the mountains and in irrigated districts quite across the continent. In the Far West it is too valuable for hay to be grazed much; sheep would destroy it.

Buchloë dactyloides is the celebrated Buffalo-grass, one of the most nutritious grasses of the West. It is short, curly, and increases by runners as well as by seed and root. In Texas it is sometimes called "Vining Mesquite." It belongs to the dry and elevated plateaus from the Rocky Mountains to Kansas, and from Mexico to British America.

Bouteloua oligostachya is the true Grama-grass, so called from Texas to Arizona, Southern California and Colorado, but known under other names north of Colorado, as far up as Montana. *B. hirsuta* is the Black Grama, with about the same range as the above, but not reaching California. Both are of the highest value, especially in Southern New Mexico and Arizona, where the "Grama Belt" is celebrated for fattening stock like grain.

Andropogon scoparius, Sage-grass, Broom-grass, forms twenty per cent of the pasture in parts of Kansas, and ranges east of the Rocky Mountains, as far south as Texas. *A. furcatus*, Blue-stem, or Blue-joint, constitues forty per cent of the pasture in some places in Kansas. These and others are good fattening grasses.

Stipa occidentalis is a common bunch-grass in the Sierra Nevada, and *S. comata* in Montana. Both are valuable for sheep.

Munroa squarrosa is a low, nutritious "Buffalo-grass," in the north, and a "Grama-grass" in Texas. It originally covered tracts of thousands of acres together on the northern plains.

Festuca is a very large genus, of much value, following sheep and cattle wherever they go. *F. ovina*, Sheep's Fescue, is common in the cooler parts, from California and Montana to the Atlantic; it is one of the best of grasses for sheep, and has followed them around the earth, even to New Zealand, Australia and Tasmania. Among the native species of value are *F. occidentalis*, in Oregon, and *F. scabrella*, from California to the Rocky Mountains.

Hilaria rigida, known as "Gallotte" or "Galletta" (perhaps the same as the "Gietta-grass" of Arizona), forms a large part of the pasture on the semi-deserts of San Bernardino County, California. It is a hard grass, but valuable.

Calamagrostis Canadensis is perhaps best entitled to the name "Blue-joint," of all the various grasses so-called. Its range is rather northerly and mountainous, from California to the Atlantic. It is a favorite grass for hay, and it stands so erect in winter that it is one of the chief supplies for sheep in a deep snow.

Eriocoma cuspidata, the Sand-grass of Utah, is a very nutritious, valley bunch-grass.

Sporobolus airoides, known in Utah as "Vilfa," is a lowland grass, remarkable for its power of taking up alkali, which gives the whole plant a salty taste. Cattle are injured by it at first, but sheep not so much.

Aira cæspitosa, a red-topped grass, is found in the arid region, surrounding the small lakes and tarns, sometimes at an elevation of eleven thousand feet; it forms a continuous sod, and is a very beautiful and valuable grass.

The chief grasses of the elevated timber tracts belong to the genus *Bromus*; when young they are tender and good, but with age they become tough and worthless.

Agrostis vulgaris, common Red-top, follows cattle and sheep in the cooler regions; always valuable.

Atropis tenuifolia is one of the most valuable bunch-grasses from San Diego to Oregon, and Colorado.

Hordeum murinum, the odious Squirrel-grass or Foxtail-grass of California. When the heads ripen they break up and the barbed seeds and awns work into the wool, and even into the flesh of sheep, in some places killing many lambs; they get into the eyes and injure or destroy the sight. It is one of the worst vegetable pests of California.

Muhlenbergia gracillina is another of the species commonly, but incorrectly, called Buffalo-grass; it is a low, nutritious grass, common on the plateaus from Colorado to Texas.

Of clover (*Trifolium*) there are five species east of the Mississippi, while west of it there are some forty, of which twenty-five are found in California alone. Only one (*T. Andersonii*), is not eaten by stock. The burs of the bur-clover (*Medicago denticulata*), furnish sheep a large amount of rich, oily seed, but they work into the wool and compel flock-masters to shear in the fall to get rid of them. Indeed, it is believed that sheep have thus been the vehicle for its dissemination in regions where it was unknown before.

Alfileria or Pin-clover (*Erodium Cicutarium*), is a valuable forage plant which follows cultivation. It furnishes a heavy swath on lowlands; the hay is black and becomes much broken and chaffy, but is very nutritious.

Over many of the drier sections of the interior, various shrubs form a notable feed in the winter. Prominent among these is the celebrated "White Sage," or, as it is sometimes called, "Winter Fat" (*Eurotia lanata*), which ranges from the Saskatchawan to New Mexico, and from the Sierra Nevada to the Rocky Mountains. After frosts come its quality is improved as is true of other shrubs of the same order (*Chenopodiaceæ*), and it is a valuable winter forage in many places in the Great Basin. Other species are here and there called white sage, but this is the one *par excellence*.

The name "Greasewood" is applied to a considerable number of plants. The most common ones, however, are the *Sarcobatus vermiculatus* and *Abione canescens*, both more or less thorny shrubs and looking most unpromising as forage, but which nevertheless have considerable value. *Purshia tridentata* is also widely known as greasewood, and is eaten by stock, and so are a number of other species less common and of less value.

The Mesquite (*Prosopis juliflora*) grows as a shrub or small tree on the dry slopes and mesas from Texas to California, and produces a crop of sweetish pods, four, six or more inches long, and each containing numerous bean-like seeds. Both the pods and the seeds are eagerly eaten by stock, and are very nutritious.

"Sage" is a name given by the early mountaineers to the shrubby species of *Artemisia* found so abundantly from the plains to the Pacific. There are many species of this genus, bitter, strong smelling, and belonging to dry regions. But the name has come to have a wider use among stockmen, and besides the white sage we have yellow sage, red sage, black sage, rabbit sage, etc., applied to various species of shrubs, some of which are eaten by stock in extremity, others more willingly, but taken as a whole there is not much dependence upon browse feed, except with the white sage, although in many places it forms an element not to be entirely ignored.

CHAPTER VIII.

A MUTTON MERINO.

One of the great needs of American diversified agriculture is, a general-purpose sheep. There are few farms in the United States which would not be the better for having some sheep upon them. They eat the refuse feed, and they manure the ground. Sheep manure, on account of its richness in silica, will make wheat grow stout and short, with heavy heads, where other manures produce long, soft straw, and not so solid heads. It is also excellent for corn. Where clover can be started and pastured by sheep, or fed to them, almost any worn-out land can be reclaimed in a few years.

The English sheep in America stands on a Merino platform. The sheep of the United States are ninety-five per cent Merinos; and they sustain the great wool and mutton substructure of the country, on which the British sheep can stand and show an extraordinary profit from the sale of early mutton lambs. He who sells lambs cuts the throat of his flock. The people of England for several years past, have had to abstain from lamb-mutton, in order to rebuild the wasted flocks of the Island. Relegating to the Merino the great foundation-work of sheep husbandry—mature mutton and wool—the breeder of the British races in this country, working on a vast body of cheap Merino ewes, devotes himself to the exceptional and necessarily suicidal industry of rearing early lambs, and makes exceptional profits.

But what does the British sheep do when standing on his own

platform? America is a Merino country (and always will be for that matter), while England is a mutton country; yet, according to that excellent authority, Mr. H. Stewart, the average annual revenue derived from a sheep in England is $1.17; in the United States, $2.16! This, too, in 1866, before the bars of the tariff were put up.

The Cotswold and the Downs are summer sheep, but the Merino is a dry feeder. Prof. J. W. Sanborn fed twenty-three Cotswolds and eleven Merinos for fourteen days, at two different times. The first time the Cotswolds ate three per cent of their live weight, daily, and the Merinos 3.9 per cent. The second time, the Cotswolds ate 2.8 per cent of their live weight; the Merinos 3.5. In a letter to myself, he says simply: "Both gained similarly." This does not matter, however; the greater gain and value of wool in the Merinos doubtless compensated for their larger ration. The main point is—the Merino is the better dry feeder.

Mr. W. D. Crout, a feeder of many years' experience, says in the *Ohio Farmer:* "I feed different classes of sheep almost every winter, and find that no other sort takes to feed so kindly, and fattens so rapidly * * * * * If I have long-wooled sheep to feed, I invariably turn them off early in the winter; but I believe I have never been fortunate enough to escape having some culls from coarse sheep."

In the winter of 1882-3, Hon. William G. Kirby, of Kalamazoo County, Michigan, fattened about one thousand wethers, of which a small number were Merinos. They were shorn in April and sold for the English market. The Merinos brought the following amounts per head:

13¼ pounds wool @ 33 cents..................... $4.54
130 pounds mutton @ 6¼ cents..................... 8.12
                                                 ―――――
                                                 $12.66

The account, in the *American Sheep-Breeder*, simply adds: "The weights attained by the Merino wethers as given above, though exceeded by the larger mutton breeds shearing comparatively light fleeces, were heavy enough to bring the top price, and in Mr. Kirby's opinion were grown and fed at a greater profit than any other of the one thousand head, which numbered equally choice specimens of the mutton breeds."

In the winter of 1874-5, Mr. O. M. Watkins, of Onondaga Co., N. Y., fed two hundred and ninety sheep, of which one hundred were Merinos, and ninety were Cotswolds. During

January a record was kept; the Merinos gained seven and one-quarter pounds each, at a cost of 8.4 cents; the Cotswolds five and one-quarter each, at a cost of 11.6 cents per pound. These animals all received the same amount of feed per day, and it was all dry feed.

The explanation is, that the English sheep are best adapted to the damp climate, the juicy turnips and the shade-cured hay of England; the Merino to the hot, dry climate, the oily corn and the sun-dried hay of America.

No one disputes the remarkable precocity of the English breeds. A Hampshire-down lamb on its native grass near Salisbury, has increased eight-tenths of a pound daily, for a good many days together! But the breeding of early market lambs is an exceptional, extravagant and necessarily suicidal industry. Only one man in a thousand can afford to eat spring lamb; the vast majority of mankind who eat mutton at all, must be content with the mature flesh. And for nearly half the year in America, if not the whole year, mutton can not be made so profitable in the large way (body and fleece taken together), from the English breeds, as from the Merino.

We want the English breeds near our cities to furnish spring lambs, and long, coarse wool, and root-fed or grass-mutton, for export to England; but the Merino will never cease to supply most Americans with their corn-and-hay-fed mutton.

The assertion that first-rate chops and roast can not be cut from any but an English carcass, is old and wornout; and, moreover, wholly unwarranted. There is only one genuine mutton-sheep worth considering, and that is the Southdown, whose wool is comparatively fine. The coarser the fiber of the fleece, the coarser the grain of the mutton. The heavy, loose-wooled Cotswold and Shropshire produce mutton, as Lord Summerville says, "fit for such markets as supply shipping and collieries"—ham-fat and thick on the rib.

The mature American Merino, with its fine-grained flesh, when it has been properly fed and butchered, yields chop, boil or roast, second only to the Southdown, if, indeed, it is at all inferior. The superiority of the Southdown, if it has any, consists, not in the sweetness and tenderness of the flesh, but in the thickness of the hams and the "marbling," or the distribution of fat among the lean.

The idea that the wool gives taste to the flesh, either by its growth before butchering, or by its touch in the butchering, or after, is a very old one, but it is erroneous. The flesh of the sheep partakes of the flavor of its feed more than does that of the steer or the hog; and the milk still more perhaps. But all the apparatus of glands and tissues, for the manufacture of wool, is situated in the skin, and all its deposits are made there, without affecting the flesh.

The disagreeable "sheepy" flavor is imparted to meat by age, by bad feeding (or no feeding at all), by wrinkles, and by delay in the removal of the viscera. Let a plain sheep be properly managed from birth to butchering, and the entrails be taken out with neatness and dispatch, and the carcass may be wrapped in the skin without detriment, barring the uncleanliness. From the enormous preponderance of the breed, the much-decried "Merino taste" is the scapegoat for all the bad feeding and worse butchering of the country. A sheep may yield the best flesh of all the domesticated animals or fowls—or the worst.

A cry comes up from the Territories and from Texas, that they must have a larger carcass, "more mutton and more wool on fewer legs." These men do not correctly perceive what is wrong with their Merinos. It is not size they lack, so much as quality. The sheep of Texas "kill red," as the butchers say. Then they "cook red;" they will not brown in the oven; they are the despair of the French *chef*. The sheep that "rustles" is muscular, he is gamy, though not necessarily "sheepy." He is never fat enough for thoroughly good eating, even when feeding on the best Montana Bunch-grass or the famous Grama of Texas. And when he is forced to live awhile on the Black sage of Nevada, or the Nopal cactus of Texas, or the Broom-sedge of Georgia, what can we expect?

Then, too, in the Far West, the value of sheep heretofore as a wool-producer, has oftentimes caused the flock-master to keep his wethers until they died of old age or abuse; whence has arisen the delusive maxim of that section of the United States, "old sheep for mutton." But the tables given by Mr. Randall, referred to in a previous chapter, show that Merinos, both ewes and wethers, whether for wool or mutton, or for a combination of the two, have passed their meridian of profitableness at the age of two years. In other words, young sheep produce not only the best mutton, but also the best and, proportionately, the most wool; so that the producer of both articles can give his customers the quality they want as to age, and at the same time promote his own interests. And when to the requisites above noted is added the others, **viz: that the animal shall be free** from wrinkles, and be slaughtered under four years of age, we shall have Merino mutton of an unexceptionable quality.

In the still, deep pastures of Ohio, or fed on corn, oats and bran, with bright fodder, hay and straw (add roots if you will), the flesh of the American Merino has juice and flavor. It will take on the color in the oven which is the delight of the gourmet; and it will enrich with gravy the plump, brown potatoes which encircle its base, round about. Mutton is as much responsive to culture as music. What the flock-masters of the Territories need is, to round out the fattening of their muscular, gamy "muttons" with a few months', or at least weeks', feeding on hay and grain, in a field or corral. The leggy wethers of the plains would have to be broken to quiet gradually; but a reasonable period of rest and feed will develop in their flesh that fat and juice, which the constant walking of their previous lives had dried up.

A fat, smooth carcass, weighing eighty-five or ninety pounds, will sell only an inconsiderable fraction, if at all, lower than the one of one hundred and ten, one hundred and twenty or one hundred and thirty pounds' weight, in the Chicago Stock Yards. In England, and also in New York, with its English tastes and prepossessions, a difference of about a cent and a half per pound of live weight is made in favor of Southdown, over Merino; sometimes even two cents. But in liberal, cosmopolitan, (or, rather *American*) Chicago, there would usually be a difference of only three-eighths to one-half cent a pound between pure one-hundred-pound Merino wethers and one-hundred-and-twenty-pound Southdowns. In a letter to me, Mr. A. C. Halliwell, of the Chicago *Drovers' Journal*, says: "* * * * * * In this market it may be well to state that the demand for light and heavy sheep varies; largely owing to the season. When lambs and choice 'handy' carcasses are scarce, it often happens that Merino grades sell higher than coarse-wools."

The reader is referred to the report of sales in the *Drovers' Journal*, of Chicago, quoted in a previous chapter, showing how almost completely quality rules the market, instead of size. I will add at random a few comments made by the reporter of the above journal:

"These Western sheep are among the best mutton sheep that now come to market. A few years ago, Oregon and Nebraska sheep were among the most objectionable that came. The reason was, they were large-framed, weighed about one hundred to one hundred and ten pounds, and were seldom more than half fat; now they have the same frames, but twenty to thirty pounds more of good, solid meat on their bones, put on with good western corn and hay.

"This market could use a very large number of fat Texas sheep every day. In fact, there is very little danger of crowding the market with fat sheep suitable for making mutton.

"Ship fat sheep as fast as they can be gathered; and do not ship lean, scrawny lots at all.

"Buyers do not show any partiality * * * * * They look at the condition of the animal. If heavy, round and fat, they pay top figures, no matter what part of the country the sheep hail from, or what breed or mixture."

Both the Merino fleece and skin are superior to those of the Southdown; so, if both were sold in the wool, in Chicago, there would be practically no difference between them.

But, if the butchers or the fashion should insist on having a carcass as large as the Southdown, the American Merino is capable of supplying it. A direct descendant of the famous ram Sweepstakes, in Fulton County, Ohio, weighed when in high condition, two hundred and five pounds. No ram is admitted to record in the "Victor-Beall Delaine Merino Register," weighing under one hundred and fifty pounds, and no ewe under one hundred pounds. Mr. H. R. Pumphrey, of Licking County, Ohio, had at one time twenty-one full-blood American Merino yearling ewes, which weighed one thousand eight hundred and ninety-five pounds, or a trifle over ninety pounds apiece; and twenty-four ewe lambs weighing one thousand six hundred and seventy-five pounds, or a little less than seventy pounds apiece. E. M. Morgan, of Champaign County, Ohio, reared thirty-eight lambs from thirty-five ewes, and at the opening of winter, these lambs averaged seventy-six pounds. His yearling wethers, "including three dry ewes," averaged one hundred and twenty-two pounds. A lot of full-blood wethers, two years old, were sold in 1881, in Jefferson County, Ohio, which averaged one hundred and twenty-two and a half pounds with the wool off. The Merino wethers of Michigan and of Washington County, Pa., often average one hundred and twenty to one hundred and twenty-five pounds with their fleeces off.

Butchers and buyers seem to have a sort of prejudice in favor of a sheep that will reach the even hundred pounds; and I do not think the breeder of pure-blood Merinos will find his highest profit in aiming much above that figure. In a Shorthorn steer, we seek the greatest amount of beef in a single animal;

but with the sheep, which yields as a collateral product that "staple of endless values and mysterious shrinkages," we must have surface to grow it on.

In advising new beginners what sheep to buy, the Texas *Live Stock Journal* says: "Purchase a flock which will turn out a good hundred-pound mutton. Then see that these sheep will shear at least above the average of four pounds of wool. With a flock shearing five, six or seven pounds of wool, with a flock turning out hundred-pound muttons at two years old, with sheep used to the country and the system in vogue, with a good range of mixed grass to run them on, any intelligent man can make a good living profit on his investment at present prices for year's clips of wool."

No right-thinking breeder of American Merinos will ever seek to grow early mutton lambs. But a cross between the Merino and Southdown produces admirable results in this direction. The Merino-Southdown, from a unilateral cross, is probably the nearest approach that will ever be made to an animal yielding, at the same time, the best staple and the finest quality of mutton.

It is as a producer of mature grain-fed mutton, and a choice delaine or combing staple, that the American Merino has before it a great future.

It is probable that wool of equal value for the manufacture of worsteds, or other fabrics requiring a true and strong fiber, can be produced in no other way, as the conditions to which the sheep is subjected when being judiciously fed for mutton are especially favorable to the growth of a fiber of this character. Life upon the range, with its attendant exposure to extremes of weather and alternations of plentiful and scant feed, can never with certainty produce this class of wool.

Hon. John L. Hayes, Secretary of the National Association of Wool Manufacturers, said to the Vermont breeders in 1884:

"The French early saw that combing could be applied to Merino wool; and that gave rise to the French Merino—a large-bodied animal growing a long wool. The French invented various fabrics, merino, etc. The English did not comb any wools, except those of long fiber. Up to 1865 we followed the English practice and our wool manufactures urged the growing of long wools. Then the processes in vogue in France were adopted here, and in 1869 the first worsted coatings were made in this country. The wool used was Merino. In 1876 there was an astonishing exhibit of worsted at Philadelphia. Since 1869 the manufacture of worsted has increased from $320,000 annually to $45,000,000, and this has come from your system of breeding, which has furnished the necessary material."

The great demand for wool created by the fostering of the protective tariff, led to the development in the United States of the greatest wool-producer the world ever saw; but it was a one-sided animal. It is a sheep whose anatomical formula may be comprehensively stated thus: Fifty pounds of carcass, fifty pounds of "drift," six pounds of wool, twenty pounds of yolk —a total of fifty-six pounds of wool and flesh, to seventy pounds of waste. Such an animal cannot stand the test of a many-sided civilization. From the production of wool there has set

Fig. 6.—A RAMBOUILLET MERINO.

in an extreme reaction to the production of mutton from English breeds. It would be wiser for us to do as the disciples of Daubenton did in France—create out of our own excellent material, the American Merino, an animal yielding a three-inch delaine staple, and sixty pounds of clear, ripe mutton.

The French Merino was brought to this country under unfortunate auspices; it came from a high culture to a raw, pioneer civilization; and this fact, together with the rapacity and swindling of its introducers, loaded it with obloquy. But with our American soil and climate, superadded to the laborious and searching care which the French farmer is willing to give, something very like the French Merino would be the nearest approach to a perfect general-purpose sheep.

As a suggestion, and perhaps a desirable model for our American Merino breeders to follow, I will give a brief outline of the Rambouillet Merino. When brought from Spain to France, in 1786, the rams weighed in full fleece from one hundred and thirty to one hundred and forty-five pounds; the ewes, ninety to one hundred and five; the rams' fleeces, ten and three-quarter pounds; the ewes', nine and one-third; the staple of the rams' fleeces was 2.18 inches long in the crimp; that of the ewes', 2.14. At the date of the Paris Exposition, 1878, the rams weighed in full fleece, one hundred and ninety-five pounds; the ewes, one hundred and forty; the rams' fleeces, nineteen and three-quarter pounds; the ewes', thirteen and one-half; the staple of the rams' fleeces was 2.57 inches long in the crimp; that of the ewes', 2.30. The staple was about of the same fineness as when they were brought from Spain, and the fleeces yielded about the same percentage of pure wool, that is thirty to thirty-three per cent. Their fleece weighed about ten per cent of their live weight, where our American fleeces reach twenty per cent.

The fluctuations of fashion sometimes operate disastrously on the wool-grower. Since the year 1869, there have been three revolutions in woolen manufacture: alternations between the long-haired wool used in making the stiff, "full-luster," British fabrics, such as alpacas and brillantines, and the Merino wool used in the soft, fine, clinging stuffs, such as cashmeres and chillis. To say nothing of the heavy losses occasioned to wool-growers by these changes, I will mention the case of the Pacific Mills, of Lawrence, Mass. This great factory lost over $2,000,000 as a result of a single one of these changes, that amount of capital being sunk in machinery, thus rendered worthless.

But the superiority of American wools, especially those from the Eastern agricultural regions, in soundness, length and strength of staple, gives our manufacturers an advantage of great value. The clips of Ohio, Pennsylvania and West Virginia are acknowledged to be the best produced in Christendom, not wholly on account of a superior adaptation of soil and climate, but also because the Merino has fallen into the hands of the best race of farmers of civilization.

The wool product of the United States for 1884 is set down at about 337,000,000 pounds, against 320,000,000 for 1883. Our imports for 1884 were a fraction short of 73,000,000 pounds, against 84,000,000 in 1883. Subtracting the amount sent over to Canada, we have about 68,000,000 pounds as our import for home manufacture. Only about 6,000 bales—2,500,000 pounds—were imported from Australia, to be used in the finest cloths of our manufacture; the greater part of the remainder imported was carpet wool. Now, it is estimated that about ninety-five per cent of all the sheep of the United States are Merinos, or Merino grades. So then we find that five per cent of our home-grown wool and perhaps ten per cent of the imported—certainly not over seven per cent of the whole amount of wool used in our manufacture—is from the English breeds. And the tendency is toward a reduction of even this trifling percentage.

The faint-hearted flock-master who may be disposed, in a temporary depression of values, to sacrifice our National race, the American Merino, for a coarser breed or a mongrel makeshift, ought to bear in mind that the Merino furnishes, and in the long run will continue to furnish, ninety per cent or more of the wool required for the clothing of the American people! All the British wool that will ever be grown in the United States would sew but a brilliant patch of color upon the Merino fabric of this Nation. In the markets of a cultivated people the coarse, showy cloths manufactured from British wools can not permanently or long compete with the dainty, soft, fur-like flannels which the Merino yields.

Hon. John L. Hayes, speaking to the wool-growers at Philadelphia, said: "By making your sheep fat in the shortest possible time, which you can do best with the English races, and killing them as soon as they are mature, you make the best and soundest wool. It will not only be young, but healthy; it will have no tender places in it. Aiming for the best mutton, you will be certain to get the best wool, which will always sell, no matter what race it belongs to."

There is some confusion of logic here, and a statement only true in a general way; but the converse proposition is equally true; that the farmer who produces the best wool (a three-inch, fine delaine staple, true and sound, and uniformly distributed over the surface of a rather plain sheep), will develop also the best mutton.

The watch-cry comes fitfully over from Great Britain to the American flock-master, that he must make wool the collateral product. It is a mischievous maxim. With the overworked and underfed millions of an old civilization, "what ye shall eat" is of prime consequence; but in this new empire of ours, still in the pride and strength of youth, the people have money to buy the wherewithal they may be clothed.

The spring lamb which feeds the gentleman, clothes the hod-carrier in Kersey or jeans, and we can not develop it to anything better; but the gentleman, after he has obtained his broadcloth from the Merino, would relegate the carcass to the hod-carrier. In other words, the British sheep caters to the two extremes of a dense population, wealth and abject poverty; but the Merino ministers to that independent class which is the boast of our country.

Our factories are rapidly acquiring the secrets of peculiar and popular foreign styles and fabrics, and even improving upon them, and inventing new processes and textures. Fancy cassimeres were, until recent times, of foreign production. Now the world-famous establishments of Sedan and Elbœuf are equaled or distanced. A bit of M. Boujeon's goods, taken from the inside of the collar of an overcoat worn by a gentleman from Paris, was the inspiration of the Crampton loom on which fancy cassimeres are now woven, not only in the United States, but also in several countries of Europe. These goods were at the Centennial Exhibition, and the Swedish judge, Mr. Carl Amberg, a practical manufacturer, in his admiration, said to Hon. John L. Hayes: "You know that the best fancy cassimeres in the world have been made at Sedan and Elbœuf, in France. If these goods were placed by the side of the Elbœuf cassimeres, you could not tell one from the other; and the goods could not be bought at Elbœuf for the prices marked here." These goods were made from American Merino wool.

The worsted coatings, differing from the fancy cassimeres in being made from combed, instead of carded wool, are a recent triumph of our manufacturing skill; they obtained distinction at the Paris Exposition of 1867. As an incidental result of this, another industry has been created, the combing and spinning of worsted yarns. Of these an exhibition was made at Philadelphia, by companies representing $1,500,000 of annual production; and they obtained an award showing them to be superior to yarns from the best Australian wools, being "kinder, more elastic and stronger."

In flannels, America has already surpassed Europe, because the goods are as well made and of better material. For a quarter of a century, European flannels have been driven from our markets, and we now export them to Canada. The yarns from these flannels are more closely twisted, the fabric shrinks less, and it is more highly finished and smoother in face. Even opera flannels are now made here from American Merino wool, which are softer than those manufactured from Australian fleece. Commendable progress has been made in competition with France, in the finer styles of ladies' dress goods, such as delaines, serges and merinos.

I addressed a number of questions to a prominent manufacturer and expert of Conshohocken, Pa., to which he replied under date of May 6, 1883:

"* * * The Australian wool is finer in blood than our American Merino wool, but it is not as strong, and for all purposes I would prefer our high-blood American Merino wool, making a stronger thread; and for nearly all goods made, would give the American Merino the preference to Australian at same price. If we could get the farmers to be more careful in putting up their wool, free from trash and other refuse stuff, [it] would bring a better price. We can now purchase Australian wool at seventy-five cents for the scoured pound, which to the American manufacturer comes much cheaper than the American Merino wool, being free from heavy string, and containing no trash whatever. My own experience is, that Australian wool is much finer, and varies in staple and soundness; and I would prefer for our use the choice, fine, high-blooded Ohio Merino to the Australian, on account of its strength, and making a much stronger yarn."

Granting always that there are sometimes conditions of soil, climate and market, in which the British breeds would be more profitable, *as a specialty*, than the Merino, let us consider the latter as one of the by-products of the farm. For, in all the region east of the Mississippi, it is chiefly as a factor in diversified farming that the American Merino will fill the measure of its great possibilities. As an element of average mixed farming, a few choice high-grade or pure-blood Merinos could be kept at a profit on land worth two hundred dollars an acre, whereas, if kept to the exclusion of other farm-products, they might be unprofitable on land worth over twenty-five dollars an acre.

Rejecting all conjectural statistics, I will give some well-attested, actual experience. I condense the following from the *American Sheep-Breeder*:

"Mr. T. W. W. Sunman, of Spades, Indiana, Nov. 1, 1878, weighed six Merino ewes, aggregating five hundred and ten pounds, or eighty-five pounds each, and put them by themselves in a yard containing a good sheep-house, where they were strictly confined during the winter. They received timothy and clover hay, mixed, and cut three-quarter-inch long; of this they had all they would eat, weighed out every day. They never exceeded eleven pounds a day, a trifle under two pounds apiece. In addition they received a quart apiece of ground feed, consisting of one-quarter oats, one-quarter corn, and one-half wheat-bran. They were kept on dry feed from Nov. 1 to March 15, and from that to May 1, on green rye. As the rye which they grazed, when cut and threshed, yielded as much grain as that part on which no sheep were kept, they were charged nothing for this six weeks' grazing. From May until late in October, the six ewes and their five lambs, grazed on a single acre of well-set pasture, containing good shade and water, for which they were charged forty cents apiece for the season.

DEBIT.

Six ewes @ $10	$60.00
Three-quarter-ton hay	7.50
Seven hundred and twenty pounds ground feed	7.20
Six months' pasture	2.40
Salt	.25
Total	$77.35

CREDIT.

Six ewes @ $9.00	$54.00
Eighty-four pounds wool @ 28 cents	23.52
Five lambs @ $2.50	12.50
Total	$90.02
Profit	$12.67

Another case I will cite out of my own experience. In 1882, I purchased eleven pure-blood ewes for two hundred dollars. The first winter they slept in a partitioned corner of the sheep-house, in which they were confined at night; while by day, except in stormy weather, they ranged in a corn-stubble, from which they picked up the greater part of their living. They received one ear of corn per head per day. One hundred average ears yield a bushel of shelled corn. Thus they consumed, in five months, fifteen bushels of corn. In addition to the corn, they received during the winter, one-half ton of hay, mostly timothy. During the summer they pastured in an extremely rough piece of land, part orchard, part locust grove. The apples from the orchard paid for the use of the ground. The second winter was more severe; and they consumed, with the same amount of corn, twenty-two hundred pounds of hay. We have, then, the following account (one ewe having died):

DEBIT.

Eleven ewes	$200.00
Twenty-seven hundred pounds hay	13.50
Thirty bushels corn	12.00
Salt	.50
Total	$226.00

CREDIT.

Ten ewes	$140.00
Twenty lambs	50.00
One hundred and eleven lbs. wool @ 20 cents	22.20
Two hundred and five lbs. wool @ 22 cents	45.10
Total	$257.30
Profit	$ 31.30

It will be seen that I have given the results with the utmost fairness. A small profit was made even on sheep bought at the "stud-flock" price of twenty dollars apiece, though their lambs and, of course, the wool are credited at only ordinary wool-flock prices. Plenty of high-grades could have been purchased at eight dollars a head, which would have given equally good fleeces and lambs, worth as much as the above, in the estimate. At eight dollars a head the profit for the two years would have been forty-one dollars and thirty cents; for one year, twenty dollars and sixty-five cents; or about two dollars and fifty cents a head. This is a clear profit more than double the "average annual revenue" from the English sheep in his own country.

CHAPTER IX.

LAMBING.

USE OF THE SHEEP-HOOK.—Probably not one flock-master in a hundred uses that invaluable labor-saving implement, the shepherd's-crook, or sheep-hook. It is surprising that practical, shrewd, inventive Americans will take upon themselves year after year heavy and unnecessary labor, when the sluggish, Oriental shepherd contrived a way to escape it in the earlier stages of his art, by the employment of the crook. Every experienced flock-master knows that unless a bunch of sheep are closely huddled, or quite tame, it is very laborious work to catch them, one by one, by the hind legs—for it is an outrage to catch them by the fleece, which the master should never tolerate. And if it is very tedious and tiresome to catch a hundred or more strong sheep when they are crowded together; how much more so is it to single out and chase down one after another in a roomy yard, where the master wants them well scattered, so that he can readily detect any points or marks he is looking for. He wants them to run by him, or to go circling around him, always keeping them with the painted side toward him, so that he can catch the mark readily.

Then again, when a foolish young ewe is standing guard in the field over her first lamb, she faces the intruder constantly and backs slowly away, stamping and snuffing, the lamb following her up; until finally in despair the master makes a sudden spring, and his fingers probably just graze the wool.

With a sheep-hook in his hand, his task is greatly lessened and simplified. If the ewe is to be caught in the sheep-house, it is barbarous to create an uproar and set all the other ewes and lambs to running about, trampling down the weakest, in a mad chase after a recusant ewe. Instead of that, let him quietly reach out the hook around the corner or between the slats of a hay-rack, and seize her by the leg—either a fore or hind-leg— preferably the latter, and no disturbance is created. I have seldom failed to capture the wildest ewe in a twenty-acre field, with the hook, at the first pass. There is no bending of the back; no foundation laid for rheumatism in after years.

Deftness with the hook, of course, can be acquired only by practice. It is best to let the sheep which is to be caught get somewhat disentangled from the others, then thrust out the hook and clap it on the leg just above the hock-joint, where the sheep can not readily kick it off. Draw it carefully back, and at the same time lift upward to keep the hook in its place.

When the sheep is within reach, seize it by the hind-leg, or throw an arm around in front of the brisket. The gentleness and even tenderness with which the fancy breeder lays hold of his show-sheep in the pen, sets it on its rump for inspection, then assists it to its feet without allowing it to struggle, and carefully brushes the dirt off the fleece and straightens out the disordered locks, may seem to some bordering on the sentimental; but it puts to shame the barbarous roughness of the novice. If a sheep is to be lifted and laid on its side or carried, it may be taken up with the left arm under the neck, and the

right arm grasping the right flank (not by the wool); or with both hands joined under the brisket, the animal being held perpendicularly, or with the left arm around the brisket and the right between the hind-legs.

MAKING A HOOK.—Any blacksmith of ordinary "rumble-gumption" can make one. Take a five-sixteenth-inch rod of spring steel, weld it to the socket of an old hoe-handle, for the insertion of the handle; bend it into a hook, as figured here-

Fig. 7.—CROOK.

with, about four inches long, an inch wide on the inside at the bulge, seven-eighths of an inch wide at the neck (to spring open and close again on the leg): flatten it at the point, and turn it out one-half inch or so, and back with a roll or a knob to prevent laceration. Insert a wooden handle six or seven feet long.

FIXTURES AND PREPARATIONS.—One of the most necessary fixtures about the sheep-house in lambing-time, is a set of portable panels for the construction of pens for milkless ewes. These panels should be about four feet high, by three feet long; they may be made of light lath, with spaces narrow enough to prevent the passage of lambs; or, better still, closed up entirely near the bottom, to prevent the lamb from seeing and smelling ewes outside of the pen. These panels can be tied together at the corners with twine to form enclosures. They are better than permanent boxes or pens, as they are not needed except in lambing time, and are more easily laid away than boxes.

It is taken for granted that the ewes occupy a stable, more or less, during this critical season; the size and arrangement of this will be considered elsewhere. All crannies and crevices ought to be stopped before lambs begin to arrive; a very young lamb, attracted by the light, perhaps, or moved by that instinct which teaches it to seek refuge and warmth, is very apt to wedge itself into a narrow place and get chilled. If the building has stone foundations—which are objectionable for this reason—they ought to be covered deep with litter; a lamb, while still damp, is almost certain to lie down on the stone and become fatally chilled.

GENERAL MANAGEMENT.—When a rain is coming on, look out for a shower of lambs; a falling barometer generally portends an increased activity in the sheep-stable, and indicates the necessity of greater watchfulness. The first thing in the morning, of course, the shepherd will go through the stable and look carefully for newly dropped lambs. As soon as convenient, the doors ought to be opened and the flock allowed to drift leisurely out into a yard (not to receive feed, as then they will rush out too rapidly), to allow the ewes with lambs dropped during the night, to become separated from the others. If any irregularity appears, if any ewes have abandoned their young, careful search must be made through the flock for those which give indications of having been recently parturient. There may be twins; they may be separated; one may have been adopted by a strange ewe, herself on the point of yeaning, and she may now be paying attention both to the stranger and her own lamb; or she may (such is the extraordinary stupidity of which young Merino ewes are capable), even have neglected her own lamb in her devotion to the one previously adopted.

When a ewe is seen to remain apart and take no notice of the flock for two or three hours, she ought to be gently caught and examined. Young Merino ewes are apt to be troubled by a retention of the fœtus, which may be due to several causes: Scirrhous *os uteri*, firm adherence and abnormal conditions of placenta and uterus, loss of power of expulsion by the uterus, paralysis, deformities, torsion of the uterus, and others. The first of these causes is most likely to be present, and it may induce a labor so protracted as to make the ewe disown her lamb. Let the operator, having laid the ewe carefully on the left side, sit at her back, and with the forefinger of the right hand, feel for the fœtus *per vaginam*. He should rest satisfied with nothing short of the fore-feet, or head, or both. If this can not be had, the mouth of the womb is evidently closed, but a patient search will seldom fail to reveal a very small and tightly corded orifice. If this can not be discovered, one must be fretted away with the finger-nail, or with the point of a knife-blade closely pressed against the finger. After this has been gradually enlarged so as to admit one finger, a second finger may be inserted, then a third.

Delivery can be successfully accomplished in three cases: First, when there is a presentation of the hind-feet; second, of the head and fore-feet; third, of the head with one or both of the fore-legs doubled back, though in this case the labor is difficult. All other presentations must be corrected; some person with a small hand should thrust the fœtus back and endeavor to turn it in such a manner, as to bring on one of the above presentations, preferably that of the head and two forefeet. Such interference as this is risky, still it is always best to resort to it promptly, as soon as it is ascertained that there is a false presentation, for protracted labor is apt to result in the strangulation of the lamb and the eversion of the uterus. More than that, it frequently disheartens the ewe, and makes her indifferent to the lamb. From the time the head distinctly emerges from the mouth of the womb, the labor-pains may be so assisted by the operator as to complete the delivery in twenty minutes.

With the fore-finger hooked in the under side of the jaw, and the remainder of the hand grasping the fore-legs, the operator may draw gently, in unison with the pains, gradually increasing the draught. If the pulling is distributed equally between the legs and the jaw, it may reach twenty or twenty-five pounds without injury to either ewe or lamb. It is far better to employ whatever force may be necessary, even to the fracture of the lower jaw (this may occur, and yet the lamb survive and recover), than to allow the ewe to linger for hours in agony, in a hopeless effort to expel the fœtus from a womb which has an insufficient exit, or none at all.

After the lamb has drawn a few breaths, the umbilical cord may be severed a foot or more from the lamb, which should then be laid under the ewe's nose. If she falls to licking it, all well; but if the parturition has been too painful, she may take no notice of the lamb. But if confined with it in a very small pen, where she can see no other sheep, she will generally own it in a few hours.

FOSTER MOTHERS, SUBSTITUTION, ETC.—When a good milker loses her lamb, her services are not necessarily lost; there are various ways of rendering her useful in the flock. If she is extremely attached to her lamb, and lingers about its dead body, she may be made to adopt a stranger by clothing it in the skin of her own, but this ruse will not deceive a sharp ewe. Let the skin be taken off without the head, but with the fore-legs to serve as sleeves, and with the tail, for it is at the root of the tail that the mother always seeks the scent by which she recognizes her offspring. The skin should be removed within twenty-four hours, else it will putrefy and sicken the lamb.

A ewe may be induced to adopt almost anything if, immediately after parturition, her own lamb is taken away before she smells it, and another, after being rubbed in her *liquor amnii*, is laid under her nose. A little salt rubbed about the rump may persuade her to fall to licking it, and thus develop a fondness for it. But all these substitutions are extremely hazardous; the master may have to keep the foster-mother alone with the lamb, and contend with her for weeks, whipping, scaring her with the shepherd dog, etc., to accomplish the desired result. If a ewe owns her lamb at all, and has milk, however little, with a prospect of giving more, it is far better to leave the lamb with her and supplement her supply with the bottle. A lamb once taken away from the mother is a source of infinite "pottering."

RESUSCITATION OF CHILLED LAMBS.—It is surprising to the novice, how near death a lamb may pass and yet be brought back by the help of man. If the thumb and fingers tightly clasped on each side of the chest, discover the faintest throbbing of the heart, it is worth while to attempt to restore it, if the lamb is a good one. (Even in a well-bred flock there are sometimes lambs so puny and flaccid—generally covered with minute pellets of wool, tightly curled down, plainly revealing the skin and prophesying a poor shearer—that they are not worth much exertion). The quickest way, and in extreme cases, the only way to recover it is, to plunge it up to the neck into water as hot as the hand can bear. But this should be only a last resort, for there is great danger that the water will obliterate the scent at the root of the tail by which the dam recognizes her own. For the same reason, it is dangerous to carry the lamb away at all, especially if wrapped in malodorous carpets, or the like. It is better to bring out hot flannels and wrap up the lamb, leaving the head out for the mother to smell occasionally. A very good way, when the case is not desperate, is to fold the legs neatly, and hold the lamb between the ewe's hindlegs until it is warmed enough to suck. A lamb once severely chilled must be closely watched for several days afterward; it is liable to a relapse unless highly nourished.

If dropped in cold weather, a great many lambs would never succeed in getting the teat, unless assisted. It is an extremely vexatious task for one person to attempt to hold a struggling ewe on her feet, and teach a very young lamb to draw. It is best always to cut the matter short by laying her on her **left** side, the lamb on its right. Then with the thumb and finger of the left hand, hold the jaws apart and milk a little into the mouth. The taste of the warm milk will generally induce it to draw, as soon as the teat is introduced into its mouth. However bright the lamb may appear, it is never safe to take anything whatever for granted as to the establishment of working relations between ewe and lamb, unless the latter is actually seen to suck.

COSSETS.—It is only a very valuable lamb that will repay the master for bringing it up by hand himself, and the hired help or the children will generally feed it so injudiciously through the summer, as to render it nearly worthless. I make it a rule of my flock, whether a lamb is to be reared by hand or not, that it *shall not receive anything whatever but fresh ewe's milk into its stomach for the first day;* and the longer cow's milk can be withheld, the better. If no fresh ewe has a supply to spare, I make no scruple to draw on one that soon will be fresh. Cow's milk is too constipating, especially if not fresh. Constipation is at best the greatest bane of the young lamb's life, and it is well to allow the cosset, once a day, for a week or two, to have its fill of the freshest ewe's milk obtainable. If a young lamb is fed a few times with a teaspoon, it may be taught to suck a leather in the bottom of the trough, and thus much trouble be avoided; but some are obstinate and must have the bottle. Sucking is better than drinking; it is slower, and causes a freer secretion of saliva.

A GOOD PRACTICAL SYSTEM.—Mr. Geo. S. Corp, of Morgan County, Ohio, in the fall removes the first sixty ewes served, keeping them separate. Eight or ten in every flock of sixty will "miss" at the first service. In two weeks after the service begins, he puts "teasers" into the flock every day, as he brings them in, to discover those that require a second service; and these are drafted into the second division of sixty. So with the second division, and the third, etc. Thus, when the season is ended, he has the divisions composed of about fifty each, which is the largest number he wishes to have in one flock.

When lambing comes on, one division at a time requires attention; the first and the last dates of service are recorded, so it is known when each division is done with. Bulletin boards hung on the wall have slips of paper pasted or tacked on them for "lamb records," showing date of birth, sire and dam of each. This lamb record is to be returned to the Secretary of the Ohio Register, of which Mr. C. is a member.

Not satisfied with providing lamb pens for ewes that disown their lambs, he has enough to hold every lamb that will be dropped in two days or more. They are thirty-three in number, in two barns, ranged along the two opposite sides, about four feet square, made of plasterer's lath, each with a little hinged door, hay rack and feed box. The building, which may be called the nursery, has a row of these on two sides of it, the row on the warmest side of the house having a stove about the middle of it. This stove is fenced about with lath. There are cages of different sizes, some only half as large as a mockingbird's cage. These are furnished with handles and may be set on top of the pens or anywhere else near the stove. Lambs shut in them will be dried and warmed, and they can not wander off, as they have a propensity to do if not restrained, as soon as they begin to get warm and limber.

Now this row of lamb pens on one side of the house (they are permanent) might be labeled MILK; that on the other side, NO MILK, though this is not actually done. When a lamb is stout and the ewe has milk, both are put at once into one of the pens on the "milk" side; when the ewe has no milk or the lamb is weak and needs help, both are put at once into a pen on the "no milk" side. This saves the shepherd considerable trouble. When he comes into the stable with a bottle of warm cow's milk he does not have to pudder about, catching this ewe and that to see whether she has a supply of milk; he simply takes the row as it comes. Sometimes the ewe's milk will "come" in six hours, sometimes in twelve, sometimes in thirty-six; in a very rare case it never comes. In the interval of waiting, the lamb and the ewe require gentle and patient care and liberal feeding. As fast as lambs gain strength enough to go alone, they and their dams are removed to a separate flock until the limit of fifty is reached, when still another flock is started, etc.

The bran boxes in the pens are about six inches square and an inch deep—a lath forms the sides—and are tacked on the top of the sill. The hay racks are also made of plasterer's lath, against the wall, having a depth a little greater than the width of the sill.

After trying the patent rubber nipples of drug stores, Mr. C. threw them aside and made a little plug of soft poplar, with a bore about the size of a small wheat straw. At first he attached a cloth to this, but he presently found that the lamb would suck it as well without the cloth. He now uses this altogether.

He has two or three "lamb creeps" in different parts of the building. One of these is formed by a board placed high enough to allow a lamb to pass under, but too low for a sheep. Another has its entrance through a little hinged door, which is propped open wide enough to admit a lamb, but it will catch a sheep by the shoulder. In these pens he has little troughs containing bran or ground feed for the special benefit of the lambs.

Whether lambs are dropped in February or April, whether they are grown for wool or mutton, it is of immense importance to keep them growing rapidly. They will, in ten days, take more than their mother's milk, and the little feed bestowed in this way will prove to be the best investment the flock-master can make. They can be weaned a month or six weeks sooner, if fed this way, and still be as large as usual, if not larger. This gives the ewes more time to rest and recuperate, and the added growth and strength of the lambs are a wonderful protection against parasitic diseases.

FEEDING WITH COW'S MILK.—It is sometimes desirable, where young lambs are fed with cow's milk to keep milk warm for some length of time. This can be easily accomplished by having a double tin can made, leaving a space of an inch between the outer and inner walls which can be filled with sand. The top and bottom should be soldered securely to both walls after putting in the sand. A tube an inch long must be placed on top and open into the inner cavity where the milk is put. Once warm the sand, and it will keep the milk warm for some hours.

Lambs fed on cow's milk, like those whose mothers receive only corn and hay, are very prone to constipation, which is the greatest pest the shepherd has to contend with.

1. Never feed with cow's milk, if possible to avoid it. If used, let it be fresh, diluted one-third its bulk with water, and well sweetened with pure, white sugar.

2. If fed every hour or two, after the first three or four feeds, it is not easy to give a lamb too much. Lambs are oftener starved to death than over-fed.

3. If constipation has already set in, do not dose the lamb with black molasses, magnesia, lard or the like. Give it an injection with a bulb-syringe, *very gently*, with blood-warm water, first oiling the tube with castor-oil. If necessary, repeat the operation.

But all the nostrums, laxatives, injection-pipes, and what not, fall immeasurably behind grass-made milk in value as preventives of constipation.

DISEASES OF LAMBS.—This remark as to grass leads to a mention of the so-called "lamb cholera"—a clear misnomer, since the malady has been distinctly shown to be non-epizoötic. It generally attacks the finest, fattest lambs of the flock; indeed, almost the only strictly safe generalization which may be made on its causes is, that it does not assail an under-fed flock, or a flock ranging on the sweet grasses, and the clear, running waters of a hilly country. For this reason, Southern Ohio has been almost exempt from its ravages, and I am chiefly indebted for information to observers living on the flatter, sourer lands of Northern Ohio, among whom I may mention Capt. J. G. Blue, of Morrow County; Mr. William Cattell, of Columbiana County; and Mr. G. W. Hervey, of Jefferson County.

The lamb is taken very suddenly and violently; falls on the ground in a tremor, with spasmodic kicks; sometimes froths at the mouth and throws the head back, further and further every minute, until finally it almost rests on the shoulders; the eyes are rolled up and have a fixed, staring look. Death usually ensues in a few minutes, and dissection reveals "the first stomach full of cakes of curd; the lungs seemed full of blood, and just inside the rectum was a slimy, watery appearance, with considerable wind. No diarrhea apparent in those; but I noticed in some a discharge like diarrhea, after they were sick, but before they died." I never lost but one lamb from this disease, a hand-fed pet; it had the above symptoms, and its stomach was very acid and tightly distended with gas.

As with all ailments to which the sheep is liable, prevention is a hundred per cent better than cure; but in this case the preventive measures must be brought to bear upon the ewes. One excellent, practical shepherd recommends to take a half-gallon of tar, mix into it all the salt it will hold together, and smear the salt-troughs with it, withholding all other salt, so as to compel the flock to lick this. The lambs will soon learn to partake with their dams. Another recommends grain and dry feed to correct the flatulency and acidity of the stomach. Better than either, perhaps, is sharp wood ashes, or lime, well mixed in the salt, say in the proportion of one part lime to ten of salt.

If the lamb is seen as soon as attacked, and the shepherd is skilled in drenching, so that he can perform the operation without strangling the animal—of which there is great danger, especially when it is unable to swallow readily—let him administer an ounce of Epsom salts in a teacup full of warm water; it may save its life. Or, put a lump of tar, as large as a hickory nut, well back on the base of the tongue, and shut the mouth and hold it closed to compel it to swallow.

EXCESS OF MILK.—When the ewes are on a full feed of grass, it frequently happens that a good milker will accumulate a supply of milk so large, as to cause one or both of the teats to become swollen and tender. If the lamb is vigorous and persistent, it will generally reduce one teat to use, but there is great danger that it will rest content with that one, and neglect the other, which will then speedily become useless. The milk must be drawn gently, and the ewe confined on dry feed three or four days. Care must be taken not to let her out too soon, or the operation will need to be repeated.

FOULING.—The tail of a very young lamb sometimes becomes so firmly glued to the posteriors by the gummy excrement, that further defecation is rendered impossible. The best thing to do is, to remove the obstruction and dock the lamb at once; but if, on account of warm weather, or for other reasons, it is not deemed expedient to do this at the time, all the parts should be scraped clean with a cob and well sprinkled with road-dust, or something similar.

CHAPTER X.

CARE OF EWES AND LAMBS.

There is nothing within the compass of the art of man which will promote a flow of milk so well as grass; and there is nothing else which will set a lamb up on its legs as well as a supply of grass-made milk. In the pastoral states, grass must necessarily be the main dependence of the shepherd; but in the older East, the pressure of other branches of spring work on the average farmer (for it is chiefly as a component of diversified farming that the Merino has an assured future in the agricultural States), will probably always cause a majority of northern flock-masters to have lambing over and out of the way before much grass grows.

FEED FOR EWES.—When a Merino ewe lambs as early as February or March, it is a long time and a hard task for her to make milk on dry feed until grass comes. What little she may make will be constipating; there is great danger that the lamb will die of costiveness, after ten days or two weeks. A guide as to condition of the suckling ewe, is the softness of the fæces; they should not be in pellets.

One of the best shepherds of my acquaintance, Mr. William F. Quinn, of Washington County, Ohio, feeds his ewes regularly, mangels, cut and sprinkled with bran. He has tried pulping, but prefers to cut them by hand into longish pieces, as large as one's finger. Pieces of this shape are not liable to cause choking. He finds that his ewes take more satisfaction in chewing these pieces than in gulping down like pigs a quantity of cold, watery mash, and are more benefited by them, on account of the more perfect admixture with saliva.

Clover hay, bran, bruised oats, bright corn-fodder, fodder-corn (if cured without must), linseed meal, cotton-seed meal are all excellent. If the ration is increased gradually, there is hardly any danger of over-feeding with any of them, except the linseed and the cotton-seed meal. There are cases so well authenticated, in which linseed meal has produced abortion, that we are not at liberty to disregard them. Still, I never saw a case of injury resulting from its use. One of the best practical shepherds of my acquaintance, Mr. L. W. Skipton, employs it habitually. To accustom his ewes to it, he at first mixes it in very small proportions with wheat bran, which he generally wets into a stiff slop; at the outside, he never allows more than a gill of the linseed meal to each sheep. Probably most flock-masters would find less trouble in teaching their sheep to eat it dry, mixed with bran.

But all these dry feeds, however excellent, do not equal roots in supplying the place of grass. And of these, probably, there is nothing superior to the white sugar beet for producing a flow of milk. I would name, in the order of their general availability, the white sugar beet, the mangel, the ruta-baga, and the white turnip. If no other succulent feed is at hand, small potatoes and apples can be given with great advantage.

If the flock is small, it will be better to wash and slice the roots, or pulp them in a mill for the whole flock. If it is large, there will be a majority of them robust, and hearty enough to eat the roots whole, if scattered on the hay orts, in the racks where they will be clean. The dainty ones can be culled out in the course of a few feedings and placed in a separate flock; this will reduce the labor of pulping.

An excess of cold, watery feed is injurious to pregnant ewes, as it is likely to produce abortion. After parturition has taken place, there is little or no danger.

KEEPING THE STABLE CLEAN.—From the succulent feed, on which alone the shepherd can hope for a modicum of success with the breeding flock, there will be an immense increase of the exhalations which are so fatal to the health of sheep. The urine decomposes and gives off ammonia. There is nothing so abominable as a slippery, reeking stable-floor, from which the lambs, slipping between the slats into the hay-racks, carry filth upon the hay; they also discolor the ewes' fleeces by gamboling upon them when lying down, and they so besmear each other that they are almost unrecognizable by their own mothers! Nor will it answer merely to heap up litter, in the hope of smothering the stench. The manure must be removed, clear down to the floor—which should be of earth—every week, or oftener, if the stench can not be suppressed; and the surface sprinkled with lime, if the offensive odor is very persistent. On dry feed, the steady dribble of orts from the hay-racks, with a little addition of straw or chaff, will absorb all the urine and prevent the escape of ammonia nearly all winter; but on succulent feed this will not answer. The manure must be removed with the utmost vigilance. The sheep's nostrils are near the ground; the shepherd may perceive nothing amiss when he enters the stable, while the flock are sickening on ammonia.

LAMBING IN THE FIELD.—When the lambing season is somewhat protracted, the latter part of it will probably extend into the grass, and there will occur spells of sharp weather of some days' duration, when the ewes will have to go afield some part of the day at least. It is desirable to keep them housed from cold winds as much as possible, but they cannot be confined altogether. On such days, the shepherd should watch the flock carefully, for there is a fatality (or, rather, an explainable natural cause), which brings lambs fastest in the roughest weather. There ought to be a piece of good pasture, preferably of orchard grass, held in reserve near the sheep-house for such weather. When a lamb is dropped, unless unusually vigorous, it will rapidly chill in a cutting wind. The shepherd can decide what to do in five minutes. Let him be provided with a light wheelbarrow, a piece of soft wool-twine and a sheep-hook. Capture the ewe, lay her on her side; take a turn in the middle of the string around the lower hind-leg and secure it with one knot; draw in the under fore-leg, and secure in the same way; then the upper hind-leg; lastly, the upper fore-leg, and make fast. Lay the lamb between her hind-legs to keep it warm, then wheel them gently to the stable. Here, if assisted to suck once, it will probably do well thenceforth.

When the flock is brought in after a windy day, care must be taken that no lamb is left behind. They are apt to hide away during the day in sheltered crevices.

GOITROUS LAMBS.—Under the headings "Congenital Goitre," and "Imperfectly Developed Lambs," Dr. Randall treats at great length certain abnormal phenomena appearing in very young lambs, which in all probability are reducible to the same category as regards their cause, and that cause wholly adventitious.

Under the heading of "Goitre," I shall have something further to say in a subsequent chapter; at present it will be sufficient to note somewhat more particularly the effects of too high feeding with grain, especially with corn. Mr. H. Miller, of Delaware County, Ohio, in a communication to the *Ohio Farmer*, stated that he had had occasion to suspect that excessive corn-feeding of the ewes produced goitre in the lambs; and, by dividing the flock into two portions, and feeding one lightly and the other heavily, he satisfied himself, by the absence of the malady from the progeny of the lighter-fed ewes, that his suspicions were correct.

The flock-master of extended experience will often, in his earlier career, find himself wondering why it is that the fattest, "stockiest" ewes in his flock will occasionally yean the smallest, whitest, most puny lambs. Sometimes, however, instead of being phenomenally under-sized, the lamb will reach the average stature, or even exceed it; but it will be of a flaccid, soft, muscular development; the under-side of the hoofs very spongy; the skin pallid, especially around the lips, nose, the septum of the nose, the ciliary caruncles, and the natural orifices of the body. It is weak—the least obstruction of the *liquor amnii* or slime about the nostrils will prevent it from getting its breath: the liquor itself is colorless, a robust lamb being generally enveloped in yellow liquor. If it survives at all, it will be hours before it can stand, even under the stimulus of warm sunshine. The ewe will be subject to garget.

The probability is, that the mother of this lamb was over-fed on corn. She may have been sterile the preceding year, and consequently fat at the coupling season, and she remained so throughout the winter, to the detriment of the lamb as above described. Still, experience and observation both convince me that there is little danger of having breeding ewes *too fat* if they have *plenty of exercise* in the open air. The errors are nearly all committed upon the other side, by having them *too poor*. A fat ewe may not produce as large a lamb at birth as a thinner one, but nine times out of ten the fat ewe's lamb will be the larger when a month old.

On the other hand, a ewe sometimes enters the period of gestation in high condition, and continues so for a month or two, then begins to fall off under good feeding, and as the end of her term approaches, staggers under her burden. Parturition is accomplished with great difficulty ; the chances are that the lamb will be still-born ; she will cast her fleece as a result of puerperal fever. She has no constitution, and is valueless for a breeder ; she ought to be drafted from the breeding flock and fattened.

The simple fact that a ewe disowns her lamb, however persistently, should not condemn her as a breeder. When she has little or no milk, it seems to be a monition of instinct that she can not rear it, and she abandons it accordingly. Under more favoring circumstances next year she may prove the most affectionate of mothers. This has often occurred in my experience.

GREEN RYE FOR EWES.—When grown on dry or well-drained uplands, green rye is undoubtedly a very valuable resource in spring, for ewes in lamb ; but when cultivated on rich, moist, river bottoms, its deleterious effects are beyond question. Although only moderately nutritious and rather distasteful to stock, in comparison with other kinds of green plants, still its exceptional earliness, bringing it forward at a time when there is nothing else above ground eatable, imparts to it a high value. I find by referring to my farm journal, that I have one year mown a fair swath of it as early as March 25.

It is well known that rye is subject, especially during cold, damp seasons, to the attack of a parasitic fungus, which attaches itself to the seed in its earliest development. This causes it to grow out in a long, dark excrescence shaped like a cock's spur, whence its name, ergot. This fungus may be detected by the microscope, not only in the head, but also in other portions of the plant ; and the white *sporidia* or dust on the surface of the ergot will inoculate other plants with the disease, if scattered in the soil at their roots, or applied to the forming seeds.

The therapeutic action of ergot is so well known as to require but a mere mention. It affects the uterus, tending to accelerate labor, and after parturition, to expel the placenta. Now for its actual effect on pregnant ewes. My rye patches are sown necessarily on the river bottoms, where it grows rank and is liable to be spurred. In my earlier ignorance I allowed the ewes to run on it nearly through the whole lambing season. They gave me an unusual and (at that time) unaccountable amount of trouble. When I succeed in inducing a ewe to recognize and own a lamb, I expect no more trouble with her if she has plenty of milk. But here they acted regardless of all precedent. I would establish practical relations between a ewe and lamb, both of them strong and healthy; she would have a full supply of milk for it, and everything would go on correctly for two or three days, and then, the first thing I knew, the lamb would be going around drawn up nearly double, disowned and half starved. The ewes were in good condition and full of milk. Several of the lambs seemed to have been dropped prematurely. The dams paid no attention to them. Myself and hired man were chasing, tying up, whipping, and otherwise employing coercive measures toward refractory ewes, all through the season. It was a warm, early spring, and there seemed no excuse for such wanton proceedings. Toward the end of the season the flock was removed to a field of red clover, and in about a week the trouble ceased. It is my opinion that the fungoid spores were present already in the young plants. If there is any reason to suspect their presence, pregnant ewes should not be allowed to graze on rye for a fortnight before yeaning, and not for a week or ten days after.

DEFECTIVE TEATS.—In case a ewe has had a teat clipped off by a careless shearer, she ought not to be admitted to the breeding-flock, but if she has got in through oversight, she had better be marked for rejection next time, unless she is otherwise exceptionally valuable. The orifice will be grown up, but a new one can be created by inserting a small trocar and canula, and leaving the latter in for several days, withdrawing it every day to apply some ointment of tar and powdered vitriol, which will assist the healing process.

Sometimes a middle-aged ewe will have a teat which, though yielding wholesome milk, is enormously enlarged, so that the lamb can not deplete it without assistance. The teat must be taken in hand promptly, else it will become so engorged as to be feverish, and then it will be many days before the ewe will permit the lamb to touch it. She must be kept up on dry feed, and the milk withdrawn **several times a day, until the lamb gets hungry.** It is singular how soon **a bright lamb will get into a habit of depending on one teat, and** it is sometimes **necessary to smear the favorite one with tar for a while, to make the lamb take to the large one.**

TWINS.—With the Merino ewe, twins are **seldom desirable,** unless it is in a standard or stud-flock, where the great value of the lambs will justify very high artificial feeding. If **the ewe is large and a free milker, and the twins of about equal size, she may be allowed to retain** both, and should **be put in a small enclosure or pasture alone with them, until they become thoroughly accustomed to each other.** But **if one is conspicuously smaller than the other, the shepherd will generally be the gainer by giving the dwarf to a neighbor. If he** has a fresh, **lambless ewe, he can compel** her to adopt it, but he must make up his mind to struggle with **her for two or three weeks.**

STIFF NECK.—Lambs **running afield in damp, cold weather,** are subject to rheumatism in this form. The great cervical muscles are flattened **and rigid, the head is drawn down almost to the ground, so that the little animal, though perfectly bright otherwise, is unable to suck.** I have never given any treatment beyond assisting the lamb to suck for a few days. I had a case once which **lasted twenty days before recovery was complete.** Doctor Randall states that a cure can readily be effected by administering a teaspoonful of turpentine, in two of lard, reducing the dose for a very young lamb. It is one of the evils attendant upon lambing before the season is sufficiently advanced to afford grass-made milk and sunny weather.

CASTRATION AND DOCKING.—It is rather severe on the ram-lambs, to have to undergo castration and docking together ; yet out of many hundreds I have operated upon, I have lost less than **a half-dozen from excessive bleeding.** I attach no importance to the "signs" in this matter, and never consult the almanac beforehand ; but there are undoubtedly certain times in the month when bleeding will be more prolonged than at others. If the reader is skeptical on this point, before he scoffs at the "superstition" of an old shepherd, let him make the experiment for himself. Instead of consulting the almanac, then, let the flock-master perform an operation (docking is the best test, as castration causes **little bleeding anyway), on two or three lambs, and if they bleed profusely, he had better defer the operation a few days.**

If castration and docking could be performed without assistance, it would be best, every way, to attend to them both before the lamb is a week old, but it is not very convenient to do it without an attendant, and for this reason, most shepherds will always continue the practice of going through the whole flock at once. Still, if settled warm weather is to be expected before lambing is over, it is best not to wait for that, but set to work promptly, and then finish the stragglers in a second batch.

Castration ought always to be performed before docking; it requires a finer and sharper blade than the latter. Cool, cloudy weather is best. Let the flock be driven up in the evening, without heating or worrying the lambs; then, during the night, they will measurably recover from their stiffness and be ready to follow the ewes afield in the morning. Let the catcher bring forward one lamb at a time, and hold it perpendicular before him, head uppermost, back against his breast, a fore-leg and a hind-leg grasped in each hand, but not drawn together so tightly as to make the belly concave and draw the testicles back. The operator seizes the end of the scrotum and cuts it off well up—the closer up it is cut the less cod there is left to hinder the shearer. Then he takes the scrotum in the right hand, works the testicles down, seizes one at the time between the thumb and fore-finger of the left hand, jamming the thumb hard down (it takes a powerful grasp and a stiff thumb to draw the testicle of a robust lamb), and draws it out with a steady pull. Care should be taken to slip the skin, the fat and the interior pouch, or membrane up, so that nothing shall be grasped but the naked testicle. The knife should not be used except to cut off the lower end of the scrotum, nor is there any particular virtue in any other method of severing the spermatic cord, except by pulling. It is wholly unnecessary to scrape it off. Let not the operator be alarmed if the spermatic nerve (the white, glistening cord), is drawn out at considerable length, and three or four inches of it left dangling from the pouch. It is bloodless, and will not attract flies; it soon dries up and will give no trouble. In all my experience, I have never found it necessary to apply any ointment to the pouch to keep off flies, if only the stump of the tail is well protected. Blood is apt to trickle from this down the whole length of the legs, and it needs careful attention.

The catcher still holding the lamb in the position above described, the operator takes another knife and severs the tail. He should first carefully ascertain by feeling with the thumb where the joint is, for the bone is hard to cut in a lamb of some weeks' age. The length of tail to be left, is not the unimportant matter the careless shepherd might think it; too long a stump is inconvenient to shear, and promotes fouling; too short a one detracts greatly from the beauty of the animal. An inch in length on the under side is enough; this will round out in fair and seemly proportions, the horseshoe-shaped "escutcheon," which is a feature of much importance in the outward make-up of a handsome Merino.

If the weather is likely to remain cool and cloudy for several days, until the blood dries up, no application will be needed. Otherwise, a half teaspoonful of fish-oil should be well worked into the wool at the stump of the tail, so that it may dribble downward. Tar is objectionable, it smears the wool. As fast as the lambs are docked they ought to be dropped outside the building, where they will not be disturbed in the further pursuit of operations.

The Scotch method of burning the tail off with a red-hot iron (several of which are kept heated close at hand), is a very good one. The operation is instantaneous, and the cauterization prevents bleeding.

A writer in the *Ohio Farmer* contributes the following:

"There is a better way of docking lambs than to use a chisel and mallet. The writer has used for two or three years a pair of toe nippers (the same as used for trimming the hoofs of sheep). The writer's plan is to take the toe nippers with him when he goes to look after the lambs each day, and dock and castrate all that are two or three days old. By the time the ewes are through lambing and ready to wash and shear, the lambs will be healed and exempt from trouble with flies."

I am inclined to believe that this would be an excellent method. One man could do the work alone. With the lamb held between the knees of the operator, as he sits on a box, the left hand should work the skin of the tail toward the body, so that when the tail is severed, there may be a hood or flap of skin to cover the bone and assist in healing the wound. With a stout pair of sheep-shears, castration might be performed at the same time; for, rough as this method may seem, when the lamb is very young, the pouch and testicles may be severed at one stroke close to the belly, with the best of results.

RE-DOCKING.—Sometimes it happens that a lamb needs re-docking after it has attained a growth of some months. In case it is a ewe lamb, it is better to re-dock than to suffer an unsightly stump to remain. It may be done safely, but it would be advisable to sear the wound with a hot iron, or put on a pinch of powdered blue vitriol.

CHAPTER XI.

TAGGING, WASHING ETC.

NECESSITY FOR TAGGING.—Where a very small flock is kept, and they run afield all winter, getting more or less old grass every day, they seldom scour when grass begins to grow green, and tagging is not necessary. But flocks which are confined through the winter, no matter how healthy and how high their condition may be, are sure to contain a certain percentage, which, when turned out, will become polluted about the vent before the time for shearing comes on; and the results of a neglect of tagging are so abominable, that no self-respecting farmer can afford to neglect this precaution.

The utmost care should be exercised in handling pregnant ewes while tagging. Blakely's sheep chair is a good thing to hold sheep in while this operation is being performed. It is just high enough for the operator to stand up, leaning to the sheep a little, in a comfortable position to work. It is adjustable so that it can be let out or taken up to conform to the size, and is adapted to all sheep. To tag sheep rapidly and well, the operator must be handy with the shears, gentle with the sheep, and have a mechanical eye. Many cut off twice as much wool as is necessary by not cutting in the right place, and leaving it where it should be cut; and before shearing time comes around, some of the sheep are as bad as ever. Tag one sheep and let it go, and take a look at it when it is going from you, and you can tell if you have sheared in the right place to escape the falling dung. By thus observing, you can, by shearing a small area in the right place, thoroughly tag a sheep by cutting off a small quantity of wool. Where more wool is cut from one side than the other, it makes the sheep look one-sided, but done in a workman-like manner, it adds to rather than detracts from the looks of the sheep.

One year we clipped the tags from about seven hundred and sixty sheep, which, after being washed three times in warm soap-suds, weighed one hundred and eleven and one-half pounds. I calculate that at least one-quarter of this amount would have been lost, if it had remained on the sheep until the regular shearing, because it would have become formed into dung-balls or clipped off in fighting maggots. Say then that we saved thirty pounds; at only forty cents a pound, it is worth twelve dollars. At one and a half cents per head for tagging, the operation cost ten dollars and ninety cents. The saving in wool paid for the labor, to say nothing of the avoidance of that most odious task that falls to the lot of the shepherd, the ridding of sheep from maggots. The ewes were heavy with lamb, but with careful handling, not one of them suffered any injury. Thus the udders were freed from wool, and the lambs all have clean, white faces, instead of the miserable, dung-smeared heads, which too often disgrace a flock.

TAGGING WETHERS.—As wether lambs are apt to become fouled about the pizzle before shearing-time, and consequently fall a prey to maggots; it is well to tag them in this place also, as well as about the vent. Even if no flies attack them, the wool becomes so clotted and heated with urine, as to create a festering sore. But I have never found it necessary to tag wethers under the belly after the first year's shearing.

CLIPPING THE HOOFS.—In all Merino flocks there is a certain percentage which will require to be caught and have their hoofs shortened twice a year; sometimes a few will need it nearly every month. There is nothing better for this purpose than Dana's toe-shears; and I have never found any better time for the operation in spring, than at the tagging. One man can catch and clip the toes, while another does the tagging. The operator, having caught the sheep, sets it on the buttock, with its back toward him, jams the left thumb between the hoofs to hold them well apart, and turns the shears in such fashion as to cut each segment of the hoofs from within, the inside being softer than the outside. If too thick and flinty, the hoof must be set on a solid block or board, and shortened with a chisel and mallet. Hoofs are always softest in wet weather.

POLICY OF WASHING.—A vast majority of experienced, well-informed shepherds are agreed upon three points:

1. Washing is an injury, both to the sheep and to the washer.
2. It is a benefit to the fleece.

3. On account of the long-established, but more or less unequal and unjust "rule of thumb," enforced by wool-buyers, by virtue of which unwashed fleeces are subject to deduction of one-third from the price of brook-washed clips, the keeper of an ordinary, out-door, wool-bearing flock, must wash his sheep or suffer pecuniary loss.

As to the first point, it is hardly worth while for the opponents of washing to assert that it is injurious to the sheep of average robustness and accustomed to run out-doors; it is sufficient to claim (what can be truthfully asserted), that it would be difficult to find a flock in such high and uniform condition, that some one or more would not suffer detriment in washing. And for the sake of these few weaklings, it is a pity that the flock could not escape the ordeal of being forced into the water, which the sheep, above all other domesticated animals, dislikes. By this I mean sudden and complete immersion. Every practical shepherd knows that hardy and well-fed sheep, even thoroughbred Merinos, will stand tranquilly through long winter rains, until their fleeces are saturated, when ten steps would carry them under shelter; nor will they be injured a particle thereby.

In the course of my experience, I have washed every year from five hundred to seven hundred sheep. Some years, when the weather was favorable, not only have I not lost any, but I could not discover that a single sheep was damaged. Other years, when cold rains supervened quickly upon washing, I have lost one per cent or more—a loss directly attributable to washing—to say nothing of the falling-off in scores or hundreds more—undoubtedly a greater total loss than I should have suffered from selling the clip unwashed.

In the hilly region of Southern Ohio, in which my experience with sheep has been cast, foot-rot is extremely rare; and I have never known, or heard of a case, where a flock contracted it on the road to the washing-pen, or in it. Yet, I am not in the least disposed to be-little the dangers from this source, which are urged by the opponents of washing as an argument of capital force and importance against the custom, where this malady prevails.

As to the washers themselves, there are plenty of men, young and hardy, who could wash sheep all day, once a year, for a generation, and not perceive in themselves any falling-off in physical vigor; yet none the less surely and inexorably they will suffer loss of vitality in the end. I have never known one to incur anything beyond a trivial chill or cold, which disappeared under mild treatment. The old-fashioned practice of taking whiskey on this occasion, only made matters worse. Alcohol is specifically determined to the brain under natural conditions, and when, in addition, the blood is driven thither by the chill of the lower extremities, drunkenness is tolerably certain to ensue.

All sheep are injured by washing, indirectly; that is, by the necessity of wearing their fleeces in hot weather, while waiting for the water to get warm enough for washing. This is true of dry flocks, and still more so of suckling ewes. In the earlier years of my shepherding, I often wondered why lambs, after springing forward rapidly for three or four weeks in the last of April and first of May, would then experience a decided check, becoming before shearing time, slightly "pot-bellied," less rangy, exhibiting those unmistakable, but often hardly describable changes of form, by which the trained eyes of the shepherd detects stunting. I am now satisfied it was caused by the drying-up of the ewes' milk, as a result of the fevered blood produced by carrying heavy fleeces in the piece during the early heat of our American summer. Proof of this was furnished by the continued progressive thrift of lambs running alongside, whose mothers had been shorn before settled hot weather commenced.

As to the second proposition, it is hardy necessary to state to a practical man that a washed fleece is easier to shear and easier to do up than an unwashed. "Shearing in the dirt" is hard work, and the fleece falls to pieces in a vexatious fashion on the table, especially if the sheep has been fed for the shambles, or is naturally very yolky. It is essential to the ready and accurate sorting of a fleece into the half-dozen or more grades to which it is ultimately assigned, that it should be kept well together, and washing is a great aid thereto. For instance, if shoulder-wool is worth fifty cents a pound, and the belly-wool fifteen cents, and they are mingled together, the sorter will invariably classify the mixture below its real value, if he does not incontinently consign it all to the lower grade of the two. This does not concern the farmer, except in a large way; he would not probably be paid by the wool-buyer a cent more for the well-folded fleece than for the jumbled one; the growth and condition being the same in both. But it does react upon him unfavorably in this way: It prejudices the manufacturer against unwashed wool, and he makes it the subject of hostile discrimination.

It is not worth while to go to the length of some advocates of washing, who assert that the buyer can not classify wool as correctly when unwashed as when it is washed. A genuine expert, though he may declare that he likes the feel of washed wool better, can pronounce upon the actual merits of a clip as well in one condition as the other. If not, he has mistaken his calling.

As to the third proposition, it is not worth while here (for this work is addressed, not to wool-buyers, but to wool-growers), to enter upon an extended discussion of the justice or injustice of the "one-third rule." I shall confine myself principally to the statement of a few practical facts, which will furnish the reader with a ready-made commentary upon this time-honored practice of the buyers.

At my request, Mr. A. F. Breckenridge, of Brown's Mills, Ohio, who is one of our best breeders of full-bloods, furnished me a transcript from his flock book, which is of interest as bearing on the question of washing. In 1877 he washed fifty-eight sheep and sheared them about June 1; in 1878 he let the same sheep go unwashed and sheared them April 30. Both years he recorded the weight of every fleece to an ounce. I give these tables entire. It will be observed that I have worked out the percentage of loss for the first table.

SIX-YEAR OLDS.				FIVE-YEAR OLDS.			
Sheep's No	Washed.	Un-washed.	Perc'tage lost.	Sheep's No.	Washed.	Un-washed.	Perc'tage lost.
	Lbs. Oz.	Lbs. Oz.			Lbs. Oz.	Lbs. Oz.	
41	8 02	9 02	11	26	7 00	8 00	....
45	6 09	7 02	10	29	5 10	7 08	....
46	6 00	7 01	14	30	7 08	9 00	....
49	6 00	7 00	14	33	7 02	7 10	....
50	6 04	7 08	17	39	5 04	8 00	....
52	5 01	7 08	33	42	6 08	8 00	....
53	6 07	8 00	20	43	6 00	6 10	....
54	6 04	8 00	21	56	5 00	6 08	....
55	5 14	8 06	30	58	6 13	7 13	....
57	6 11	7 01	8	59	6 13	8 06	....
60	5 12	6 12	15	63	5 10	7 00	....
61	6 10	8 01	18				
62	7 12	9 10	20				

FOR WOOL AND MUTTON.

Sheep's No.	Four-year olds Washed. Lbs. Oz.	Un-washed. Lbs. Oz.	Perc'tage lost.	Sheep's No.	Three-year olds Washed. Lbs. Oz.	Un-washed. Lbs. Oz.	Perc'tage lost.
65	5 08	8 04	....	81	10 00	10 00	....
66	6 12	7 08	....	82	9 00	9 03	....
67	5 14	8 12	....	83	11 00	11 05	....
68	6 08	9 04	....	84	10 08	11 00	....
69	10 00	12 00	....	85	9 12	10 12	....
70	11 00	11 12	....	86	7 12	10 00	....
71	8 00	9 12	....	87	9 01	9 03	....
72	7 11	10 00	....	88	8 12	10 00	....
73	6 11	10 02	....	89	8 01	9 03	....
74	7 00	10 08	....	90	9 03	10 08	....
75	6 02	8 00	....	91	8 02	10 00	....
76	10 13	11 08	....	92	7 08	9 00	....
77	5 12	7 02	....	93	8 12	10 04	....
78	6 12	9 06	....	94	6 11	8 12	....
79	5 00	8 02	....	95	8 12	10 00	....
80	7 00	9 12	....	96	8 02	11 00	....
				97	8 00	10 00	....
				98	8 02	10 00	....

Of course these tables do not make a thoroughly fair exhibit, since the unwashed fleeces had only eleven months' growth. In the computations herewith following, I have in every instance added one-eleventh to the unwashed weights, though that gain is doubtless somewhat inaccurate, since the fleece would grow less than an average during the twelfth month. But it is probably as near an approach to correctness as will ever be attained until some experimenter who can afford it shears a flock at the same date for several years, alternately washed and unwashed.

I find that the three-year-olds suffered an average loss of one pound, fourteen and two-thirds of an ounce per fleece, or nineteen per cent. on an average fleece of eleven pounds.

The four-year-olds suffered a loss of three pounds and one ounce, or twenty-nine per cent. on an average fleece of ten pounds and six ounces.

The five-year-olds suffered a loss of two pounds and one ounce, or twenty-four per cent. on an average fleece of eight pounds and eight ounces.

The six-year-olds suffered a loss of two pounds and one ounce, or twenty-four per cent. on an average fleece of eight pounds and eight ounces.

In looking over the above tables, the reader will perhaps be surprised to observe that generally the heaviest fleeces, which doubtless lost most in the scouring-tub, lost least in the brook or washing process. The explanation of this fact is, that where very yolky sheep are housed, the yolk becomes more or less inspissated, so that it does not yield to the solvent action of cold water. In confirmation of the showing of the tables in this respect, I will adduce a remarkable experience which was given me by another breeder of full-bloods, a perfectly trustworthy gentleman. He had a ram which he had shorn three years, and his heaviest fleece in that time was twenty-two pounds and twelve ounces. The fourth year he took him into the water, and with his own hands washed him thoroughly. After a lapse of two weeks he was shorn and yielded twenty-four pounds and four ounces!

There were occasional disturbing factors which produced apparent discrepancies in the above tables, as, for instance, the suckling of a lamb one year and not the other; but, as a geologist would say, the extent of the plateau is so considerable that the location of the "fault" is scarcely discernible.

Now, the average shrinkage on these four lots of sheep was twenty-four per cent. Reasoning from this fact, the farmer would probably arrive promptly at the conclusion that the deduction on unwashed wool ought to be twenty-four per cent., (say, one quarter), instead of one-third required by the manufacturer. But this view is erroneous.

The fallacy lies in the fact, that the basis of all calculations as to the value of wool is the *scoured pound*, in other words, clean wool. This is the foundation of all reckonings. The manufacturer simply ascertains what the scoured pound is worth in the markets of the United States—X, XX, XXX, or picklock, quarter-blood or common, medium, or whatever the grade may be. Then he glances at a table in which are given the average rates of shrinkage of washed and unwashed wools, of the different grades, from different sections of the country; with the value of each per pound.

To illustrate, let us take the general average for the United States. Messrs. David Scull, Jr. & Bro., wool commission merchants of Philadelphia, in a letter to myself, stated that the rate of shrinkage in scouring is sixty-seven to seventy per cent. on unwashed Merino wool, and forty-eight to fifty-two per cent. on washed. (My friend, Mr. W. M. Brown, Superintendent of the Beverly Woolen Mill, gave the figures as fifty-nine and forty per cent. respectively; but the clips he was accustomed to handle were lighter and drier than the average of the United States).

That the figures given by Messrs. Scull are approximately correct, is shown by the following table, which gives the results of the scouring of seventeen fleeces sent by the Missouri Wool Grower's Association, from Sedalia, to Messrs. Walter Brown & Co., of Boston, and scoured by a professional scourer at Walpole, Mass.:

BREED.	Age of sheep.	Age of fleece.	Weight of fleece when shorn.	Weight of fleece when sorted.	Clean wool obtained.	Shrinkage.	Amount of proceeds obtained.
	Years.	Days.	Lbs. Oz.	Lbs. Oz.	Lbs. Oz.	Per ct.	
Merino ewe.....	3	372	15 5	14 13½	5 9¼	62.32	$4.04
Grade ewe......	1	365	8 10	8 4	2 14	65.16	2.86
Merino ram.....	2	372	28 4	27 11	7 14	71.09	5.45
Merino ewe.....	3	376	17 8	17 0¼	5 13¼	65.69	3.90
Merino ewe.....	3	376	16 3	15 15	5 6	66.78	3.55
Merino ram.....	3	370	28 14	28 13	7 15¼	72.40	5.75
Merino ram.....	7	360	21 1	20 11	6 12¼	67.15	4.66
Merino ewe.....	1	360	13 7	13 1½	5 0	61.82	3.57
Merino ram.....	1	365	12 6	12 4	4 11¾	61.36	3.45
Merino ram.....	2	360	25 7	25 3	7 12¾	69.05	5.44
Merino ewe.....	2	358	18 1	18 0	6 3¼	65.05	4.33
Merino ewe.....	4	365	17 10	17 5	5 6¼	68.78	3.46
Merino ram.....	2	371	25 13	25 6	7 4¾	70.82	4.69
S. D. ram......	1	365	8 0	7 13	3 7¾	55.39	2.25
S. D. M. ram....	1	365	6 5	6 3	3 3	48.05	2.14
Cotsw'd ewe ...	3	365	16 5	16 0	11 3	30.03	4 3
Cotsw'd ewe....	1	365	12 8	11 10	6 11¾	4..07	2.97

We may accept therefore the figures sixty-seven to seventy as a fair percentage of loss, with this reservation, however, that a greater part of the unwashed wool of the Eastern States, as is evidently the case with that above tabulated, was taken from stud-flocks, housed sheep; and that this percentage would be too high for ordinary Merino flocks.

It may further be remarked, incidentally, that this table shows an invariable loss in the fleece between shearing and sorting.

Now, let us suppose that the buyer purchases one hundred pounds of washed wool at thirty cents a pound. The shrinkage in scouring is forty-eight per cent. This leaves him fifty-two pounds of clean wool, costing thirty dollars. Now, suppose he wishes to purchase one hundred pounds of unwashed wool, how much must he pay per pound so that it shall cost him the same per scoured pound as the first lot? The rate of shrinkage here is sixty-seven per cent. That is, one hundred pounds of unwashed wool will yield thirty-three pounds of scoured. This gives a simple problem in the "double rule of three." If fifty-two pounds cost thirty dollars, how much ought thirty-three pounds to cost? 52 : 30 : : 33 : 19.03.

That is, the hundred pounds would have to be bought for nineteen dollars and three cents, or nineteen and three-tenths cents a pound. This would require a deduction of a little over one-third from the price of washed wool.

The rates of shrinkage in washing are varied a great deal by methods of feeding, by housing, by individual peculiarities, by modes and degrees of washing. These variations appear conspicuously in Mr. Breckenridge's flock, ranging from eight to thirty-three per cent. If such differences exhibit themselves in the flock of a man who is a very careful breeder and feeder, and who has studiously sought after uniformity of type, what may we not look for in the flocks of a whole county? How wide will be the differentiations in a State?

In view of these facts it seems little less than a truism to assert that the "one-third rule" is a very raw and crude principle on which to conduct the purchase of wool.

SHEARING WITHOUT WASHING. — Col. F. D. Curtis writes vigorously in the *Country Gentleman* on this subject:

"I had an illustration of the differences between sheep shorn and unshorn, as to comfort and growth of sheep and lambs, this spring. A number of my flock were sheared in April, and the rest not till the last of June, and then before most of my neighbors. Those sheared in April are fat, while those which carried their fleeces did not gain at all in condition. I am satisfied that sheep should be shorn without washing, and that they should be shorn by the first of May, or in any latitude before they can be turned out to spring pasture. Of course this early shearing should not take place where there are no provisions made for sheltering them. Where they can be kept within a comfortable enclosure, they will do better without the hot and debilitating fleeces. Their lambs will get more milk; the sheep will be more active, eat more, and have more vigor. I have sheared sheep the first of April (latitude forty-three degrees north), and had no trouble. They must be kept out of the wind and wet.

"When shorn so early and before going out to pasture, except in rare cases, it will not be necessary to tag them before the whole fleece is removed. There would be less trouble with ewes if they were sheared before having their lambs. The little lambs could get at the teat much better, and with a careful man to shear there would be no risk in handling the sheep. Lambs are often born after the ewes have been both washed and sheared, and do well. * * * * * * A great deal of bother can also be saved, in trying to keep the sheep dry for the shearer, who very likely does not come. The annual bloating and starving of the sheep at shearing time can be dispensed with, as the sheep can be taken from the winter quarters, and when shorn returned to them. I am so impressed with the advantage of early shearing in this way, that I shall make it a rule to do so."

In this section, Mr. M. Palmer, and Mr. J. Chadwick—the former after an experience of more than fifty years; the latter with one nearly as long—have discarded washing, though both keep grade Merinos. Mr. Geo. S. Corp, owning about six hundred grade Merinos, has shorn them unwashed, about April 1; two years. He experienced a loss, one year, of nine dollars, and a gain the next year of seven dollars, saying nothing of the great gain in the condition of the sheep from being shorn early, without wetting. He will adhere to the practice.

MODES OF WASHING.—As we wash in the river, we are obliged to wait rather late for the water to become warm enough. The extreme dates which I find, on referring to my farm journal, are May 15 and May 29; the average will range between these two. When the washing season is at hand, I watch my barometer. I want to have reasonable indications of a spell of rising, clearing weather, and I want to put the sheep into the water in the beginning of that spell, so that they may have as many days of steady sunshine as possible in which to dry off. Washing in the river, as I have said, involves the necessity of waiting, but it is a great convenience, especially if one can have a clean, gravelly beach on which to land them. In most places the bank is so sloping that it is necessary to build three sides of a pen, but we are favored with an overhanging ledge of limestone. Jutting up against this, we put up three lines of portable fence, making pens which front directly on the water. We crowd the foremost flock down the bank and keep them huddled against the upper pen, while three or four men catch and dip, and then pass them around the projecting fence into the first pen. They are now passed into the second pen, and the men at once proceed to washing, or else dip a second flock, bringing these latter into the first pen. In either case the sheep are well soaked before the washing commences.

I frequently catch the flock of breeding ewes myself, as I am unwilling to have them abused. (I should have stated above that we catch all the sucking lambs out and leave them behind in the sheep house.) There are men who, when they stand in a line in the water and pass the sheep from man to man, will do nothing but swing the sheep to and fro. I watch the washers, and instruct them to squeeze out every part of the fleece. A man need only go out deep enough to float the sheep off its feet, then by taking the wool between his forearms, he can squeeze out considerable sections at once. They land on a stony beach and no dirt gets into the wool, if they do flounder about. It is true, they travel home by a dusty road, but the dust which settles on them does not amount to anything. We turn them on a clean sod pasture to dry off.

I have several times had suckling ewes come home from washing, hungry, go on a white clover pasture, eat greedily, and die of hoven in a few hours.

Managing in this way, four men and a boy or two will pass seven hundred and fifty sheep through the water, take up the fence, and get home by four o'clock in the afternoon, and the sheep will be washed clean; that is, as clean as cold water can make them.

I have seen an arrangement which would commend itself to the farmer living at a distance from any large stream, since it can be employed on a mere mountain run; it consisted of a plank box ten feet long, six feet wide, and deep enough to swim the sheep. The stream was dammed up, some distance above the place, and while the reservoir thus made was filling, a hole was dug out and plank fitted in, as described above. Steps at the end led to the bottom of the box, for taking in the sheep, and when washed they were let out on an inclined plane made of rough boards, with strips nailed on to prevent slipping. Here the water was squeezed from the wool, and the sheep passed out upon clean sward. Some arrangement of this kind is better (for a small flock), than to drive them a long distance, over a dusty road, and expose them to the danger of contracting the foot-rot or some other contagion from "scalawag" flocks.

Some writers recommend a waterfall and a spout, under which the sheep can be held and washed by a man, without getting into the water himself. If a man is afraid of a wetting, he can pursue this course; but the sheep will not be nearly as well washed as it would be if taken into deep water. If the washers can keep their bodies dry, they are not nearly so liable to receive injury from their prolonged wetting.

CHAPTER XII.

SHEARING AND DOING UP WOOL.

LENGTH OF TIME BETWEEN WASHING AND SHEARING.—How long a time should be allowed to elapse between washing and shearing, is a question which must be determined by circumstances. If the washing was done with thoroughness, the fleece was deprived of that modicum of yolk to which it is in fairness entitled to impart to it luster, elasticity, and, in general, a good style; and the farmer has a perfect right to allow his flocks to linger in the pasture until the sunshine has brought out the oily exudation, and until capillary attraction and the motion of the fibers one upon another have distributed it to the extremities. More than this honesty does not permit. When the wool has reached that condition of oiliness which may be found in a fine, healthy head of hair, on which a daily brushing has kept the natural oil distributed through its entire length, then, and not before, it should be shorn. What then shall be said of those flock-masters, who both keep such gummy flocks and so imperfectly wash them, that at shearing time the yolk may not only be seen glistening along the fibers in pellucid globules like glycerine, but even coagulated in yellow, pasty masses? Much depends on the weather after washing. If the sun is hot, ten days will be a long enough interval; if the weather is cool and cloudy, two weeks will not be too long a period.

GENERAL MANAGEMENT.—It is extremely convenient to have a pasture close at hand, from which the sheep can be brought up in small flocks as needed by the shearers. Thus they will keep full bellies, and the shearers will be troubled with fewer wrinkles. If a shower is threatening, of course the sheep will have to be closely housed over night. In " catching weather," we have frequently had to keep them confined until they became very hungry and hollow. Of course, also, the sheep-house will be kept well littered. This is essential throughout the house, but is especially important in the limited space where they are crowded in, a few at a time, to be caught. If the litter is replenished here every few hours, it will clean off their feet so that they will not foul the shearing-table. We generally fence off this small space simply with hay-boxes, and suspend an empty barrel in the passage-way through which the shearers enter. My shearing-table is about five feet wide, by fifteen feet long, and is supported on trestles which bring the table about up to the shearer's knees. Now and then a shearer prefers to take his sheep right on the floor.

The reader may or may not be familiar with a contrivance for holding the sheep fast during the operation of shearing. It consists of a large wooden bowl, in which the animal is set on its buttock and which prevents it from kicking. To this bowl is attached a frame like a chairback, both bowl and frame revolving on a pivot in the centre of the bowl. A strap passes diagonally across the frame, by which the sheep is lashed to it. This relieves the shearer of the strain of holding the sheep in position. By unbuttoning the strap, the sheep can be reversed for the other side to be shorn.

I observe that our best shearers proceed with the fleece as follows: Beginning on the brisket, they shear down past the arm-pits, and then from right to left clear across the belly in successive clips or strips, until the whole belly piece is taken off and left hanging on the left side of the fleece. Then they open up the neck, and beginning at the ears, shear neck and body to the rump on the left side, running the shears round to the backbone, and holding them in such a position that the clips or flutes left by them are parallel with the ribs, not only on the body, but on the neck. Then turning the sheep over, they shear the right side in the same manner. When clipped in this way, the sheep presents a zebra-like appearance, which is commendable for its regularity and workman-like neatness.

Much depends on the manner of shearing. The wool should not be cut twice, as this injures the appearance of the fleece when done up, also lessens its value to the manufacturer, as there will be more or less waste in the combs and cards. The shearer should keep the fleece together, not parting it on the shoulder as some do. I have seen shearers open the fleece on the right shoulder, running up the neck from the middle of this shoulder, and shearing to the middle of the left shoulder, and by the time the fleece was off this part was in pieces. I would give all such shearers the "go-by." Both shoulders should be left whole, as here is the finest wool of the fleece. Neither should the shearer cut a second time the portions clipped over in the spring in tagging; the wool is so short here that it is of no value, and if little locks of it are seen about the fleece they give a suspicion of chipping or mincing.

SORTING AND MARKING SHEEP.—Now is the time above all others in the year for the flock-master to subject the sheep to a critical examination, with a view to determine whether it is worthy of being longer retained. I find either of the following methods good: Have a rope hanging before each shearer with a strap at the end of it, which can be buckled around the sheep just behind the fore-legs in such a manner as to allow only its hind feet to touch the floor. This will keep it from escaping until the master, who is supposed to be occupied tying the fleeces, can find leisure to inspect the fleece and put such mark upon the sheep as he may wish. But a better way would be for him to turn over the tying to an experienced operator, and give his attention altogether to the inspection and marking of the sheep. I have a small grocer's scale on a low table, with a platform of light boards attached, on which the shearer can deposit the fleece. A single glance will reveal whether the weight of it comes up to the required standard or not, and a mark can then be affixed to the sheep accordingly.

The average flock-master, who does not care to go to the expense of having his flock entered in some one or more of the fashionable Registers, will scarcely find it worth while to follow any complicated system of record, such as is recommended by Dr. Randall. If he wishes to observe a system of numbering, he will hardly find anything better for the purpose than Dana's ear-labels. If the breeding flock is so small as to require only one ram, the owner has no option, and will not be required to institute any very fine discriminations among his ewes. But if it is large enough to demand the services of several rams, it will then be advisable to record in a book a few points, as "length of staple," "yolkiness," "density." etc., with a view to assigning each ewe to such a ram as shall be most likely to correct her deficiencies.

After trying several different plans of marking, I have adopted substantially the following: I employ red lead, or Venetian red, with linseed oil; tar is highly objectionable, since it makes a lasting clot which has to be clipped off before the fleece can be used by the manufacturer. First, I affix the letter of ownership—on the left hip for a ewe, on the left shoulder for a wether. It is important to mark all sheep on the same side, so that the eyes of the master can catch the mark readily as they circle around him. In addition to this, I stamp on the right hip the letter O, denoting that the animal falls below the standard and is to be drafted. I use the same letter during the lambing season, to designate a ewe which has shown herself unfit for further service as a breeder. The selection of two-year-old ewes for the breeding flock next fall should be guided more by the form than by the fleece, but the latter is important, and unless the breeder keeps a book record of each member of his flock, he ought to affix at shearing some mark to denote an extra shearer.

FOLDING THE FLEECE.—There should be near the wool-press a table or platform of ample size, on which fleeces may be deposited and spread out for folding. No fleece ought to be divided, however large it may be, for the sorter wishes to have the whole fleece before him, in order that he may divide it correctly into the different sorts. But it is permissible to detach the belly-piece for convenience in shearing, if it is folded into its proper place in the fleece.

TAGS, ETC.—The best course for the farmer to pursue in respect to that bone of contention, the tags, is to sort out carefully all very thick "sweat-locks," and the tags which are hard with dung, and wash them separately. Then the cleaner portions of the tags can be washed by themselves very much in the mode and measure of the wool on the sheep's back. The sweat-locks and the most objectionable tags should be put to soak in soft water for twenty-four or forty-eight hours, then washed out two or three times in warm soap-suds, and wrung out with a clothes-wringer. By this means they can be rendered white; whereas, if washed all together with cold water, the whole mass has a greenish cast, which is very objectionable to buyers. Tags washed thus thoroughly are perfectly entitled to be put inside the fleeces, a handful in each.

Dead or pulled wool should be kept separate, because all parts of the fleece are mingled together, can not be sorted, and consequently grade about on a par with the lowest; for this reason I prefer to remove the pelt from a dead sheep, as this retains every sort of wool in its own position. With the pulled wool may be put all the bits from the shearing-table which are worth picking up at all, for the hairy locks clipped from the legs are fit only for the manure-heap.

No wool which is damp with maggots, dew, rain, urine, or dung, ought to be rolled up in the fleece; it will heat and impart to it a dark color and an offensive odor. Tag-locks which consist mostly of dung (being different from "sweat-locks," which are entitled to go with the fleece), are worth only four cents a pound, and ought to be excluded. "Cots and common" form the coarsest grades of wool; the hard-matted locks have to be broken up by machinery before they can be used, and are then fit only for the lowest kind of goods. On these fleeces the buyer will probably insist on a reduction of at least five cents per pound. The hard clot-bur ought to be pulled out (if the farmer is so negligent as to allow this to grow and get into the wool, he had better remove it before shearing), but for the beggar-lice (*Cynoglossum Morisoni*), there is not much help, though it injures goods by specking them.

It is the folder's task to spread out the fleece on the table, weather side uppermost, clip off all the dung-balls, gather it as nearly as possible into the shape and density which it had on the animal's back, and then fold it for the press. The breech is folded over first, next the flank, then the neck, and lastly the flank to which is attached the belly-piece (the belly-piece ought to be where the sorter can find and remove it before he unfolds the entire fleece). The fleece ought now to be about square. Across the middle of this square the folder lays his left arm, and with a dextrous motion of the right, folds (not rolls), one half upon the other. Working an arm under each end of the fleece, he lifts it from the table with the two edges of the fold against his breast, and lays it in the press.

WOOL-PRESS.—In my own practice, I have been able to do up wool most satisfactorily with what may be called a rolling press (in contradistinction to a flat press), shown in figure 8. The outline dimensions are as follows: The table is two feet six inches high, two feet two inches wide, and four feet long. The leaves are four feet long and one foot wide. The box inclosed between the leaves is eleven inches wide. The head piece *c c c*, which is concave on the inside to adjust itself to the circumference of the fleece, is six and one-half inches high. The side-pieces of the table project far enough beyond the end to support the roller, *e*, which is three inches in diameter at the thickest part, tapering slightly toward the ends. The drop leaf, *b b*, is hinged, and falls forward toward the operator. When the

120				THE AMERICAN MERINO

fleece is placed in position the drop-leaf is raised to a perpendicular, where it is held by the upright, *f*, which works on a roller. This roller might be placed in the top of the table legs, instead of being a few inches from them, as in the engraving.

The fleece being now in the box, the leather band, *d d*, (six

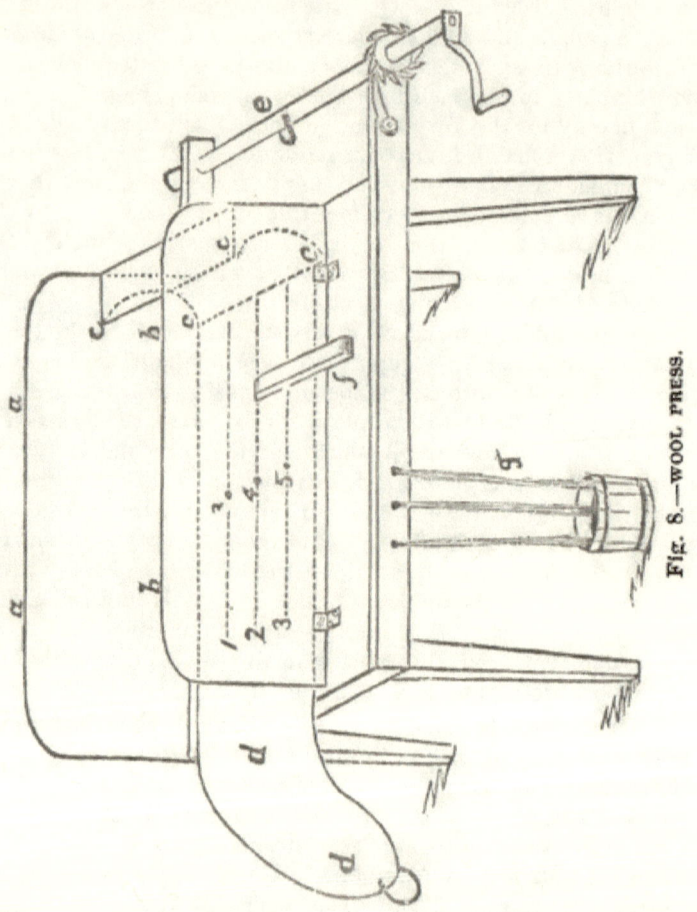

Fig. 8.—WOOL PRESS.

feet long and eleven inches wide) is carried forward over it, and the loop in the end is fastened to the roller by a little iron hook. One end of the band being fastened to the head-piece, when the other end is wound up on the roller it draws the fleece down into a tight drum-shaped package. The strings, *g*, entering the three holes in the table frame, pass up through three others in the bottom of the box at 3, 4, 5, and so along under the fleece to the head-piece, being fastened in creases at the top of it. The leather band has three slits in it, through which the ends of the strings can be reached with the right hand, while the left brings up the slack, and the knots are tied on the top of the fleece.

The ratchet is now lifted, the roller runs back, the band is detached and thrown off, the drop-leaf is let down, and the strings cut with the knife, which should be kept lying at the foot of the head piece under the fleece. The strings are now drawn up and fastened in their creases, and the press is ready to receive another fleece. I find it an advantage to let the twine pay out from the inside of the ball instead of the outside.

If the fleece was properly folded according to the above directions, it will come out from the press a cylinder (a better shape than a cube), and it will be so bound in one part by another that bulging or bursting is almost impossible.

STORING.—A wool-room need not necessarily be ornate, but it should be of ample size, convenient to the shearing-room, and made of dressed lumber. It ought to be furnished with windows, and yet so made that it can be shut up perfectly dark and tight enough to exclude bumble-bees, mice and rats, which are fond of burrowing in wool. If on the ground floor, it should be so high that rats can not bank up the earth underneath to touch the floor, as this will cause the wool to mold. But if space on the ground is considered too valuable to be appropriated to a room which is used only for a few months, it may be constructed overhead, and the fleeces pitched up one by one from the press; or the sheep may be hoisted into the second story by an elevator and be shorn there, as in a sheep-house to be described hereafter.

The important point in storing wool is, to have the pile of such shape that the buyer can, if he wishes, inspect every fleece without moving it from its place. The best contrivance I have seen for this purpose, is one which is employed by Mr. C. C. Smith, of Waterford, Ohio. This consists of a double row of upright studding, running across the room nearly to the wall at each end. These studs are framed together into something like a corn-crib, the width of which is only sufficient to accomodate one average fleece. This frame-work consists of smooth, light slats, stretching across between the studding, far enough apart to prevent a fleece from slipping between, and all of them inclined inward like the slats of window-shutters. This inclined position allows the fleeces to settle smoothly. The slats can all be taken out down to the floor, and then slipped into place one after another as the fleeces are piled up.

Fleeces stored this way will lose from two to three per cent. in weight in the course of six months, while a large pile close to the ground will shrink principally in the outside fleeces, and those in the interior will retain sufficient moisture to keep the shrinkage of the whole somewhere near one per cent. As a matter of course, very yolky wool will shrink more than the dry and light. I once had a pile of ram's fleeces lose about two and three-quarters per cent. in sixty-four days.

SPEED IN SHEARING.—The neatest shearer in the county in which I reside, once sheared for me fifty-eight head inside of ten working hours; they were about three-quarter bloods. I challenge any man to leave a sheep in better shape than he does. In his prime, he averaged forty-five to fifty a day. Another noted shearer in this county, has sheared over seventy Saxons in one day—seventy-seven, if I remember rightly. But of such grades as are generally found in this county, an average good shearer, working by the head, will clip thirty-five to forty in a day. The practice of "paying by the head" leads to racing between the shearers and a slighting of their work. Leg-wool is of no value, it is true, but a shearer who does not trim it off neatly, as well as that from the body, should be dismissed. It is best to employ capable and conscientious shearers, and pay them by the day. Six cents a head, or two dollars a day is commonly paid east of the Mississippi.

WHERE TO SELL WOOL.—The average farmer will almost invariably find it to his interest to sell his clip in his own wool-room, unless the amount of it is so small that he can transport it to and fro in a wagon. Warned by the example of neighbors, I have never shipped a clip to a storing-house or commission-house. After his wool has once passed from his sight, the farmer is practically powerless; he had better make up his mind to accept without complaint whatever is tendered. But in his own wool-room, especially if the clip is thoroughly good, he is independent.

QUALITIES AND GRADES OF WOOL.—It may be well to give here a brief extract from a little book on wool, issued by a wool commission firm of Philadelphia, Messrs. W. C. Houston, Jr., & Co.

"In any section or State all the wools are bought at about the same figure, whereas one clip will often be worth five cents per pound more than another, on account of growth and condition. By growth is meant the length, strength and elasticity of the staple, the working properties of the wool, and whether it is healthy and of good grade, or weak, coarse and of wild and 'frowsy' character. By condition is meant whether the fleeces are light and bright, or heavy with grease and dirt, or dark in color. Condition relates chiefly to shrinkage in scouring for goods. The more a fleece loses in scouring, the less it is worth to a manufacturer, on account of the smaller percentage of clean wool it yields. It can readily be seen that poor or heavy condition may overcome the advantages of good growth. Wool may be of good growth like some breeders' clips of well-bred Merino, but heavy with grease, and therefore poor in condition. And similarly, wool may be light and bright (in good condition), but having a wild, coarse or weak staple, will be of poor growth. If a fleece is wild and poorly grown, it will go into low-priced goods, no matter how light it is, so that poor growth may be counterbalanced by good condition.

"Good growth (sound, healthy staple) and good condition (a light, bright fleece) make up the first requisites of good wool. The growth and condition depend on care and intelligence in breeding, and also, considerably, on the locality where the wool is grown. In wild or prairie sections, the wool is apt to be 'brashy' (weak staple and of wild growth), and is generally discolored by the soil; while in localities more under cultivation the wool is apt to be of better growth and brighter. This is one of the reasons why Ohio produces better wool than Wisconsin or Minnesota. As the land of a section is brought more under cultivation, the wools improve. But this must be supplemented by proper attention to breeding; for we have received some lots of unwashed from Iowa, that were better in grade and condition than shipments from Indiana and Illinois.

"The terms 'growth' and 'condition' being understood, we pass to grades. *Fine* is the full-blood Merino. In well and high-bred washed wools, fine is sub-divided into X and XX, according to the fineness of the fiber. *Fine delaine* is the elastic and long staple fiber, of about two and one-half inches in length throughout the whole fleece. *Medium* is a three-eighths to one-half blood Merino cross. The proper crossing of full-blood Merino on a coarse-wool sheep produces medium grade. *Medium combing* is the long staple of the medium grade, about three and a half inches in length. A cross of Merino and Leicester gives medium combing—the Merino giving fineness of fiber and the Leicester length of staple.

"*Quarter blood* may be called a mongrel wool; like a cur dog, it has no defined characteristic of breed. It is generally wool of common sheep, that don't contain enough Merino blood to class as medium. Quarter may be a run-out medium, or a coarse sheep not yet sufficiently graded up with Merino blood. It is of a wilder and not so close a growth as medium. It is difficult for our western friends to make the distinction between medium and quarter blood. In the West all wool between the fine and extremely coarse fleeces is classed as medium; whereas here that range is split into a medium and quarter blood, the bulk of the wool sometimes going to the latter grade. We can hardly make the difference more clear than we have, except to add that in a real medium the Merino blood can be distinguished in the fiber of the wool; whereas in quarter blood the Merino characteristic has entirely died out, if it ever was there. Quarter is a wild, coarse wool, as contrasted with medium, which is a closer and finer growth, approaching Merino. *Coarse or quarter combing* is the long staple of quarter blood grade. *Common* is the rough, hairy wool and cotted or matted fleeces. It is often a run-out Cotswold, and this grade is found mostly in coarse sections and in flocks that are run out. *Common combing* is the long, hairy wool, on the order of full-blood Cotswold and Canada.

"The grades fine, medium, quarter and common, apply to all wools. In unwashed wool there is little or no difference in the price of combing and its corresponding grade of clothing, and the only advantage of taking out the combing is that it can be run a little lower than clothing. For instance, medium clothing and combing sell at the same price, but as what is known as medium combing, is made almost a grade lower than medium clothing, we can sell more medium wool by making a medium combing; the same holds true of quarter clothing and combing. In well-bred washed wools, the combing is worth more than the clothing, because it grades up better. Fine delaine is practically never taken out of unwashed wools; and in fine washed, that is not well-bred and well-grown, there is rarely any to take out. Neither combing nor delaine are made out of dark, heavy, or poorly-grown wools, because the staple is generally weak, brashy, or not of sufficient elasticity."

SACKING AND TRANSPORTATION.—If the clip is too small to justify the trouble, or has to be transported only a short distance to market, it may be hauled tolerably well on a hay-rack, if care is used in stowing the fleece. It is better, however, to sack it. Custom requires the buyer to do this on the farmer's premises, unless it is otherwise stipulated in the bargain; indeed, few farmers have the appliances necessary in sacking. For convenience in sacking, it is well to have the wool-room on the second floor; and in the floor a circular trap-door, two feet in diameter. The sack is hung down through this, swinging clear of the floor beneath, and supported by its edges lapped under an iron hoop with an inside diameter the same as that of the trap door. After five or six fleeces have been thrown down into the sack, a man descends into it, arranges and treads on them, and so continues until the sack is filled. It is then raised a little with a lever underneath, the mouth secured with clamps, the hoop removed, and the sack is then lowered to the floor and the mouth sewed up with twine. Cobs placed in the corners of the sack at the bottom furnish convenient handholds. The implements required are, a canvas-needle, two wool-boards, with a half-circle cut out of each (for use in case the sacks have to be suspended between joists or timbers, or a temporary frame-work), iron clamps with leather straps, and a hoop of half-inch, round bar-iron.

CHAPTER XIII.

SUMMER MANAGEMENT.

SHEEP AS SCAVENGERS.—When old fields have become overgrown with briers and bushes, and the farmer desires to extirpate them, sheep will do the work for him better than any other stock, but they will sometimes require assistance. If brier-clumps are very thick or very high, the flock cannot do the work unaided. The bushes must be mown and burned, or, if well filled at the bottom with dead leaves and grass, they can be fired in a dry time, and, if some pains are taken to beat down the green ones as the fire is burning, the whole clump can be consumed. The young shoots which sprout up in the ash-heap will be eaten off by the sheep much more thoroughly than those growing where there are no ashes. I have found it one of the best ways of renewing old moss-bound pastures, to fire them in a dry spell in the spring, when there is dry herbage enough on the ground to carry the flame ; then let the sheep have the range of them through the summer. They take a great deal of satisfaction in grazing, sleeping and stamping in the burnt district ; and, as above stated, they will take much more pains to crop off the sprouts here than they will in unburned territory. The ashes must give them a relish ; probably it is the greater percentage of potash they contain, since sheep are noted for their fondness for and need of certain mineral ingredients in their feed. I have often observed their relish for these ash-fertilized plants ; they return to them again and again, cropping them down close to the ground, where they would scarcely taste them if growing in the open field.

Every observing shepherd has noticed that sheep have their decided preferences in a rolling or hilly pasture, generally choosing a southern or eastern slope. Old farmers will tell you it is because the grass on these poorer, thinner exposures is shorter and sweeter. Probably this is one reason, but I cannot help thinking there is another. These southern slopes are nearly always wind-swept and sun-burned, and receive no staying deposits of forest leaves ; hence the bed-rock is close to the surface, and frequently crops out in shelly ledges. This character of the soil gives the grass a more mineral and earthy quality than is possessed by that growing on the north slopes ; for on these the soil is generally red clay, and strong with the humus or vegetable mold resulting from the rotted forest leaves of centuries. And the fondness of sheep for mineral ingredients in their feed was above alluded to. Hence they linger on these naked, wind swept, southern slopes, nibbling the already scanty grass into the very ground, and neglecting the rich, rank feed on the northern slopes until they are fairly "starved to it," often to the wonder and annoyance of the shepherd.

In general, sheep are so nice in their tastes and preferences that a pasture of any considerable extent, especially if it has a diversity of soils and exposures, is apt to become patchy if left entirely to the sheep. They are fond of knolls for stamping-grounds and sleeping-grounds, and will manure them to excess if they have their own way.

There are various ways of regulating these matters. A portable fence might do good service here ; I never tried it. A few young cattle with the sheep will give their attention to the north slopes and the rank pasture spots, while the sheep are grazing on the shorter feed. The sheep themselves will depasture these northern slopes in the fall when feed grows scarce; but meantime much grass has grown up and died, so going to waste; and the briers make their whole summer growth unchecked.

I have found it an advantage to run a permanent fence between the north and the south slopes, so compelling the sheep to divide their time between them. Still, they will hang along the fence for hours, sleeping by it, waiting and watching for a chance to get through. So, as a still better measure, I generally keep one of my flocks in ignorance of the existence of certain south slopes, by never turning them on them; thus, when it comes their turn to occupy the contiguous north slopes, in the rapid rotation which it is my policy to keep up during the summer, they graze there quiet and contented.

I am always more careful to keep the large briers and shoots cut on the north slopes; I salt the flocks there whenever practicable; and burn all brush and trash which may accumulate there.

All burs of whatever description ought to be cut, dried and burned before they get ripe enough to part from the plant. Burdock and Thistle burs are worse than Cockle burs, if possible; they burst asunder and fill the wool with the most odious prickles and filaments, while the hard burs can be removed whole. No words of condemnation can be too severe for the farmer who allows burs to grow and ripen and get into the fleeces.

NUMBER OF SHEEP PER ACRE.—T. W. W. Sunman, of Spades, Ind., gives in the *American Sheep-Breeder* the following experience: "We took six head and put them on an acre of ground well set in grass containing some white clover, well watered and good shade. They were turned in somewhere about the 12th or 15th of April, and remained there until along in October without any additional feeding, when they were turned to early sown rye and pastures saved for fall pasture. The acre furnished all the pasture the sheep required and to spare. In the spring of 1880 we turned eleven head of one and two-year-old ewes upon this same acre of ground, and they remained there from May to October, receiving no additional feed, and had plenty of grass all the time.

"In 1881 we took in one-half acre more land, making in all one and one-half acre; upon this we pastured seventeen head of one, two and four-year-old sheep, consisting of fifteen ewes and two rams. There was all the pasture the sheep wanted and to spare, and we believe would have furnished pasturage for four or six more, but this was a good year for pasture."

But this is an exceptional case. When the shepherd, in going over his pastures, finds an occasional grass-tuft pulled up by the roots, he may know that he is over-pasturing. I have kept twenty-three sheep in good condition on three acres nearly all summer.

NECESSITY OF WATER.—If the nights are cool and there is a heavy deposit of dew every night, sheep will do well for a long time without water, if they have constant access to salt, so that they do not eat too much at any one time. Otherwise they ought to have water within reach all the time. A flock of ewes with lambs at heel, ought always to have free access to water, summer and winter, without regard to weather.

WORKING OFF THE CULLS.—With a flock of considerable size this is one of the most difficult operations connected with its management. There is no profit in grain-feeding old ewes or the long-legged, short-wooled, ungainly culls, into which a large flock, despite the most careful management, is continually "tailing out." Occasionally a batch of them can be sold to a neighbor who, having a fresh run, and wishing to keep only a small flock, can make something out of them when segregated into smaller bands; but usually the only method practicable is to fatten them as quickly and cheaply as possible, and sell them for what they will bring.

An old, toothless or splintery-toothed crone of a ewe, is an extremely poor piece of property. Scarcely better is a younger one yielding a short, dry fleece, or a short, yolky one which collects into hard, yellowish blocks, that almost require a hammer to soften them; or with a bare belly and long, bare legs; or with a tail set on low, and a weak, drooping neck. Of course, ewes that are in service will produce lighter and thinner fleeces from year to year, and some deficiency in this regard may be tolerated in one of exceptional excellence otherwise; but if these faults appear in a younger sheep, it ought not to be retained after the first shearing. It is a capital mistake to allow an inferior sheep to drift into the breeding flock, for then there will be two culls instead of one.

Shippers commonly say they do not care how old a sheep is if it is only fat. But that condition which the ordinary farmer calls fat may be only "grass bloat," or it may be fat enough to make fairly good mutton for his own table; but it will not endure the long, rough ride to New York, Chicago, or Baltimore. How to make an old ewe fat enough for the shipper, is a difficult matter.

I generally succeed best with culls by putting them by themselves, young and old; feeding them all the wheat bran and corn meal they will eat (their teeth will be too sore to crack corn), and giving them the benefit of the first fresh cropping from each pasture. As soon as they have been on it a week or ten days, I pass them on to another fresh one, and let the main flock follow them up, taking each field in turn after them as soon as they leave it. By having three or four fields and swinging the flocks rapidly through them in succession, I can keep the main flock very large—much larger than it ought to be in winter quarters—without detriment to it, and even keep them improving in flesh all the while for three or four months, until the culls are ready to turn off, when the main flock can be broken up small again before frost sets in.

Old ewes and other refuse sheep ought to be pushed rapidly while the grass is tender; like an old "shelly" cow, they are a drug in the market at best, and in the fall they will be crowded to one side by wethers.

The butcher or shipper ought never to be required to take culls for the sake of getting good, straight wethers. Some shippers will not handle the former at any price; they will have to be disposed of to some "cheap John" dealer for a bagatelle; but for thoroughly good wethers the farmer can demand and obtain a good price. By all means keep the two classes separate.

TEETH AS AN INDICATION OF AGE.—It is often the case that a man will develop into an excellent practical shepherd, but without a taste for keeping a record of his sheep by books, marks, labels, etc. He will have frequent occasion to refer to the teeth as decisive of age. The milk or lamb teeth are easily distinguished from the grass teeth by their smallness and dark color. The old rule among farmers was that a "full mouth" (eight grass teeth), denoted a four-year-old, each year bringing forth two new teeth; but in the modern improved breeds, unless ill-fed, the grass teeth make their appearance about as follows: The first pair at one year; the second pair at eighteen months; the third pair at twenty-seven months; the fourth and last pair at thirty-six months, or three years.

A LEG OF MUTTON.—A fat young ewe affords the best ripe mutton; next, a young wether. The sheep selected for mutton should be kept quiet and cool in a dark place, twenty-four hours, without anything to eat, but with all the water it will drink; above all things it should not be worried and heated. The neck being laid across a block, may be severed at a blow with an axe, and the flow of blood should be made as complete as possible by the butcher seizing a hind-leg and gently pulling and pushing with a foot on the carcass. The disemboweling and skinning should be quickly dispatched. Let the sheep be hung up, ripped, and the bowels removed; then the skinning can be performed afterward. Immediately after the sheep is hung up, if a hole is made between the hind-legs and the abdomen filled up with very cold water, it will assist in preventing the "sheepy" taste.

When Daniel Webster said he learned in England the secret of good mutton, namely, that it improves with age, he must have meant that it grows better each day after it is butchered. The longer it can be kept the better, within decent limits. If the farmer wishes to avoid surfeiting his family on mutton, let him convert a part of it—the legs preferably—into smoked "mutton hams" or corned mutton; then hang two or three good roasts down a deep well, and proceed with moderation in all things. The advice of the old English "quarter-of-mutton chant" to the cook is: "Let her boil the leg and roast the loin, and make a pudding of the suet," and the advice is sound. The roasted loin is always a juicy piece; but the shoulder-blade, gently browned, with onion sauce or baked tomatoes, runs it close in the favor of gourmets, who will also generally be found to prefer a neck chop to one from the ribs, since in a coarse-grained sheep oil has a tendency to gather there.

Charcoal or vinegar will remove what the Scotch call the "braxy flavor," if it exists, though it should not be noticeable after the above precautions in butchering have been taken. The old English fashion of cooking before an open wood fire, as directed by Dean Swift, was very good; but an intelligent cook can prepare just as choice a roast in a modern American stove-oven. If the sheep was young the piece may be put into the oven at once; otherwise it ought to be macerated by boiling awhile, with the amount of water so gauged that when tender, it will be "done dry." Then let it be put into the oven, with this remnant of juice, and nicely browned; and the gravy should be thickened with flour and water previously stirred together without lumps, and poured into the pan about ten minutes before it is taken out of the oven.

MAGGOTS.—Mr. E. J. Hiatt, the editor of *The Shepherds' National Journal*, and himself a shepherd of long experience and excellent judgment, gives the following :

"Sassafras oil and alcohol, one-fifth of the former and four-fifths of the latter, mixed, will destroy maggots on short notice; this is a safe and sure remedy and is particularly valuable to destroy maggots when they are located where it is difficult to get at them. They may be destroyed without shearing off the wool.

"Turpentine has been used, but this is injurious to some sheep and cannot be used with safety when the sheep are allowed to run in the rain, and it is also unsafe in cases where the sheep is fevered and reduced in strength from being unnoticed or neglected, until its life was in great danger. Water should not be used as it only increases the danger of a second attack. There is much less danger of trouble with maggots when sheep are kept from the rain.

"A liquid is sometimes used, which is made by boiling or stewing the bark or stalks of the Elder. This is more troublesome, but could be used in the absence of something better."

I had come to the same conclusion as Mr. Hiatt respecting the use of turpentine, also benzine—both being too severe on the sheep in most cases. I salt twice a week until shearing-time, and carry to a field with me, besides the salt, the crook, some tar, and a pair of shears. If a sheep is seen to stamp and twitch its tail, catch it on the spot. When your suspicions are found to be correct, shear off close all the wool infested by the vermin, clean them off and apply tar thoroughly. If they have established any considerable footing, scrutinize with the utmost thoroughness the wool adjacent, for colonies of them will migrate around about and begin operations afresh.

TICKS.—It is an impeachment of the shepherd's care and vigilance to have these abominable pests on his sheep, at least for any length of time, since they are liable to get into any flock through purchase. In the early summer is the time above all others in the year to give them the slip. After shearing they will disappear in two or three weeks from the shorn sheep, and part of those on the ewes will take refuge on the lambs. The grown sheep will need no more attention if they are kept in good growing condition through the summer, but unless the lambs are treated in some way, the vermin will survive through the summer, some will return to the ewes before weaning-time, and the remainder will be ready to begin their deadly work through the winter, as they seldom do much injury in summer. Ticks never flourish on fat sheep. Indeed, this rule holds good in reference to nearly all ovine parasites; but it is almost an impossibility to get lambs in good condition when infested with ticks. It is not advisable to dip them in cold weather, but in summer it may be done with safety and benefit. Some shepherds recommend Eady's Sheep Dip, others carbolic acid, etc.; I have tried kerosene, snuff, sulphur (rubbed into the wool), and tobacco water and a solution of arsenic (as a dip). I think, all things considered, the tobacco-water is best, if the material is readily obtainable, though if applied strong it has a tendency to color the wool and make it harsh.

Twelve or fifteen pounds of refuse tobacco and chopped stems; or six pounds of white arsenic, will make a solution sufficiently strong for one hundred lambs; though with either one, a little of it should be tried on a few ticks before the dipping begins. A few gallons of water will suffice for the boiling, then the decoction may be diluted with about a barrel of cold water. The keeper of Merinos ought not to be troubled with ticks sufficiently (they are more troublesome on the British breeds) to justify the expense of making special dipping apparatus. Two wash-tubs or large iron kettles will answer the purpose.

A person whose hands have no abrasions of the skin need not fear to plunge them freely into either the tobacco or arsenic decoction. One hand should grasp the lamb's mouth and nostrils (to prevent it from getting the liquid into them), the other the fore-legs, while an assistant holds the hind-legs. The lamb should be lowered, back down, into the liquid and held there until it thoroughly pervades the wool nearly up to the eyes and the roots of the ears. Then let it be placed on its feet in the other tub, and the wool squeezed out. Unless this dipping is very thoroughly performed, some of the eggs of the ticks will escape, and in two weeks the operation must be repeated.

In cold weather, as above remarked, dipping is not advisable; but the ticks may be so held in check by means of sulphur mixed in the salt that they will work the lambs little or no injury until shearing-time comes. Indeed, some very good practical shepherds of my acquaintance assert that they destroy or prevent ticks altogether by the use of sulphur, putting three pounds of sulphur to five of salt, and giving about a handful of the compound twice a week to forty or fifty sheep in their feed. In the summer they are not molested by them, and in the fall, if any are discoverable, they renew the sulphur. When feeding sulphur, I am careful to keep lambs housed from storms.

My experience with sulphur has been so satisfactory that I should never bother with kerosene, snuff, mercurial ointment, or any other substance to be rubbed into the wool.

SALTING.—I take it for granted that every flock-master who peruses these pages never denies his sheep salt, unless it may be from occasional negligence. By keeping it in a covered trough and taking account of the quantity consumed during a series of weeks in early summer, I ascertained that an average sheep requires about one-eighth pint per week. During a protracted drouth, or late in autumn, when the grass has become dry, sheep consume less salt than in the spring when the grass is washy. Strong, healthy sheep, well cared for otherwise, may flourish for an indefinite period without any salt; but every flock-master of extended experience, who has turned that experience to account, is well satisfied that salt is very beneficial to sheep, and that the money it costs is well expended in warding off disease.

In a journey through New Mexico several years ago, I conversed with a resident wool-grower, Mr. Anton Lippart, who stated a remarkable fact in his experience. One winter during a severe and protracted drouth, he lost about twelve hundred sheep, while a neighbor similarly situated lost less than a score. His neighbor saved his sheep *with salt and water!* The liberal supply of salt so toned up and stimulated the sheep, that they consumed the coarsest feed and turned everything to account.

It is wasteful to salt sheep on the ground, even in the cleanest places, but this system has its compensating advantages in that it compels the flock-master to see his sheep once a week, which he might otherwise neglect to do. By scattering the salt in a circle of handfuls, he can count and inspect every member of a large flock. I never found it worth while to provide a covered trough, except in one case, and that was as a receptacle for salt and copperas as a preventive of Paper-skin in lambs. (See Chapter on Diseases). The salt-trough in the pasture serves another useful purpose in accustoming lambs to eat from a trough as a preparation for weaning.

THE DUST BATH.—Some writers and practical men recommend tar, smeared in the salt-trough, and thence attaching itself to the animals' noses, as a repellant of the gad-fly and a preventive of the deposition of its eggs. In a close-fenced and cleared pasture, with no shade, except that beside the fence, tar, or whale oil, may be rubbed on their noses with good effect. I attach great importance to shade and dust. If on the top of some commanding hill or knoll, there is a clump of trees under which the breeze draws cool and refreshing, here the sheep will always be found congregated in the heat of the day, and here each one will wear out, by stamping, a little circular depression for himself in which, with evident satisfaction, he will lie down and get up many times a day, paw, turn round, and otherwise raise a dust into which to thrust his nose. He will lie for an hour or more with his nose close pressed against the ground, inhaling the dust. It is an instinct; he seeks in this way to escape his enemy.

The gad-fly is more apt to trouble lambs and tegs than older sheep, and I deem it a matter of importance to provide for these, if possible, an enclosed building as a refuge during the heat of summer. Even a shed with only one side, if it is somewhat dark and cool, is a better protection against the fly than the open field or a thin coppice.

WEANING LAMBS.—If they are thriving as well as they ought, lambs need not run with the ewes above four months. They will be more quiet if left in the field they are accustomed to, with the ewes removed out of sight and hearing.

If there are shade and water in the field which they know where to find, they will help themselves. If not, they ought to be driven to water every day; and it is a good plan to fetch them to the stable before the sun gets very hot, to prevent them from rambling aimlessly about the field, panting in the sunshine, or crowding into the fence-corners.

The lambs should have a fresh rowen or an upland pasture, if one is available, well stocked with June grass, Red-top, or some other short, tender, nutritious grass. There should be strips of forest in it, with shady knolls for stamping-grounds, where they may find an abundance of the dust which is so essential to their health during the dog days. An old ewe should be left with them for a flock-leader. If they are accustomed during the summer to a stationary salt-trough, the task of teaching them to eat feed will be reduced to a trifle, as they will approach the troughs freely. A mere dusting of salt should be sprinkled on their feed for a few days (being withheld from them otherwise); after that it may be left in quantity in the trough appropriated to it, or sprinkled on a clean sod. It is of the highest importance that lambs and yearlings should have daily access to salt, summer and winter, at least in a humid climate. I will give a brief description of my mode of making a salt-trough. For the supports take two equal pieces of one-and-a-half-inch plank, fifteen inches wide, and saw notches in the top deep enough to receive the trough. Make the trough V-shaped, sixteen feet long, of boards six inches wide, using for end-boards the pieces sawed out of the plank. Let the supports be about eighteen inches long, and nail to them, one on each side of the trough, upright standards. Across these standards at the top nail two V-shaped pieces to support the roof, which is made like the trough and turned bottom up. The standards must be high enough to allow the sheep to insert their heads freely between the roof and trough, which requires a space of about nine inches.

For a safe, nutritious, healthy, universally available and everywhere procurable feed for weaned lambs, there is nothing which is comparable to wheat bran. I find it profitable to enrich it by the addition of a little shorts or oil-cake meal. In default of this, let a small proportion of oats be introduced into the ration when the frost falls, and some corn when the snow flies. Buckwheat bran is too coarse and rough for lambs.

TAGGING LAMBS. — Merino lambs four months old should have wool of considerable length, and in the heat of midsummer this renders them liable to the invasion of those detestable vermin, the maggots. Out of a flock of one hundred and thirty lambs, I have lost over twenty in less than two weeks from this source alone. Of late years I have invariably tagged at weaning all the ewe-lambs, and as many wethers as showed signs of fouling about the pizzle. In very hot, muggy weather sheep will sometimes become fly-blown anywhere about the fleece if there is the least fetor attaching to the animal, around the hoofs, the head, the wrinkles, or the natural orifices of the body. The most rigid cleanliness must be maintained to carry lambs through the dog-days in bad years. Four hours' work in tagging may save ten times that amount of the most odious drudgery the shepherd has—fighting the maggots.

SUMMER HOUSING AND FEEDING.—Some very good shepherds, indeed a great majority of the keepers of stud-flocks, give their sheep a little hay all summer. It is only a very little, and that of very sweet hay. A still smaller number give lambs and choice rams a daily ration of grain, generally consisting of wheat bran and oats mixed in about equal portions. It is claimed that this dry feeding in summer steadies the animal's appetite, acts as a corrective of acidity and flatulency, a preventive of colic and scours, and a general tonic to the system; this more especially when the weather is exceptionally wet and the grass slushy. To the breeders of high-priced standard sheep there is undoubtedly much force in this argument; they find profit in the course above indicated; and, conducted within the careful, reasonable limits implied in the foregoing statement, it affords no just ground for the odious charge of pampering.

Neither have I any quarrel with the veteran shepherd who chooses to house his flock every day in the year, and who would suffer a load of hay to take a shower rather than a dozen favorite sheep. It would argue the height of folly to assume that he does not know his business, and that this policy is necessarily incompatible with common honesty. Our countrymen who breed fine stock may be trusted to discover ultimately those methods which will develop that stock to the acme of symmetry and beauty. And it cannot be denied that a Merino systematically housed and blanketed is much more pleasing to the view than one which exposure has rendered rough and shaggy. The soft, moist feel of the exterior, devoid of clots or indurations; the rich, dull luster of orange or gold revealed in the deep clefts between the blocks when opened; the fibers glistening, when held up separate, with a pellucid, semiliquid unguent—these are eminently satisfactory to the admirer of fine sheep. A fleece which has been housed for some time and is then exposed to the rains, bleaches out dirty-white, yellowish, yellow-gray, brown, or remains black, according to the consistence of the yolk; the latter has its stratifications destroyed and is washed down into the wool and into disfiguring masses like the drift along a stream, etc. A frost on a fleece is considered even more injurious to its appearance than a rain. I appreciate the artistic perception which delights in the full and fat exterior; the soft, flannel-like fleece, which yet offers a firm and thick handful where grasped; the eyes closely walled about with wool; the silken white nose and ears; the comfortable, buttoned-up chin and cheeks—the perfect presentment of hearty and well-fed opulence.

All these things may be fair and honest, they may be matters of legitimate pride and art. Everything depends on the master's motive in this summer feeding and housing.

These practices will be found only in stud or standard flocks. And when it becomes necessary for the farmer to bring sheep down from the high level of the stud-flock to the *niveau* of the plain, out-door, wool-bearing flock, he will find—such has been my experience—that hardly any amount of summer-housing will unfit the sheep for a gradual, progressive and judicious initiation into the ways of a working flock, but that irreparable mischief may be wrought by high feeding.

My father once bought a ram for four hundred dollars, which soon developed goitre and partial impotency, and died when he should have been in his prime. It was a mystery to him at the time, but subsequent investigation revealed that he had been grossly pampered.

I paid a high rent for a ram one year, and out of seventy-five ewes served by him, a great part came in heat a second time, and less than forty bore lambs of his getting. He was a large and powerful two-year-old, but in less than a year he died suddenly and mysteriously. He had undoubtedly been over-fed, but not intentionally, as his owner made honorable restitution.

Over-feeding and excessive fatness are the cause of some barrenness among Merino ewes, and, as indicated in a previous chapter, of weakness, under-size and lack of constitution in lambs. But the unscrupulous men who practice pampering on their show-sheep and their sale-sheep are well aware of this fact, and do not allow themselves to be losers from their disreputable doings. A friend informs me that, during a visit to the farm of a noted breeder in Vermont, after looking long and with undisguised admiration at the various flocks paraded for his inspection, he inquired in some surprise where his breeding flock was. He was told that they were "not in good condition to be seen," but, on insisting somewhat, he was conducted to a stony, rugged hill-pasture, where they found the ewes literally "roughing it"—a shaggy-looking lot, but rosy-skinned and hardy, the very picture of health and thrift!

The Merino is tolerant of much abuse, and when well-fed it will submit to the most rigid imprisonment for a long time with impunity and with apparent thrift. Indeed, for animals fattening for the shambles, destined to be butchered in a few months, this confinement is probably conducive to the highest profit; but stock sheep subjected to it will go to pieces in the end.

Exercise, *labor*, WORK, is the law of all being; and a violation of it will inexorably entail the penalty at last.

CHAPTER XIV.

FROM GRASS TO HAY.

SHEEP IN CORN.—In seasons when there is not much wind and the corn stands up well, it is frequently advisable to turn flocks of young sheep into the standing corn a week or two before cutting it begins. There are many leaves on the lower portion of the stalks which are never harvested, besides weeds which impede the labor of cutting, all of which sheep will consume for a change. It is best to alternate the flocks, shifting them every few days. To one not accustomed to the experience, it is surprising to see how clean and tidy a flock will clear up a corn-field—what an immense amount of trash they will consume. But it is necessary to be on the lookout for the equinoctial storms. I once had a flock caught in a two-days' rain, and they bogged down to the middle in the plowed ground, so that we had to carry some out a-shoulder.

IN ORCHARDS.—Sheep are better scavengers in a bearing orchard than hogs, notwithstanding they will bark small trees. Even if ringed, hogs will exterminate most grasses in a small lot, but orchard grass will flourish under the trees and under the hardest gnawing of the sheep. Besides that, sheep will eat up all the windfalls, no matter how small, bitter, astringent or rotten, with a more unquestioning appetite than swine; hence they protect the trees more effectually against insect enemies. It is mainly old suckling ewes that damage the trees, and these only in the spring when herbage is scanty. They may be prevented from gnawing the bark by an application of coal tar, kerosene, tar, or a wash prepared by mixing one quart of soft soap, one quart of lime, one quart of pine tar with three gallons of sheep, cow or hen manure, stirring in a sufficient quantity of water to make it about the same consistency as ordinary whitewash. Apply to the body of the trees with a whitewash brush, splint broom, or with the hand well protected with a heavy cloth mitten. This wash will protect the trees against injury from sheep, except the rams' horns, and is also conducive to the growth and health of the trees. It is valuable in preventing the damages so frequently done by insects, worms, etc.; for this purpose apply as near the roots as possible, and as often as it is washed off by the rain from the body of the tree.

But most farmers in the busy season will forget to renew the application, and at best it will not prevent damage by the rams' horns. Hence I have found the best practical protection to be stakes; locust stakes will last from six to ten years or more.

A few sheep may be kept in an orchard which does not afford enough herbage for their support; and, if fed on pumpkins, turnip-tops, apple pomace, salt-hay, brewers' grains, sweet-corn fodder, or fodder-corn, they will rid the orchard of every weed, down to yellow dock, burdock, elder, poke, and even stunt the thistles if salt is thrown around them. But they incur some risks; I once had a valuable ewe choked by a clingstone peach.

SOILING SHEEP.—Green feed soon becomes stale in a rack; it is necessary to feed sheep "little and often." With the mutton-breeds what may be called out-door soiling, or hurdle-feeding on roots, rape, mustard, etc., is often found profitable; but it will seldom repay the labor to soil Merino sheep in the ordinary meaning of the term, except as above suggested, in an orchard or some small lot which it is desired to free from weeds and briers.

MAINTAINING AN EVEN CONDITION.—I wish to impress strongly upon the mind of the inexperienced flock-master the necessity of keeping up an even, uniform condition, a progressive growth in his flocks, throughout the year. Not only do the horse and steer give quicker note of a falling-off (by their hair beginning to stand out straight and other indications), than the sheep whose carcass is deeply hidden from the master's eyes in a voluminous fleece; but the horse and the steer, by reason of their stronger muscular and vascular systems, will also more easily recover from a temporary decline.

After the fairs are all over and the show-sheep turned out, the ribbons laid away as trophies, the busy farmer—busiest now of all times of the year—is apt to neglect his flocks, and they enter upon the down-grade. But all the while he is driving his fall work, or perhaps chatting at the corner-store, there is a secret recorder that is every day, like the priest behind the wall in the Inquisition, laying up secret evidence against him, jotting down its own note and comment, which the expert may open and read.

What is this mysterious spy? It is the fiber of the wool. Let the sheep be neglected a few weeks in the late autumn and lose condition, let it fall sick, let it even be violently chased by dogs for twenty minutes, and the fiber will be "jointed," there will be a weak place in it which will cause it to break in the cards or the loom. The reader may puff out his cheeks at this as a mere bit of sentiment; but there is a case on record where a Boston expert told the much-wondering farmer that he had moved his flock from a wooded to a prairie region, and informed him in what month he did it—all from the simple evidence furnished by the fleeces.

Eternal vigilance is the price of good wool. The perfect Merino fiber of Ohio, Pennsylvania and West Virginia—true and sound, of a uniform diameter throughout its whole extent—admirably typifies the ceaseless care and the untiring industry of the true shepherd; while the staple of Australia, thin at one end, thick at the other, with perhaps one or more attenuations between, fitly represents a slipshod, "feast-and-famine" system of husbandry.

FALL CARE.—All the flocks, especially the lambs and the breeding ewes, should be vigilantly watched at this time of the year. As soon as heavy frosts begin to fall the sheep ought to be housed at night, and not turned out in the morning until the frost disappears, as they will frequently wander around an hour or more, doing themselves no good and the pasture much damage. Wherever they touch a frosty clover-leaf or other tender herbage, it is ruined, whereas if it had been allowed to thaw out untouched, it would have been uninjured. I never lost any sheep from frozen clover, but it will physic young sheep and put them in ill condition to enter winter quarters.

If the autumn has not been too rainy, second-growth clover, cut and cured, will be excellent feed to shade off on from grass to hay—for the ewes ought to have a little dry feed in their mangers while waiting in the morning for the frost to melt. But in a wet season, clover rowen is not fit for hay; it will "slobber" anything except hogs. I have had sheep killed by it. The farmer can easily tell whether it will be safe to harvest it by testing a horse with it while green.

Very rank clover, grown on river bottoms and cut while green, will sometimes cause ewes to "slink" their lambs; even the first cutting has done this, to say nothing of the second; at least, such has been my experience. Yet I should not hesitate to give upland clover to pregnant ewes without stint.

FALL FEED FOR LAMBS.—One year my pastures were much curtailed by a severe drought, and I was somewhat puzzled how to provide for my lambs a supply of that succulent herbage which is so necessary to their thrift. The cossets running about the house had access to a turnip-patch of two or three acres, and, observing them cropping the tops, I conceived the idea of turning the entire flock into the patch for a limited time each day. The plan worked admirably; in course of time the lambs had completely stripped off the tops, thus saving me the most onerous part of the labor of harvesting turnips, and they had only here and there taken a mouthful from a turnip, not impairing them in the least for use the following spring. It supplemented the fall feed admirably, and carried the lambs into winter quarters in excellent condition. A slight tendency to scours developed itself after the tops were severely frosted, but it was easily corrected by lessening their daily run on the turnips and increasing the ration of hay and bran.

Pumpkins are good feed for lambs in autumn (see Chapter on Paperskin). They will eat them tolerably well if broken up on a very clean and close sward; but it is better to provide flat-bottomed troughs with compartments, each being large enough to receive the half of a pumpkin split in such fashion as to lie flat, with the inside uppermost.

Acorns are a valuable resource for grown sheep, but I have not had favorable results when I allowed lambs to run freely in an oak forest. The acorns have almost invariably been productive of scours.

One thing is certain—lambs must be grained liberally, or else they must have a very choice reserve of green feed to wind up the grazing season on, or they will lose ground and go into winter quarters on the down grade. I feed my lambs more grain in November than in January. In January they are well established in their winter habits and have an abundance of the best and sweetest hay; whereas in November they are in a transition condition, gathering up under protest the leavings of the summer grass which the frost has weakened. I mix one part oats to two of bran, and of this I give about a bushel and a half a day to one hundred head.

AT THE END OF THE SEASON.—Sometimes an inch or two of snow will fall on the grass before it is time to bring the flocks into winter quarters, and lie a few days; or it may be desirable for other reasons to keep the sheep out a little beyond such time as the pasturage, unaided, would keep them in good flesh. I have found it advantageous under these circumstances to carry out, say a half bushel of shelled corn to the hundred grown sheep, and sow it broadcast on a short, clean sod. This enables all to share equally. On the north hillsides grass nearly always grows ranker than elsewhere, and the sheep will pass by these strong-growing patches all summer. Late in the fall they can be made, with the help of a small ration of corn, to depasture them down and so leave the pasture uniform. These tussocks would otherwise afford a winter harbor for ground mice. Sometimes I have found it advantageous to keep a few young cattle with a flock; they will graze these north hillsides, while the sheep will keep on the south slopes.

CHAPTER XV.

SELECTION AND CARE OF RAMS.

CONSTITUTION.—"A steep rump and a crooked leg," is one of the shepherd's catch-words. A crooked leg generally means also a "cat-ham," and a cat-ham is usually a sign of weakness. Still, however objectionable these points may be, they are not to be compared with flat nostrils (almost invariably accompanied by catarrh and a disgusting accumulation of mucus in the nares); weak pasterns, causing the animal to walk somewhat flat-footed, plantigrade, or bear-fashion; a straight, thin, ewe-nose; and a fine ewe-fleece—all of which denote a poor constitution.

The test of supreme importance is the bright, rosy skin. A ram may have excrescences; yet if he has this, he possesses vigor. Mr. G. B. Quinn's "Red Legs" had a shambling anatomy, thin shoulders, and steep rump; yet he had great power. Mr. C. C. Smith's "Silver Horn" was excessively wrinkly, as the annexed measurements show; still he had sufficient vigor.

"Silver Horn," live weight .... $138^{1}/_{2}$ pounds.
Length .... 3 feet 7 inches.
Total length (including wrinkles) ...... 9 " 5 "
Through shoulders .... $7^{2}/_{12}$ "
Through hips .... $9^{6}/_{12}$ "
Height .... 2 " 2 "
Length of neck .... $11^{3}/_{4}$ "
Girth (about the heart) .... 3 " $1/_{2}$ "
Width of loin .... 6 "
Width of escutcheon .... 7 "
Length of nose .... $8^{1}/_{2}$ "
Length of nose not wooled .... $1^{3}/_{4}$ "
Depth of flank wrinkle .... $6^{3}/_{4}$ "
Escutcheon wrinkle overlaps .... $2^{7}/_{12}$ "

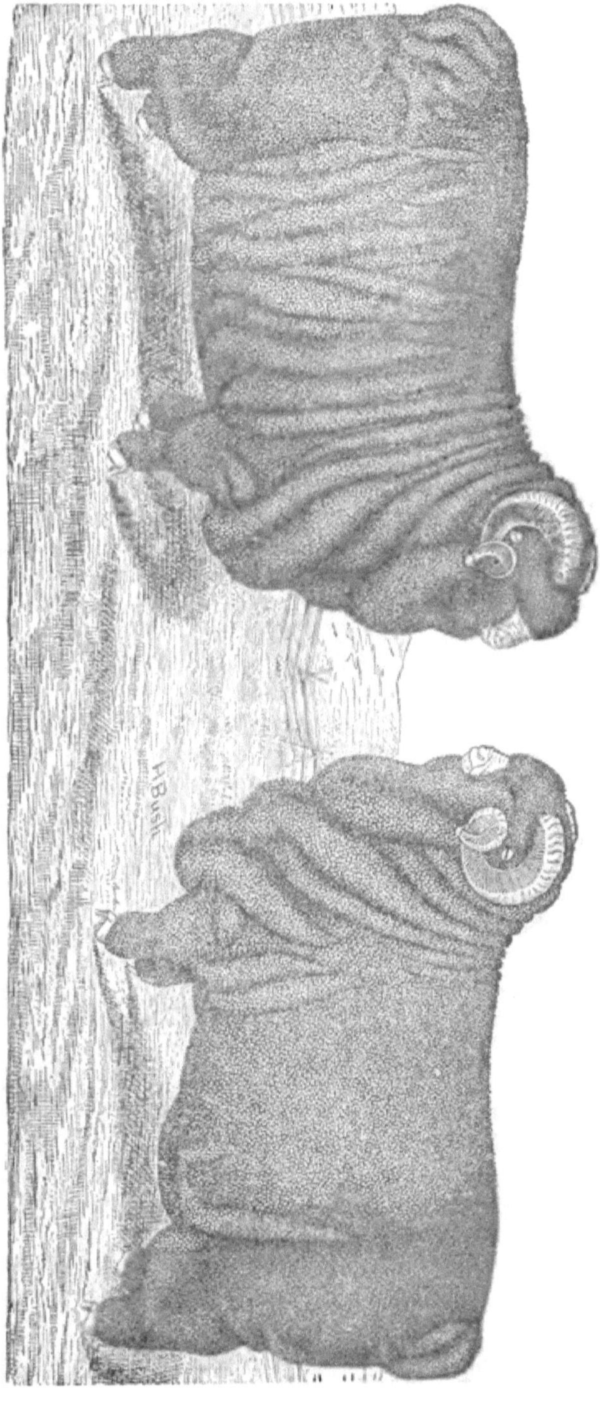

FOR WOOL AND MUTTON. 143

Fig. 9.—TWO RAMS, BELONGING TO MESSRS. THOMPSON & MOORE, WASHINGTON CO., O.

POINTS OF A GOOD RAM.—Let him have clean, short, shining hoofs, which never require the toe-clippers; a round barrel; a good diameter through the hams and shoulders; a neck well set on, thick, powerful, devoid of the feeble Saxon droop just in front of the shoulders; a nose held nearly perpendicular, arched, reddish, covered with fine corrugations, and in mature age, having two deep channels running from the inner corners of the eyes slanting down athwart the face; nostrils round and well-opened; eyes large and brilliant; horns, when grown, making one turn and a half, close to the head, spanning clear across the forehead, deep, with a sharp, cutting edge underneath, and with clean, clear-grained wrinkles, thickly set together. Let his ears be hot, so that blood will flow freely from a cut. A cold-eared, cold-blooded animal is of no value. Such a sheep does not possess sufficient animal heat to keep his yolk liquescent and diffused to the extremities of the fibers. The scrotum should be well covered, the wool joining on to the belly; the spermatic cords thick and large, and the investing skin of a bright, ruddy color. A long, pendulous scrotum with small cords betokens a weak constitution. I like to see the neck swelling into voluminous folds. especially a liberal apron; the body plain; the stifle and ham slashed with two or three obliquely transverse wrinkles free from gare. But best of all is a broad, horseshoe-shaped escutcheon, a tail nearly as wide as a man's two hands, with the skin at the sides folded and tucked under, which indicates, in my opinion, generous breeding and generous blood. The Hiatt Bro.'s ram, "Ohio," had the finest escutcheon I ever saw on any sheep.

As to fleece, so far as my observation goes, the more vigorous the ram, generally, the whiter the wool he produces. I know full well the beauty of those fleeces which, as the animal's body bends a little to one side, reveal deep rifts of a rich reddish-yellow, like the color of California gold; but they are not so hardy generally.

A ram should be sought that has a short and broad head, and powerful jaws, the lower one spread well apart. Between the lower jaws and under the tongue are the salivary glands, and if the jaws are well spread these glands will be large and afford a good supply of saliva, a very important ingredient in digestion. When the head is long and the jaws lacking in width, these glands will be small and not yield sufficient to carry on digestion with a force always assuring the animal's good condition.

OPPOSITES TO BE MATED.—Another important point is to se-

FOR WOOL AND MUTTON.   145

lect none but those that appear full of life, wide awake, with eyes not partly closed, but wide open. An active temperament is always indicated by bright, sparkling eyes and the two set well apart. A ram with the right form and temperament when crossed with ewes unlike himself, will give an increase, carry-

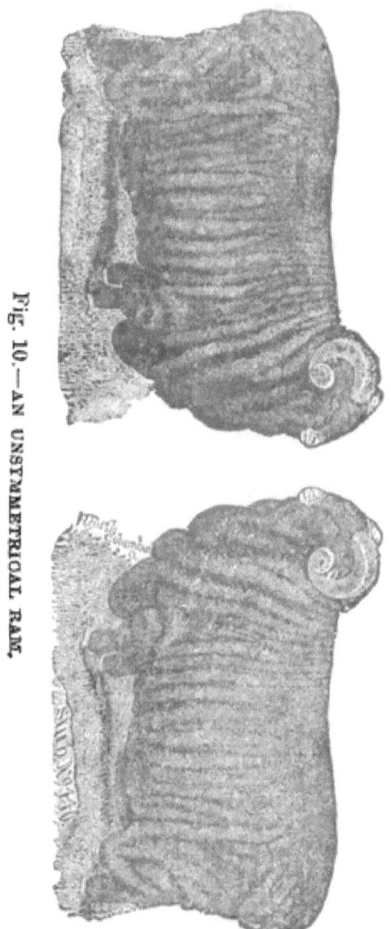

Fig. 10.—AN UNSYMMETRICAL RAM.

ing heavier fleeces than sire or dam. In chemistry it requires two distinct properties to produce the third; it takes two distinct gases to make a drop of water, and two opposite winds to blow the misty vapor together to form rain drops; and in generation two opposites are required to produce strong and healthy issue. It looks as though the power that governs the universe had a great aversion to perfect sameness, for there are no two things in nature exactly alike; and few animals of the same family and line of breeding are so near alike as not to be easily distinguished one from the other. And because this is so, it is hard to tell if ever a point could be reached beyond which no improvement could be made.

The question of in-and-in breeding often comes up for discussion among the best breeders of all kinds of stock. It is fully settled to be safe, to a certain degree, but in all such experiments as these, a full knowledge of the traits and qualities peculiar to both lines of ancestry must be possessed by the breeder, or serious mistakes will be made. In sheep breeding it rarely occurs that any chance need be taken in this particular. Near relatives may be coupled with better results when there is a sufficient distance existing between, than can those that are too much alike, when there is no relationship existing.

CORRELATION OF WOOL AND YOLK.—It is a common remark of the keepers of stud-flocks, that the rams which scour the most wool shear the heaviest fleeces. This may be set down as the major premise in a favorite line of argument, while the minor premise would be, that the heaviest fleece is what the wool-grower wants. Another common argument is (to use a homely comparison), that yolk is the peculiar sustaining or nourishing element which creates wool, very much as "mother" is the sustainer and nourisher of vinegar. (I shall, in another place, refer more at length to this theory).

The essential fallacy of this theory consists in ignoring, or overlooking, the fact that the keeper of the stud-flock seeks one object and the wool-grower another. In the wool-flock a ram is desired in which the oil-follicles are so developed, correlatively, as to insure the highest possible development of the wool-follicles—but no higher. In the stud-flock a ram is required in which there is the highest possible activity of the oil-follicles, because it it is his function to mate with the native, say, of New Mexico, in which there is no development of the oil-follicles at all. The farmer should carefully observe this distinction.

Let me illustrate: The famous "Patrick Henry," owned by L. P. Clark, of Vermont, sheared thirty-seven pounds and scoured nine pounds and ten ounces; that is, his fleece lost in the scouring-tub seventy-four per cent. A ram shorn at Sedalia, Mo., clipped twenty-eight pounds and four ounces, and his fleece, scoured by Walter Brown & Co., of Boston, cleansed seven pounds and fourteen ounces, a loss of seventy-one and sixty-nine hundredths per cent. Another one sheared twenty-eight pounds and fourteen ounces; scoured seven pounds and fifteen and one-half ounces, a shrinkage of seventy-two and forty hundredths per cent. Now take other rams, shearing a medium-weight fleece, and we find the shrinkage is not so great in percentage. For instance, one shearing twelve pounds and six ounces, in the same lot, showed a percentage of loss of only sixty-one and thirty-six hundredths. Others ran along in the same vicinity. The point I wish to make and to emphasize is, that the heaviest shearers are the heaviest losers. It is a common saying and a truthful one, that it is the extra five bushels of wheat per acre which makes the profit. This principle will not apply to the excessively yolky fleeces, but rather that other one: "The last straw breaks the camel's back." The great Vermont ram had to produce eight pounds and twelve ounces of yolk to beat his Missouri competitor one pound and twelve ounces in wool. Such an animal might be, and doubtless was, highly valuable for stud-flock purposes; but he would not have to the ordinary wool-grower (unless his ewes were exceptionally dry-topped), an increased value at all commensurate with the increased percentage of yolk in his fleece. Of course, this celebrated ram possessed a peculiar aptitude for the secretion of yolk (though this can be greatly augmented in any Merino by very rich, copious feeding); but it is only a truism to assert that yolk is valueless, except in so far as it involves the production of wool. And surely no one could be found to believe that the bushels of rich feed required to produce eight pounds and twelve ounces of yolk, were not worth more than one pound and twelve ounces gain in wool. I do not deny that such a ram is a prize in a stud-flock, but it is only because the monstrous extreme of yolkiness in Vermont is matched against the monstrous extreme of dryness in New Mexico.

MANAGEMENT IN SUMMER.—It is best to have a thoroughly experienced workman to shear the rams, and pay him his price, even if it is a dollar a head. There are very few shearers who will give proper attention to shearing closely around the horns. Oftentimes quack shearers will only half shear, in their hurry, because of the inconvenience and labor of getting behind and around the horns. It is only with some effort on the part of the shearer at this point, that the job is completed in a workmanlike manner. There are rams that need their horns "slabbed." That is, their horns grow so near their heads, as they circle, as to come in contact with the jaw bone, and if not removed often cause death. At shearing-time this should be attended to ; let one man hold, and with a sharp saw you can soon remove a wedged-shaped piece that will answer the requirement. I have always used for this purpose a small **tenon-saw, and the same will answer** for removing the rudimentary, **re-entering horns** which sometimes give trouble to wethers.

Before the ram is dismissed to the pasture, it is well to **give him** a very light smearing of tar close around the base **of the horns;** the **fetor** which prevails there is apt to attract flies, and maggots will **result.** It is surprising how quickly these abominable vermin will destroy a powerful ram if he is not promptly **taken in hand.** They soon invade the ears, and spread and multiply with amazing rapidity, until they invest the whole neck and breast; a disgusting **stench arises;** fever is created, and the **wretched** creature perishes in agony.

Some shepherds fastidiously object to the **tar** being smeared around the horns. There is no necessity for it if the ram is housed, or kept close to the house where the owner will see him every day through the summer; but if he is at a distance, it is best to employ the tar. If very lightly put on it will not damage any wool which is of value, nor injure the animal's appearance; it is lasting in its effects, so that it will not have to be renewed more than once during the summer, while fish-oil or whale-oil will evaporate in a fortnight.

To PREVENT FIGHTING.—The ram must have some company during the summer, and a little bunch of calves or hogs will answer all purposes, if he is kept out of sight and hearing of other sheep. Or he may be placed with a few refuse wethers which it is desired to fatten and sell. In whatever company he is kept, it is best to have him not very distant from the house. There is no other domestic animal so restless and liable to escape, especially as autumn approaches, as the ram, and generally the more valuable he is as a lamb-getter, the more restless and pugnacious he is.

If two or more rams are kept together, they are liable to fight; first, in the spring when freshly shorn; second, toward autumn when the coupling season is coming on. When freshly shorn they sometimes fail to recognize each other, and toward autumn the awakening procreative instinct renders them quarrelsome. At this latter season it is important to keep other sheep at a distance. I have had two rams, which had lived peaceably

FOR WOOL AND MUTTON.

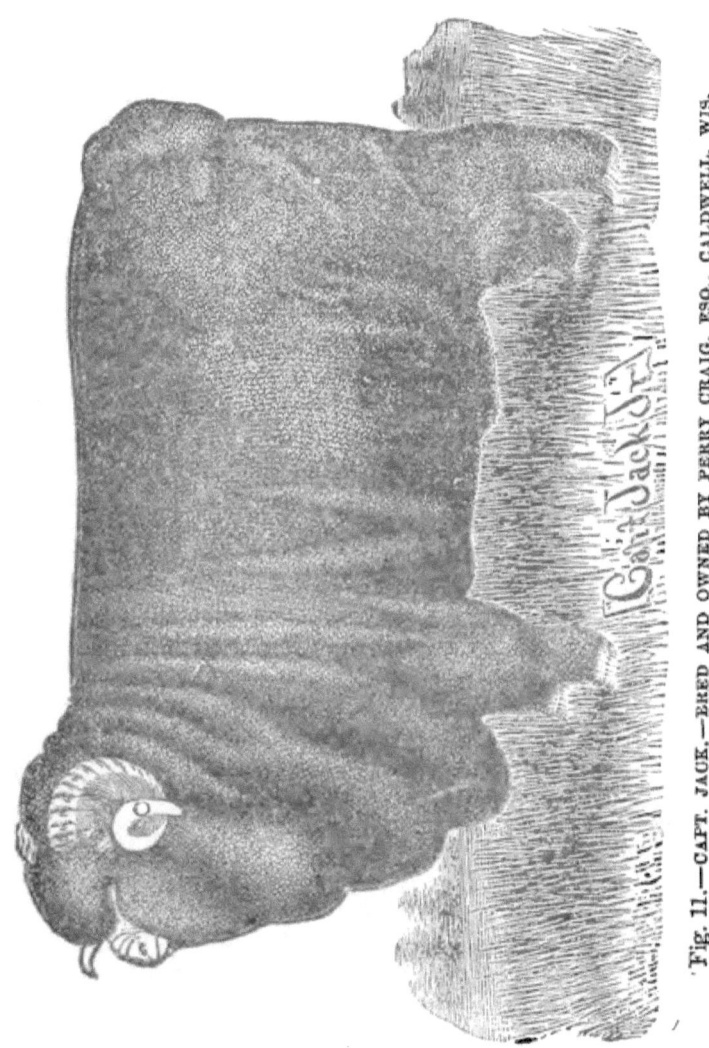

Fig. 11.—CAPT. JACK.—BRED AND OWNED BY PERRY CRAIG, ESQ., CALDWELL, WIS.

together all summer, become so excited by a flock of lambs that were driven by, that they fell foul of each other and the less vigorous one was well-nigh killed.

If they are housed at night, they may be put into a tolerably small apartment; by keeping thus closely together they do not have room to harm each other, and will soon become sufficiently acquainted so that they can be driven to pasture with but little fear of fighting. Should there be one or more that feel disposed to continue their combativeness, drive them to the barn, procure a piece of leather about seven inches square—an old boot top will answer—then with a sharp knife cut as in figure 12. The upper part of this cap is placed on top of the head, between the horns; then tie the two points on each side together, around the horns. A little practice will enable one to fit a cap in this manner as nicely as a shoemaker will fit a boot to the foot. If necessary the cap can be drawn tight to the nose by making holes, and tying from the sides underneath the jaw. This cap will entirely destroy a front view, and at the same time give a side view, enabling the animal to travel about where he chooses. This will stop the fighting; at least it will so confound the rams that they can not deliver effective battle.

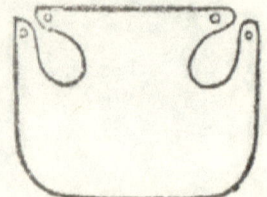

Fig. 12.—BLINDER.

How to Tie a Ram.—During service it is necessary to keep the ram shut up a greater part of the time, if not constantly. He will get little enough exercise at best, and will generally vent his impatience by butting. I make no particular attempt to curb him, but rather set up some springy boards that he can not damage and let him practice on them. To restrain a ram at all from his natural liberty during his service is a necessary evil, and it should be mitigated every way that is possible.

Rams are so restless under confinement, that where a number of them are in one apartment it is necessary to use the greatest care in fastening them, lest one should get loose and hammer another one to death. In the first place, pierce the left horn in front with a gimlet, then insert a three-sixteenth-inch staple and ring. In this ring have a leather loop six inches long, and in this loop insert the snap of a stout dog chain, for if the snap is put directly in the iron ring, the animal will work it out in spite of all precautions. The chain should be provided with a swivel, and the T at the end should be passed through an auger hole in a board. Thus secured he is generally safe; but if he is exceptionally restless, it will be found advisable to attach a stout, leather hitching-strap to the ring, and tie him up short.

FEEDING.—I take it for granted that every progressive shepherd no longer follows the plan of turning the ram in with the flock, but rather stables him and thereby husbands his powers. Some seek to compromise by turning him into the flock during the daytime, and removing him at night, but this way is very little better than the other. The ram ought to be taken up long enough before his service begins to get the grass bloat out of him, say a week or ten days. He should be accustomed gradually to dry feed, and there is nothing better to assist him in the transit from grass to hay than sweet-corn stalks or pumpkins. Give him half a small pumpkin in his box, flesh side up, and let him scoop it out at his leisure; it will give him exercise. Furnish him all he will eat, three times a day, of the best hay on the farm, adding thereto only so much grain as may be necessary to keep him in good stock condition—a trifle lean, if anything, so that he will consume his grain and pumpkin with relish, and never leave any in his feed-box to get stale. I have given rams wheat, rye, oats, corn and bran, separately and combined, in various ways. Theoretically, the more glutinous grains are better for him, but practically I see no difference; at least not when the animal receives a liberal ration of pumpkin. I should hesitate to give so heating a grain as corn to a ram, in large feeds, unless he had with it plenty of green sweet-corn or grass, or pumpkin. With a generous supply of the latter he will eat two or three ears of corn per day, and yet refuse water for days together. I have settled down practically to corn for a grain ration; I give one average ear a day before service begins, and two during service, varying somewhat according to the size and appetite of the ram. Of corn, oats and bran mixed in equal parts, I should give three times a day what I could conveniently grasp in one hand. By all means contrive somehow to give the ram some exercise and sunlight in a dry paddock or barnyard. A ram in service requires above all things, muscle —clear muscle, not clogged or dulled with fat.

Pumpkin should not be given to a ram after it has once been frozen; it is liable to give him the scours. Neither should it be unripe or rotten, or be given with all the seeds. Small apples or potatoes are also good as a laxative.

MANAGEMENT OF THE SERVICE.— After experimenting considerably with different methods, I have adopted the following plan with the breeding flock: I drive them up in the morning as soon as the sun has warmed up the atmosphere and yard them. Then I turn loose among them the most energetic one of the rams, and follow him up leisurely with the crook in hand. As fast as he discovers the ewes which are in season, I catch them and put them into a separate enclosure, until they are all drafted out, or until enough are secured for the day's operations. Then I dismiss the flock for the remainder of the day. It only remains now to sort them and select those which are best adapted by their individual qualities to the several rams, and turn one at a time into a smaller pen with the appropriate ram. It is best not to allow but a single effective service. I have lately adopted the plan of permitting each ram to cover no more than three ewes per day, with an interval of at least two hours between the services. This eliminates the possibility of any impairment of vigor, and secures strong, healthy lambs.

Oftentimes the most valuable ram is slow and clumsy, and in this case the shepherd can save time and avoid trouble by holding the ewe by the neck until she has been effectively served. If the ram's sheath hangs too low it will be necessary to belt him up somewhat tight with a leather surcingle. Sometimes he can be materially assisted by being allowed the benefit of a little slope in the ground or of a table a few inches high.

CROSS RAMS.—When a very good ram is incurably vicious, his services may still be retained by keeping him constantly chained up and bringing the ewes to him. In this way he can never get the advantage of the shepherd. At other times he can be rendered harmless by the leather cap described on page 150. Constitution is of such transcendent importance in the sheep, that a fighting ram is likely to be exceptionally valuable, and he ought never to be killed for that fault alone. Most cross rams, if not too old, can be subdued by two or three vigorous kickings in the shoulder; let the master seize him by the horn and put in the kicks until he has enough. A small hoop-pole, with two feet of the little end slightly twisted to make it pliable, can be applied with good effect about his nose and legs. Mr. E. J. Hiatt quaintly says: "A small mallet or light hammer carefully applied to the head or butt of the horns will satisfy any ram, and we always allow the ram the privilege of deciding how frequent and how severe the application must be. We are careful not to encourage a quarrel with a ram, but when nothing else will satisfy him, the remedy should be promptly applied." The keeper of a stud-flock generally has the leisure and the opportunity to make pets of his rams, to train them up gentle from the beginning; but the ordinary shepherd can seldom find time for this.

WINTER TREATMENT.—Rams usually come from service into winter quarters more or less reduced in vitality, and require careful treatment during the winter. The grain ration given during service should not be discontinued for some time ; the ram should be placed in a clean, warm apartment, freshly littered every few days; and be allowed to have his liberty for a few hours every other day or so, though he may be tied up very short all the rest of the time without injury. It is well to keep him blanketed until spring. A suitable blanket may be made of gunny-cloth or stout muslin, by cutting it to cover the body only, with loops of strings at the corners through which to pass the legs.

ONE RAM, OR MORE.—It is undeniable that greater uniformity can be secured by the use of a single ram; and when he is of known and tested power, he can be depended on to do an astonishing amount of work without injury, if his vigor is properly husbanded. The noted ram, "Fortune," owned by Mr. Solomon W. Jewett, of Vermont, used to get about two hundred lambs every year. Mr. Paris Gibson states that he had a ram which served three hundred and twenty ewes in one season, getting three hundred and fifty lambs, then sheared twenty-six pounds of wool, and the following season made an equally good record. Dr. Randall states that the "Old Robinson Ram" was believed to have gotten over three thousand lambs in his life of thirteen or fourteen years.

I said above that greater uniformity could be secured by the use of one ram than by the use of several. This would probably be the case in respect of the form of the lambs, but it might be fairly questioned whether this result could be expected in regard to their fleeces. The question as to the relative influence of the male and female in determining the external and internal characteristics of their progeny, is largely a speculative one and does not profoundly concern the practical shepherd.

But uniformity presupposes perfection and precludes progress. If the breeder is satisfied that he has a perfect flock, he will not wish to depart from the standard in any respect. But I never saw a flock, even of registered full-breeds, which did not exhibit much variability. And indeed improvement is impossible in any flock which does not. It is only by selecting those individuals which vary in a useful or promising direction, and repeating the process as often as we discern a departure toward betterment, that we can elevate the standard of the breed. And the larger the flock, the greater will be the number of promising variations, the wider will be our range of selection, and the more rapid will be our progress. Marshall, as quoted by Darwin, used to say of the sheep of Yorkshire: "As they generally belong to poor people, and are mostly in small lots, they never can be improved."

If, then, the size of the flock will at all justify the expense, it is well to have two, three or more rams; and the most obvious difference between them would be that one should be somewhat yolky, to serve the too dry-topped ewes, and the other the reverse. I have never seen but one flock approaching so nearly to absolute uniformity in body and fleece that two rams could not be employed upon it to advantage, and that was owned by Mr. Columbus Cheadle, of Morgan County, Ohio,—the work of a life-time.

"STUBBLING," BLACKING, ETC.—As a general principle, the owner of a sheep may legitimately do anything to improve its appearance, which will not injure its health or procreative powers; but, if questioned by the novice for honest information, he should honestly give it.

To shear a sheep with a "stubble" all over the body is wrong, even if it is so stated to the buyer or to the committee of a fair, because it is then impossible to tell accurately what the length of fiber would have been if shorn with ordinary closeness. This is a gross and clumsy fraud. But to "stubble" the cap—which is an almost universal practice with breeders now—to improve the appearance of the head, is legitimate, if so stated upon interrogation.

The practice of dressing the fleece with lampblack has been abandoned by most breeders, even by the dishonest. It made the fleece too black! But burnt umber is very often rubbed sparingly on the hips, the breast, legs and chin, where the wool has become frayed and whitened by rubbing, by dew or rain on the grass, or by lying down. The umber uniting with the natural yolk of the fleece, gives it a color true to nature. There is no objection to this practice that I am aware of; but if the inexperienced wool-grower asks in regard to it, a frank explanation ought to be given. The application of linseed oil, merely to add weight to the fleece, is a contemptible fraud.

It is legitimate for the breeder to put a light blanket of sheeting or gunny-cloth on a sheep during the summer, simply to render the exterior of the fleece mellow, moist and smooth to the touch. But he must take care not to over-do the matter. If it is left on more than a day or two in hot weather, the sheep may perspire freely, and the fleece will then become a muck of macerated yolk, odious to the touch, and requiring long treatment to restore it to a lively, elastic condition. The sheep's health will also be injured.

CHAPTER XVI.

THE BREEDING FLOCK.

SELECTION OF BREEDING EWES.—A great many of the characteristics of a good ram should also be sought in the ewe. The most obvious point of difference, of course, is determined and accentuated by the sexual functions. We seek in a ram a massive and powerful front, thick fore-quarters, a cluster of voluminous folds about the neck; but the ewe should be, if anything, heavier in the hind quarters, because these are compelled to carry the burdens and resist the strain of the great processes of reproduction and lactation. Many excellent practical breeders seek what they denominate a "pony sheep," but I have seldom attained the best results with short-legged ewes. It is seldom that the highest beauty of form is found united to superior breeding qualities—unless, indeed, long practical training has taught us to regard as the most comely, that figure which is found to be the best adapted to successfully sustain the arduous labors of maternity.

I have succeeded best with moderately large, strong, rangy ewes; of a figure typified—to use a homely comparison—by a wedge; with an even taper from the shoulders back to the hind-quarters. A ram in full fleece should have an almost perpendicular drop from the rump to the ground; be thick through the heart; with a girth just back of the shoulders about equal to that just in front of the hind-legs. But in the ewe, there may be tolerated a slight departure from the perpendicular, caused by a little less fullness in the ham; while the rear girth should be from an eighth to a sixth greater than the forward. In the best sucklers, especially when somewhat advanced in years, there is a deep, pendulous fold along the median line of the belly, terminating in the udder—an indication of a generous anatomy and a generous milker.

Mr. E. J. Hiatt's "Old Sue," which at the age of fifteen had shorn two hundred and seventeen and three-quarter pounds of wool, and reared sixteen lambs, had a notable development of the posterior half of the body, conjoined with plainness (both technical and actual), as she was totally destitute of "style."

It is true of sheep, as of all other animals, that those of medium size are almost invariably the surest and safest perpetuators of their race. Hence a small ewe should be avoided no less than an over-sized one.

POINTS IN WHICH THE EWE PREVAILS.—As a general rule, the ewe gives the size and the ram the form; and it is this fact which to so great an extent diminishes the danger which would otherwise be incurred by the coupling of a Merino ewe with a large English ram. This law of self-preservation, prevailing in every species, which gives the ewe the molding of the size, relegates to the ram more or less the shaping of other characteristics. So prepotent is the ram in this respect that, if a Merino ewe is impregnated for the first time by an English ram, the the chances are that some of her subsequent lambs will bear traces of his blood.

We are often asked why, in the increase of some years, one sex predominates. It is held by some to be a universal law that exists in all the different races of animals, that the natural tendency of the male is to produce the female, while the tendency of the female is to produce the male. The party in which the life principle is the strongest at the time of conception predominates. If it be the male, the issue will be a female, and if it be the female, the issue will be a male. Young rams kept in a thriving condition and bred to old ewes in low condition, will be sure to leave more ewe than ram lambs. A knowledge of this fact may sometimes be turned to advantage.

BEST TIME FOR DRAFTING.—Two-year-old ewes, which have never yet borne lambs, at shearing-time, of course, can be marked only with reference to their fleeces and their size. But ewes which have been tested ought never to be allowed to go until the coupling season is at hand before the mark of condemnation is affixed—if it is required. At lambing-time the shepherd ought to have his **stamping apparatus constantly ready**, and if a ewe is found to have an incurably deformed teat, or disowns the second lamb in succession (one season of disowning should not condemn her), or yeans a little trifling lamb, or in any other way gives proof of her unfitness as a breeder, the mark of dismissal should be promptly set upon her. In all other respects her record ought to be made up at shearing-time, because in the fall the wool will be grown long, and, if the farmer is not guided in his selection by indelible marks, or by a book record, he is apt to choose amiss.

If the farmer is tempted, in order to make out a certain number of breeders, to admit into the flock a small or unsightly ewe, he ought to bear in mind that ten good lambs are better than fifteen, of which five are inferior; and that an ungainly lamb or ewe is almost certain to come conspicuously to the front when the flock is on exhibition. The rearing of a lamb destroys for a long time the ewe's beauty of form and compactness, and makes her of second-rate mutton quality; and if fattened in the latter part of the season, she comes into a poor market and one which only good wethers will fill, principally for feeding. But if drafted now and thrown in with the flock of wethers she will by next season, after running farrow, regain somewhat her beauty of form, and also take on flesh in the early part of the season, thus enabling her to be turned off immediately after shearing. Again, the owner will not be so much the loser, as she will somewhat make up in wool for her lack in not having been bred.

CONDITION AT COUPLING.—Ewes will produce larger and better lambs if they are in good plump condition at the time of coupling; if not in fair condition they should be gaining and be kept improving until coupling, or until they reach the desired condition. They will not breed well when loaded with fat. Those which lost their lambs or failed to conceive are liable to become too fat to be sure breeders; when this is the case they should be placed on short pasture so as to reduce their weight. The use of valuable ewes is sometimes lost for a year or two by allowing them to become filled with fat; such ewes are valuable, their inclination to take on flesh readily is a good point, but requires guarding, that it may not impair their prolificacy,

PERIOD OF GESTATION. — Mr. E. M. Morgan, of Champaign County, Ohio, in a communication to the *Ohio Farmer* makes the following statements: "The first column shows date of putting ram with ewes, and the second, the date of dropping of first lamb:

          Nov. 16, 1874.............................April 16, 1875
          Oct. 25, 1875.............................March 24, 1876
          Oct. 17, 1876.............................March 12, 1877
          Oct. 12, 1877.............................March 8, 1878
          Oct. 21, 1878.............................March 16, 1879
          Nov. 3, 1879. ....April 2, 1880
          Oct. 15, 1880.............................March 16, 1881

On October 21, 1878, the ram was put with the ewes in the barn and served three within half an hour, which were caught and marked. On the 16th of March following, the first lamb was dropped by one of these ewes, the second on the 22d, and the third on the 27th of March, making a variation of eleven days in the time between first and last. These three lambs were all ewes. This seems to disprove the theory that an animal will go longer with male than with female progeny. Taking the average time of all our ewes, we find it to be one hundred and forty-nine days, for the seven years we have kept record."

TIME OF LAMBING.—It is important for the farmer to be well assured in his own mind whether his circumstances favor early or late lambing. Latitude has much to do in deciding this question; likewise the size of the breeding flock, and the convenience and comfort of the sheep house, or the contrary. I have steadfastly advocated lambing on grass, because here in Southern Ohio, and with one hundred and fifty or one hundred and seventy-five ewes in the flock, it is undoubtedly the wisest policy. In a higher latitude and with a smaller flock, the case would probably be different. We know that sheep, as well as others of the mammals, are not as good milkers in hot climates as they are in cooler ones. The excess of heat interferes with the lactific functions and curtails the secretion of milk. My belief is that when a ewe does not yean until the strong heat of summer sets in, say along toward the middle or last of May, her usefulness as a suckler is seriously impaired. It is different with her from what it is with a cow. The ewe still bears the thick, warm fleece which was intended as a protection against the rigor of winter; consequently the heat operates on her with a much greater and more prostrating power than it does upon the cow. As a corollary to this proposition, it follows necessarily that suckling ewes should be shorn before the weather becomes hot. If they are left with their fleeces on, the accumulation of heat dries up their milk.

FEEDING FOR MILK.—Ewes that are to bear lambs very early must be fed for milk as much as a dairy cow. The feed must be of a character that will produce the greatest quantity of milk. This can be secured by providing plenty of clover, millet or fodder. Some very good shepherds recommend wetting these and mixing with ground feed. The finer feed can be made, the better for any stock; but wetting is unnecessary, if only an abundance of water is provided. The more feed can be masticated and insalivated, the better, and wetting hinders this. But the water should be kept at a temperature not much below sixty degrees, to induce pregnant animals to drink freely.

MAY LAMBS.—I have found among old shepherds a prejudice against "May lambs;" and this prejudice is founded principally on the belief that the burning sun of our inland American summer "stunts or wilts" the lambs. There is no denying that a May or June lamb, though it generally shoots up for a few weeks with a rapid growth, does become stunted later on and gets into a decidedly poor condition before weaning time, unless the ewe is an exceptionally good milker, or the lamb has a ration of grain through the summer. It should always be borne in mind that the ewe is not, like the cow, an all-the-year-round milker; the ewe's lactific activity is exceptional, and though very often, especially on grass, of considerable force for a time, it quickly ceases. Hence it is of great importance to bring the lamb along early enough and so rapidly that it may be well confirmed in its grass-eating habits, and may have acquired the additional capacity of stomach, required for this less concentrated food, while the grass is yet lush, tender and inviting in spring. A lamb does not take to grass so readily if it first begins to eat it in summer after it has become dry and tough.

Of course there is no foundation for the belief that the sun "wilts" a late lamb. I never give myself any concern about a May or June lamb, if I am only able to provide nourishment enough for it; for I have often abundantly proved, by rearing them as cossets about the house, that this sufficiency of aliment was all that was needed.

NECESSITY OF EXERCISE.—The Merino ewe is something like the Texas cow—not the best of mothers. A native of the desert, she still retains in her blood a remnant of nomadic, oriental wildness. An industrious, insatiable feeder, accustomed to rove widely in search of her living, not tranquil and sedentary like the large-uddered English ewe; like the ostrich, she is apt to abandon her young, to take care of itself. She needs watching, and needs a certain pressure to be brought to bear upon her too feeble maternal instincts.

Extended experience has taught me that a Merino ewe which has a copious flow of milk is seldom failing in duty toward her offspring. The first and paramount duty of the shepherd, therefore, is to pursue such a preliminary course as will best secure this desideratum. A regimen of roots, oil-cake meal, bran, fodder, clover hay, etc., will readily suggest itself; but, valuable as these are, they are not for the Merino ewe of the very highest importance. The article which, in my opinion, holds this rank is grass, and (perhaps scarcely secondary in value) the exercise which is necessary to obtain it. There is no other domestic animal which so eagerly craves and industriously searches for a morsel of green feed cropped directly from the surface of the earth. And it is this restless, vagabondizing, gormandizing propensity of the Merino which the shepherd can take advantage of and promote, to the end that he may develop the rather feeble maternal instinct. It is as profoundly and universally true of the lower animals, and especially of the pregnant ewe, as of man, that they ought to work for their living. Pasturing (that is, a daily run on a sod, whether it furnishes much or next to nothing) means work, and work means health; while roots mean cold-blooded and watery idleness. There is nothing else which so strengthens the frame and enriches the system with warm, red blood (and, by necessity of the inseparable relation between them, that of the unborn lamb also), as a frequent ramble over the pasture lot.

Even when quite sedentary, the ewe may be made to give milk with tolerable success by judicious feeding on oil-cake meal. Perhaps as good a way as any is to make it into a slop with wheat bran, a tablespoonful of oil-cake to a pint of bran per head; but unless she has frequent and abundant exercise, the lamb will be weak, and will need close watching if dropped on a frosty night. In the course of my experience, I have had large, rangy grade ewes—and a grade is popularly supposed to be hardier than a full-blood—which had been full fed and warmly housed, drop large, finely formed lambs, which yet were so flaccid and so nerveless that it would be hours before they could stand alone, and that only after the most assiduous attentions of the shepherd, warming them before the fire, rubbing them with wisps of straw, etc. On the other hand, I have had full-blooded ewes, which had roved nearly all day during the winter through a corn stubble, getting next to nothing in it but the exercise, drop lambs on so cold a night that their feet were frozen and deformed ; yet they got up, sucked, and were lively as crickets in the morning, without having received a particle of assistance from the flock-master.

The English sheep books abound in directions for the making and administering of cordials, syrups, etc., and for rendering assistance to ewes in labor ; but a few teaspoonfuls of grass-made milk are worth more than all the nostrums ever compounded. Neither is it necessary to defer the season of lambing until grass has grown green in April.

RYE FOR PASTURE.—In the latitude of Southern Ohio a very considerable growth of rye may often be had for pasture as early as March 15th. An hour's grazing on it per day will have a surprising effect in stimulating the secretion of milk ; indeed, it is best not to allow the ewes to remain on it above a half hour the first day. Rye may be sown for fall pasture as early as August 1st. If the weather should be very favorable there will be danger of its jointing before winter sets in ; this can be prevented by keeping it pastured off. The value of the crop is much injured if it is allowed to joint or head out in autumn.

The white rye yields the greater amount of grain, but the old-fashioned black rye is hardy, makes a rank growth, and is probably preferable for pasture. It should not be cropped too close in the fall, as its greatest value is in the green herbage which it furnishes for ewes and lambs before grass grows in the spring. On rich limestone soils and in low latitudes, wheat often makes such a strong growth that it will furnish a large amount of grazing for ewes and lambs in March and April ; and there is a mass of testimony to the effect that such depasturing is beneficial to the wheat itself when it is very forward.

Second in value—and on the rich river bottoms of the West, I would assign it the first rank, on account of the tendency of rye to develop ergot on such soils—is an orchard-grass rowen, reserved for this purpose, with its mixture of weather-beaten herbage above with green growth beneath.

ACORNS.—In my experience I have found that, while acorns are not only innocuous, but fattening to dry flocks, they exert an injurious effect upon ewes and goats in a forward state of pregnancy. If they feed on them for any considerable length of time while in this condition, their young when dropped will be feeble in the legs, unable to stand or walk for several days, and walking in a sort of plantigrade fashion for some time after they do succeed in getting on their feet.

RECURRENCE OF EWES.—When there is for any reason, a failure to conceive, the ewe will be in heat again, if at all, in about two weeks. To make a provision for these I manage the coupling in the following manner: As fast as the ewes are served, I affix a special mark and turn them into another apartment, which opens into a small paddock or ram-pasture kept for this purpose. If allowed to go with the flock again, they would in all probability present themselves again the next day, and so tax the ram a second time uselessly. On the following day, when the main flock is brought up, the little band of served ewes will also come to the stable, and, after the business of the day is over and the main flock dismissed, the ewes served the previous day can be allowed to go with them.

If the coupling is well managed there ought not to be many ewes that "miss." When the pasture is weak and watery, or short from dry weather, they ought to be grain-fed at the rate of a half bushel of shelled corn daily per hundred, for a week before, and all the while during the service. It is well, when they are brought up in the forenoon to keep them on the sunny side of the building; the warm rays of the sun have a stimulating effect. The ram ought to be allowed ample time to search out all that are in season, for there are always some that are backward and will never approach the ram or give any evidence of being in season. If the ram is indifferent or logy, he ought to be kept tied in the shade between-times. On muggy, sultry days, frequently twice the usual number of ewes will come in heat; this will demand increased activity, it will tax the ram to the utmost, and sometimes the shepherd will lose ground by not having an extra animal to fall back upon.

A sudden change to cold weather is also to be guarded against. A long, cold rain, followed by high winds, hinders the dispatch of business; the sexual heat is checked; some ewes may pass their season altogether, and thus two weeks will be lost. They ought by all means to be housed during such weather.

The shepherd ought to use all dispatch to push the coupling through in thirty days or less. After winter comes on, if there is a remnant of ewes not served, they will be in heat no more and they are lost. Besides, it is tedious to have the lambing drag at great length in the spring.

"TEASERS."—No well-informed shepherd ever resorts to the

"teaser" in these days. It was the clumsy device of an unpracticed age; an outrage against nature, an imposition on both ewe and ram.

AGE OF EWES.—Bringing her first lamb at three, the average Merino ewe is entitled to be released from service at seven. It is useless to cite cases—as I might do by the dozen—where service began younger and continued longer. All rules have their exceptions. As long as the ewe's teeth continue firm and sound, and she stands up stoutly under her burden through the winter, she may be retained in the breeding flock; but let the shepherd beware lest he should keep her one year too long, and before spring lose both her and the lamb, for then she dies in his debt. A ewe in a flock of ten may bear lambs two or three years longer than one in a flock of one hundred. In flocks of considerable size the crones must be weeded out rigorously, or the flockmaster will suffer loss.

FALL AND WINTER LAMBS. — "Spring lamb," like "spring chicken," has its own proper season of the year, and out of this season there will never be any considerable demand for either. In the winter the appetite calls for fat mutton, thick on the rib. But now and then some ambitious farmer dreams anew the dream (which is as old as the appetite for mutton), of growing "spring lamb" the fall before. It is a reversal of the course of nature which never can prosper except in rare instances, under peculiarly favorable circumstances and good management. In a communication to the *Ohio Farmer*, Mr. E. M. Morgan, of Champaign Co., O., gives some experience which is so interesting, that I quote the greater part of it:

"In the spring of 1882, after washing our sheep, supposing that no evil would result from it, we let the ram run with our breeding ewes (then suckling lambs dropped from March 15 to April 15), until shearing. In the fall, about November 1st, fifteen or eighteen of these same ewes dropped lambs, the result of letting the ram run with them from washing to shearing time.

"When we began feeding for the winter, we fixed a place in one end of the stable so the lambs could enter and the ewes could not, and sprinkled some bran and salt in the trough. Very soon the lambs learned to go there, and in a short time they would run for their pen to get their rations, as greedy as a litter of pigs for a mess of milk. We fed them liberally through the winter and they came out in the spring in fine condition.

Encouraged by their fine appearance, we turned the ram with our ewes again, on the 9th of May, and will try our luck again with fall lambs. At washing time this spring we washed the lambs, thinking we would shear one or two; and if thought profitable, would shear the whole lot. The first one sheared clipped a fleece that weighed exactly five pounds. Encouraged by this, we sheared the other twelve, and from the lot got fifty-four pounds of wool, which we sold along with our other wool, at the same price. * * * * * * "The ewes came through the winter in fine condition, and when I weaned the lambs they were in much better condition than I ever had ewes when the lambs were weaned in the fall, and sheared an average of seven and eleven-twenty-sixths pounds per fleece. A lot of thirty-two yearlings, wintered with the ewes, clipped an average of nine and three-sixteenths pounds per head, all nicely washed wool, and all sold at market prices. I would say to those who are prepared to properly care for fall lambs to give it a trial. My sheep are high grade Merinos."

There is no gainsaying that a winter lamb, when it is well nourished, will surpass the later comers out of proportion to its gain of time at the start; and it will keep ahead for two or even three years. One year I had fourteen lambs dropped in January by reason of a ram getting into the flock prematurely; with much labor I saved ten of them. At the age of a year they weighed sixteen and one-half pounds per head more than the April lambs, and clipped about one and one-quarter pound more wool. Judged by the eye alone, they were still nearly as much in advance at the age of two years.

EWES GETTING CAST.—Ewes are liable in the spring, when far advanced in pregnancy, to get on a little slope with their backs down-hill, in which condition they are unable to rise. The wool spreads out on the ground and prevents the sheep from rising, when without the fleece it would be able to get on its feet. Cattle will struggle a while, then rest and renew their efforts, and they generally get up; but sheep get discouraged and abandon all efforts. In a short time they will bloat and die, unless assisted. The shepherd should be on the lookout for castaways when they are in the field; and he should level all inequalities in the surface of the yard and stable where breeding ewes are confined.

CHAPTER XVII.

SHEEP-HOUSES AND THEIR APPURTENANCES.

It would be easy to fill this volume with plans and sketches of possible sheep-houses, all of which would be theoretically good. I shall limit myself to such as have been put to the test of actual use and found serviceable.

For Breeding Ewes.—The figure herewith presented is that of a building owned by Mr. C. C. Smith, of Washington Co., O.

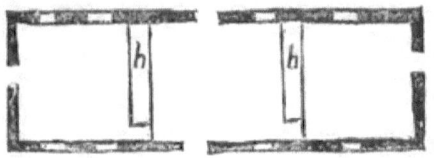

Fig. 13.—SHEEP-HOUSE OF C. C. SMITH.—GROUND PLAN.

It is fifty by twenty feet, eighteen feet high to the eaves, covered with a sheet-iron roof, two-sided, with the usual pitch. It is designed to shelter at the most about eighty sheep, and is used mostly for breeding ewes and as a shearing-room and wool-room. Hence the comparatively small allowance of space for hay should not be accepted as a guide for a general purpose sheep-house.

The lower story is eight feet high, the second ten. The upper story is divided crosswise into two equal compartments—one for wool and one for hay—with a tight partition between. Hence the hay-mow, as I intimated above, is too small to contain a winter's supply for the flock below, a point which it is always desirable to compass in the average sheep-house. To curtail this mow, twenty-five by twenty feet, as little as possible, the owner, instead of throwing a girt across between the plates to prevent spreading, put in dove-tailed braces from the top of the posts down to the joist girts, as in figure 14.

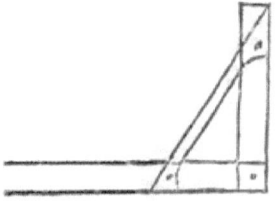

Fig. 14.—BRACE.

Across the wool-room the plates are connected by an iron rod. The floor is tight and smooth, and the sides ceiled in the same fashion. At one corner of the room, at one end of the shearing table, there is an elevator and cage large enough to hoist five or six sheep from the lower story to be shorn; the floor of the elevator to serve as part of the floor of the room. This is worked by weights. The object sought in elevating the sheep to the second story for shearing is two-fold—to avoid all dirt about the shearing table and to have the wool where it is wanted for storage, in a perfectly clean place.

The wool racks are so constructed that every fleece can be inspected without one of them being moved. When fifteen or twenty buyers come along during the season and look the clip over, it is liable to become seriously frayed and shredded if heaped in the usual pyramid in the center of the room. (By referring to the chapter on shearing, the reader will learn the construction of these racks).

The building is sided and battened perfectly tight. There are no sills; the posts stand on stones. The floor is of gravel, several inches higher than the surrounding level, and the siding reaches down within two or three inches of the floor. The bedding will be so thick as to reach up against the siding, preventing a cold wind from blowing underneath. Manure can always be removed much more easily when several inches of straw is thrown down in the fall.

Five feet above the ground are windows, sliding laterally, with four panes of glass, ten by sixteen inches each. The four doors, one on each end and one on each side midway, are double; the outer ones battened tight and opening outward; the inner ones of slats and opening inward. The slats are close enough together to exclude chickens. Thus the building can be ventilated by the slat doors, or all the doors can be closed and a draft be allowed to pass overhead through the windows. The end doors are folding-doors, wide enough to allow the manure wagon to enter. The ground floor can be divided, as desired, into two, three or four compartments, by hay-racks running across the building, each rack with a little gate at the end of it. A cistern stands midway of one side, the water from the opposite roof-slope being carried to it through the building.

The cut fig. 13, on the preceding page, shows the ground plan.

The second story is lighted by small slat-windows.

Hay is hoisted into the second story at one end of the building with a horse-fork.

There is a smaller building intended for a stove-room or lying-in hospital, being situated only a few steps from the large one above described. It is about fifteen feet square, perfectly tight, double walls with saw-dust between, divided off by light gates into eight or ten little pens, each large enough for one ewe and lamb. A stove standing in the center of the floor, well heated up at 9 o'clock at night, keeps the atmosphere sufficiently warm through the night to insure the safety of the weakest lamb arriving before six in the morning.

A GENERAL-PURPOSE HOUSE.—I give below a diagram of one of my own sheep-houses, merely premising that I have embodied in the description some changes which subsequent experience taught me would have been improvements to the building. It is forty by forty-five feet, giving (without the racks) eighteen hundred feet of superficial area, which I find sufficient for a dry flock numbering one hundred and fifty, or for one hundred and twenty-five ewes. I use it principally for the latter. It is composed of a main central frame, twenty by

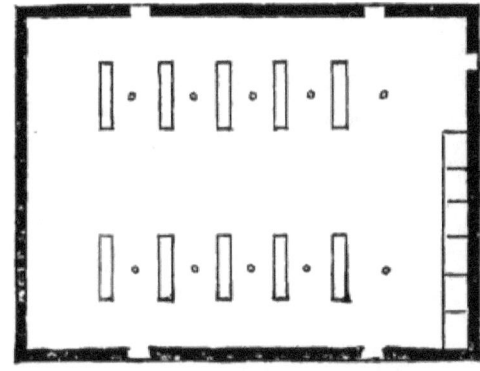

Fig. 15.—GENERAL-PURPOSE HOUSE.

forty-five, and two wings or sheds, each ten by forty-five. The main building is seventeen feet high to the eaves; this gives ten feet storage for hay, and I find by experience that a body of hay twenty by forty-five by ten, will comfortably feed one hundred and fifty sheep once a day for two months. (Hay for the rest of the winter is pitched in from an adjoining barn through a partition.) Hay is thrown down from the loft of the central building into the wings, through chutes constructed something like dormer windows, falling into racks placed as indicated in the engraving (the rows of dots denote the posts of the main structure).

At one side is a series of portable pens for ewes and lambs in lambing-time. The wings have not quite so steep a pitch as the main building, which is one-third. The elevation of the wings at the outside is only seven feet, which is simply enough to allow a span of horses to pass under comfortably in hauling out manure. There is not a sill in the building; all the posts stand on stones, which are planted on solid foundations of broken stone, let down into the ground about twenty inches. A sill is useless in a sheep-house; it is worse than useless, for it is apt to rot and let the building sag down one way or the other. By keeping all sills out, there is afforded a free drive-way all about the building, and out through the side of it wherever it is convenient to cut a door. I filled up the inside of the building with yellow loam—which packs harder than almost anything else except clay—six inches higher than the surrounding level, to prevent the interior from being flooded in winter. In place of a sill I set up thin, wide stones on edge inside the siding and leaning against the same, jointed and fitted so as to prevent the earth from touching the siding. A corresponding ridge of earth or gravel outside, tamped against the stones and sloping down as a spatter-board for the eaves (though it would be still better to have an eaves-trough) will prevent the earth from pressing the stones out too much against the siding.

There are nine windows in the building, arranged to slide laterally, so that the inside can be ventilated in muggy weather, as the siding is very tight. There are five doors, one opening into the grain yard, one into the fodder yard, and three for the ingress and egress of the manure wagon. They are sliding-doors, as I consider a swing-door on an out-building a nuisance. A one-and-a-quarter-inch strip of wood is faced with a one-and-a-half-inch bar of iron, three-eighths of an inch thick, which projects one-fourth of an inch above the wooden strip and furnishes a guide for the door rollers to travel on. This strip put on with two-inch screws, one every foot, will hold up ten times the weight of a door. The bottom of a door has to be confined with stakes, to prevent the sheep from carrying it away when they rush out in great numbers, hungry for their feed.

This building is sided with dressed pine and painted; the old-fashioned linseed-oil and white-lead paints give me better results in the long run than any of the modern ready-mixed proprietary articles. I had it covered with home-made oak shingles, twenty inches long and laid six inches to the weather. Where the material is at hand these are better and cheaper than sawn pine shingles. An iron roof is preferable to pine shingles. In one end of the building, overhead, is the wool-room; in the other end the corn-room, to which the corn is elevated by horse-power, with rope and pulleys, in two boxes which together fill the wagon-bed.

A HOUSE FOR A SMALL FLOCK.—Any sheep-house is defective which is not provided with facilities for securing perfect ventilation on the one hand, and on the other, for closing it up tight in severe weather. The Merino is intolerant, above all things, of a foul, reeking atmosphere and dampness underfoot. Inside slat-doors, as in Mr. Smith's sheep-house, are excellent ; another very good arrangement consists of doors hinged on the upper side, so that they can be dropped down during storms accompanied by wind.

So, also, is any sheep-house defective which has no hay-loft, although a mere shed or wind-break may be constructed without one. But all hay-lofts should have a perfectly tight floor. I have seen sheep going around with hay-seed sprouted and the grass growing out of the wool on their backs.

A building twenty feet wide will comfortably house two sheep for every foot in length (if not breeding ewes). Thus fifty feet in length would accomodate one hundred sheep. It should have its length running east and west, then it will make a more effective wind-break for the yard attached to it. A rack running centrally the whole length of the building, except four feet at each end, will give feeding-room for all the sheep. This rack may be connected at the top with a tight board hopper that reaches to an opening in the otherwise tight flooring of the hay-loft above. This sheep-house can be divided into as many rooms as the occasion may demand. When the hay is put into the mow some short strips or boards are laid across the opening in the floor to the rack below, and the hay is put in one continuous mow the whole length of the building. After the mow has become settled, just before winter sets in, a hay knife is used to cut a hole three feet wide down to the opening in the floor. The hay thus cut out is flung up on the mow that by this time has settled enough to receive it. The hay is put into this mow through convenient doors in the side of the building made for that purpose, and is given to the sheep by simply pitching it down into the rack. There is no wasting of hay by this means of feeding, and the flock can be fed without having to be turned out of doors into a storm.

GROUPING OF SHEEP-HOUSES. — When farm buildings are closely grouped, if one of them takes fire, all will burn. But it is better to incur this risk than to compel one's self, by distributing the buildings about over the farm, to travel on a winter's morning a half-mile or a mile in the snow or the storm. Four hundred Merinos can be wintered in perfect health and good condition on three-fourths of an acre, if proper diligence is used in cleaning out the stables and keeping down the ammonia. I make this assertion understandingly, because my experience has demonstrated the entire practicability of so doing.

Besides, it is very desirable to secure for every flock on the farm, as great a variety of feed as possible; hence it is more convenient to mass together the straw, hay, fodder, millet and the various kinds of grain at or near the farm headquarters, than it is to parcel them out in smaller lots in three or four different places. It is inexpedient to give one flock all the straw, another all the fodder, etc.; neither is it convenient to drive

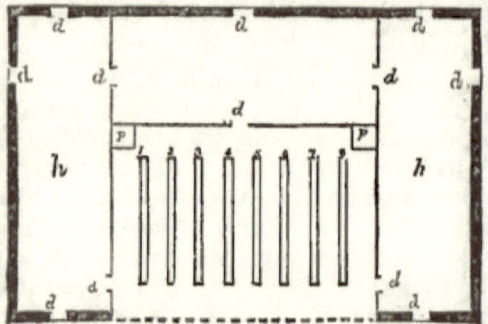

Fig. 16.—GROUP OF THREE BUILDINGS.
h, House; *p*, Pump: 1, 2, 3, 4, 5, Grain Troughs.

the flocks about in a rotation through sheep-houses, separated by considerable distances, in order to secure this very desirable alternation or variety in feed. I find it best every way to group the stables about headquarters, and then during the summer collect near them all the various feed-stuffs required for their support through the winter.

Figure 16 shows a group of buildings in which the grain-yard, open only on the east, is protected from three-fourths of the winds; it is accessible from all the stables. The troughs should be about eight feet apart, to allow a row of sheep to stand on one side of each, with room enough for others to run along between. To prevent the sheep from jumping into the troughs or over them, which they are extremely likely to do, a row of stakes must be driven along one side of each trough and two slats nailed to them; the lower slat is about nine inches above the trough, the upper one about four feet above the ground. This grain-yard will have to be lower than the floor of the three houses, else the water falling into it may run into them. The plan contemplates two fenced yards on the east side of the group—one for the right-hand the other for the left-hand stable. If any considerable amount of orts, stalks and manure accumulates in these two yards, it will obstruct the drainage of the grain-yard before spring; and to relieve the latter it will be necessary to run a drain-tile under some one of the three houses, according to the slope of the ground.

The flock on the west side of the group can be accomodated by a yard on that side, which will have to be screened from the west and north winds by a high, tight board fence. The doors are so aligned that a manure-wagon can be driven lengthwise through either of the three houses.

DOORS AND GATES.—I have found, in the course of long personal experience in feeding sheep, that a door or gate through which a hundred grown sheep are to rush, eager for their feed of grain, must be so constructed that it can be opened in the quickest possible way. A door swinging out laterally is apt to be obstructed by snow-drifts, ice, orts, manure, etc.; besides which, it is always swinging open in the wind when it is needed to be shut, and *vice versa*. Then, too, no matter which way it opens, when the time arrives for it to be opened the sheep are very often huddled against it so that it cannot be moved. If swing-doors are used at all, they ought to be folding-doors, carefully hung in such fashion as to avoid all obstacles.

After experimenting with several kinds of doors and gates I have adopted the following: Where the side of the stable is low, I have a sliding-door, eight feet long, hung on rollers and a slide as described earlier in the chapter. Where the side of the stable is high enough to admit it, I have a hoist-gate, of the same length, suspended from a pulley overhead with weights enough to balance it. It should be made of very light slats, set close enough together to prevent the sheep from getting their heads between them, as the gate is hoisted. The frame-work or guide on each side, in which the gate plays up and down, must be nicely adjusted and true, and it is well to have small pulleys let into the outer sides of the heel and toe posts, to obviate friction. An iron rod fastened to each of the posts, bent upward and provided with a loop in the middle to receive the rope, is the best attachment for hoisting the door.

Where a gate is required in a yard, through which a large flock must pass quickly several times a day, it is extremely important to have it so arranged that they can pass through without friction. The best gate is a panel of portable fence, twelve feet long and five feet high, made of slats close enough together to exclude the sheeps' heads, light but strong, and put together with clinch-nails. This can be kept closed by some simple fastenings at each end. When these are loosened the gate is thrown to the ground, and the flock rush pell-mell over it.

FEED-RACKS.—The purpose of a feed-rack is two-fold :—
1. To keep the feed in.
2. To keep the sheep out.

The first and greatest requisite toward the making of a good hay-rack I would formulate thus: Cut hay green. That is to say, if the hay is thoroughly good the sheep will stand quietly and eat it; but if it is inferior they will continually run to and fro, pulling out a little here and a little there, chewing the heads off as they run, and dropping the remainder underfoot.

With regard to the first point above mentioned, I may say that, so far as my own experience goes, when the hay is bright and sweet, I have found the old-fashioned, slatted box-rack good enough for all practical purposes. And when the hay is not good, no rack, however ingenious, will prevent sheep from wasting more or less feed. I will enumerate some general principles which ought to be observed in the construction of all racks:

1. *Portableness.* A rack fastened down anywhere, though it will undoubtedly wear longer, is objectionable. It is inconvenient to remove manure from beneath it, and it cannot be used to partition off a house into compartments of different sizes and shapes to suit an emergency. There is nothing better for this purpose than a portable rack.

2. *A tight floor.* Every sheep-house should have the earth for its floor, and if the rack has no floor of its own a great deal of fine feed will mold on the ground and be lost.

3. *Sufficient elevation.* In every flock of considerable size, no matter how well bred, there will be some leggy animals that will never be satisfied until they are inside the rack. The rack should be forty inches high.

4. *Separation of hay-rack and grain-trough.* Many of the best practical shepherds, with small flocks to care for, by various contrivances unite rack and trough together; but in my own experience, especially with large flocks, I always found these

objectionable. It is almost impossible to prevent troughs so situated from becoming receptacles for dung. In any event, they have to be cleaned out at every feed, else the grain will be mingled with orts, chaff, seed, etc. Out-door troughs collect snow and ice, it is true, but that is all, and they do not require to be cleaned half as often as troughs attached to racks. Nevertheless I have figured further on some of these combinations.

Racks are single or double; that is, the sheep reaches through between one set of slats or between two sets. There is no great gain in a rack made double, except that a place is furnished for a feed-trough, which is placed at the bottom of the hay-rack, between the outer and inner sets of slats. In a double rack the sheep is prevented from thrusting its neck full length into the hay and cannot get chaff into its wool. This is unimportant, however, unless sheep are to be shorn immediately at the close of the feeding season; if they run on pasture a few weeks the chaff will work out of the neck-wool of its own accord.

A single rack should be, for lambs, about two feet wide; for grown sheep, about thirty inches. This width should enable

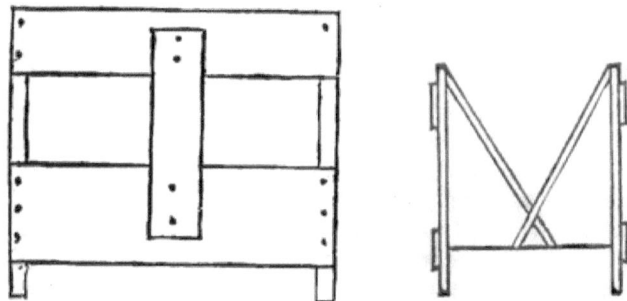

Fig. 17.—END VIEW, SINGLE RACK. Fig. 18.—END VIEW, DOUBLE RACK.

two sheep, standing on opposite sides, to reach the middle. It may be of any length desired; fourteen feet is convenient. For lambs it need not be over thirty inches high. There should be about nine inches space between the top-board and the bottom-board; the slats four inches wide; the spaces between them eight inches wide for grown sheep, six inches for lambs. This gives each lamb ten inches space to stand in, each grown sheep a foot. All edges should be rounded off to prevent tearing of wool. The corner posts, four in number, may be about four inches square. An end view is shown in figure 17.

Figure 18 represents the end of a double-rack, or an inside

V-shaped rack and an outside perpendicular-sided one. The inside rack is made of slats nailed on a V-shaped trough which is inverted and nailed down on the floor of the rack.

Figure 19 shows a V-shaped rack with a feed-trough at the bottom on each side. The rack-sticks are round; they are let into the bottom-plank by auger-holes, and into the top-boards the same way.

Figure 20 is an end view of another double-rack; two V-shaped

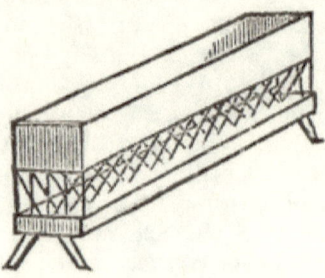

Fig. 19.—SINGLE RACK AND FEED-TROUGH.

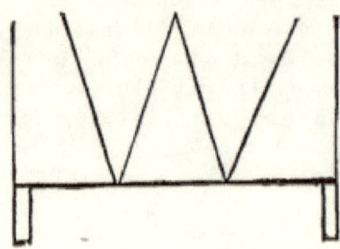

Fig. 20.—END VIEW, TWO DOUBLE-RACKS.

racks inside of one square one, with a feed-trough at the bottom of each.

CISTERNS FOR SHEEP-HOUSES.—The importance of having a supply of water in the winter not below the temperature of fifty degrees is so great, especially for breeding ewes, that nothing will justify the flock-master in neglecting it. To put in and equip a one hundred and fifty-barrel cistern, with all its appurtenances of tin eaves-troughs, spouting, etc., costing twenty-five dollars, or such a matter besides the labor, is one thing. To sink a fifty-barrel cistern, wall it and plaster it one's self, and furnish everything required about it of wood, made by one's own hands, and costing altogether not above six or seven dollars, is another and very different thing.

I do not say that every flock-master should attempt to do all this work himself, whatever his circumstances, but if he feels particularly poor, he can do it and keep the money it would cost in his pocket.

A cistern seven feet across and six feet deep will contain fifty barrels. But a deep, bottle-shaped cistern is better—say ten feet deep and five feet across. Seventy-five cents will pay for the digging. If the soil is a stiff tenacious clay, or soapstone, the brick wall need not go below the frost line. In most cases, however, I should wall it to the bottom; it is safer, and cheaper in the end. Brickbats, costing half price, will do just as well as whole bricks. One thousand of these, costing two dollars, (equal to five hundred bricks at half price) will lay the wall. A barrel of Louisville cement, two dollars, and two barrels of sand complete the bill. The bricks can be laid in clay mortar; the cement is only needed for the inside plastering.

In the lower portions of the cistern every brick (laid the long way, and the broken end chipped off somewhat square) must be jammed back firmly against the solid earth to resist the pressure of the water. The inner surface of the wall must be kept as even as possible to receive the plastering.

Observing the bottle-shape (not the square-shouldered bottle, but the sloping), the builder will begin three or four feet below the surface to draw in his wall slowly, in such fashion as to form a mouth about eighteen inches across at the surface. A wall curving in so gently as this can be laid by any farmer; a broad, bold arch would require a skilled mason. Bear in mind, the cistern under consideration is only five feet across at bottom. To build this kind of a neck, of course, the operator can no longer thrust the bricks back against the solid earth, as he did at the bottom. They must be chipped at the ends to bear firmly against each other, and laid flat, not with a pitch inward as they are when a mason is rounding an arch. Hence the last course at the surface is flat and does not need an iron collar to keep it from falling in. An earthenware elbow must be introduced near the top for the reception of the conductor; also a waste pipe. The space between the solid earth and the brick wall, down to the line where the latter begins to curve in, will have to be puddled with clay or loam.

After the brick-work has stood a few days the plastering can be done. One part of cement to two of sand is the rule. The helper will have to be trained to mix it of the right consistency. The idea that it has to be mixed in very small quantities at a time to keep it from setting is erroneous. All that is requisite is to keep the mass wetted and stirred.

The top will be finished off with a square box of oak plank, a foot deep. Against this box the earth can be banked up to protect the brick-work from injury by frost, and also afford a foundation upon which the pump can be placed.

WATERING TROUGHS.—I have tried various ways of watering sheep in the sheep-house, including plain, three-cornered wooden troughs, old iron sugar kettles, tubs, etc.; but they all proved unsatisfactory. Sheep are so anxious to get the cleanest, freshest water, to drink at the fountain head or next to the spout, that they crowd each other hard; any appliance has to be made solid to resist pressure, and of such shape and elevation that they cannot get into the water and foul it. All permanent troughs in the stable are open to objection, in that they have to be low enough to accommodate sheep, and this makes them a constant receptacle for dung. One way of remedying this is a board nailed on slanting, in the fashion shown in figure 21.

Fig. 21. — BOARD COVERING WATER-TROUGH.

The board allows the sheep to reach over and drink, and at the same time keeps out the manure tolerably well. The board cover should be eight inches wide for mature sheep, reaching two inches over the edge of the trough.

WOOL-ROOM.—No permanent shearing-room is needed, unless, which is not desirable, it is also used as a wool-room. A shearing-table can be set up or hinged anywhere against the side of the stable, on trestles about two feet high, though some shearers prefer a table higher than this, while others want a lower one. A table four feet wide will accommodate a shearer for every four feet of its length. After shearing is over this table can be turned back on its hinges and the trestles can be stored away for future use.

In the wool-room will be found the press and sheep-hook, already described; a small grocer's scales with a set of weights, and a light slat frame to be attached to one arm of the balances, for receiving the fleeces; also a set of wooden letters and paint for marking the sheep; shears; toe-shears; ear-tags; medical and surgical appliances, etc.

A very good table on which to spread and fold fleeces can be made by placing some old doors, or a table-top made for the purpose, on top of a hay-rack standing near the wool-press.

SHEARING-CARDS.—To prevent false counting by the shearers, it is well for the farmer to provide himself with a set of shearing-cards. Let there be, for instance, fifty marked A, a like number marked B, etc. In the morning each shearer takes all the cards marked with a certain letter. Whenever he deposits a fleece on the table he throws down one of the cards upon it. The cards are taken up by the wool tier, or overseer, and at night they indicate the number of sheep shorn by each man.

A Shearer's Table.—While it is not the purpose or province of this work to bring into notice proprietary articles of any kind, yet I deem it not amiss to briefly call attention to such labor-saving inventions as are of undoubted utility to the flock-master. Such, for instance, is a "Self-adjusting Shearer's Table," of which it is said: "It holds a sheep in any desired position, so that the shearer stands on both feet and has the use of both hands, and the wool when shorn is never 'kicked' or torn and is in the best possible condition for the wool box. The invention was made by a Mr. Addison, of Ohio. It is adjusted in a moment to any sized sheep, and the position of the sheep is changed by touching a spring. It will be specially 'the thing' for shearing the wrinkly, heavy-fleeced Merino, as the sheep is held in an easy position and the shearing quickly performed." I never saw this particular device, but have witnessed the operations of one very similar. It consisted of a wooden bowl, in which the sheep was placed on its buttock and strapped to a light frame-work standing up at the proper angle for the sheep to rest in while being shorn.

CHAPTER XVIII.

WINTER MANAGEMENT.

Yarding.—Perhaps the most vigorous and piquant defense of the loose-ranging system of wintering sheep, that has come to my notice, was one contributed to the *Ohio Farmer* by Mr. Simon Smith, of Harrison Co., Ohio, the owner of a flock of one hundred and fifty pure Atwood Merinos. He says: "The reason I don't house my sheep is because they must have exercise at will to keep them healthy, and must be exposed, not abused, to make the wool grow long. I do not grain them, because grain, if properly fed, makes the fat too solid (except to butcher) for stock sheep. It also produces grease or gum in the fleece, which (especially if housed) excludes the air from the roots of the fibers, which tends to shorten the staple. Some think grain-fed sheep produce more wool, when, in fact, four-fifths of the gain is grease. Sheep that are grained and housed will not live out half their days. What I consider abuse is to confine them to filthy pastures or force them to stand in a bleak wind, or impatiently to bleat at an empty rack, or gnaw the bottom of the salt box. I forgot to say that a sheep with a wrinkly, greasy or gummy coat cannot stand inclement weather."

In another place he says: "Experience has taught me that sheep will spread their own manure, trim their own tails, provide their own blankets and make their own prognostications of the weather, if managed in accordance with nature."

It is undoubtedly true that small flocks, even of full-blood Merinos, if kept on a tough sod, with bushes and bitter browse, and clumps of trees for a wind-break, with a moderate daily feed of corn, either shelled or broken one ear across another, will take the storms of winter with impunity and come through thriving, with red noses, long, clean wool, and healthy systems which will not scour a particle when the grass grows green in the spring. But with larger flocks, where the feeding of hay or fodder is necessary, the objection to the ranging system is that the sheep do not of their own accord regulate the matter of exercise judiciously. On a warm day they will rove all day, and on a cold day not at all. If the fodder is thrown out to them in an open field they will wander about the field, coming to it three or four times a day and browsing a little while, then they are off again. On a very cold day they will stand in the lee of the shed or of the fence—if nothing better offers—and lose a great deal of time when they would be eating if the feed were close at hand, and in a place not exposed to the wind. The sheep is very irresolute about breaking away from a warm sheltered place and setting out in search of feed. On an excessively cold day the sheep cannot be forced to take exercise, unless they are driven to water or something of that sort; and it is not worth while to attempt it, especially if there has been a sudden change from mild weather, for they will seldom drink the first day after such a change, even if water is offered them in the shed.

The summing up of the whole matter, therefore, is this: It is best to keep sheep in a yard sheltered by their shed, with a warm southern exposure. Let them have their regular time for exercise as much as for their grain ration or their hay. If snow continues on the ground a long time, so that they have no inducement to take exercise in search of grass, turn them into a corn-stubble from which the fodder has been hauled out and ricked. They will rove up and down in this and pick a large amount of "thimbles" from the stubs, no matter how weather-beaten they may be, which they would not eat if given to them in the yard. A sheep is grateful for the privilege of picking up a portion of his living in his own way, nibbling about in all kinds of hidden nooks with his nimble prehensile lips ; and even after they have picked the stubs over twenty times, it will pay to turn them in again simply for the sake of the exercise. The great use of the system of yarding is that it allows the master to regulate the time and amount of exercise, and also secures more effectual alimentation.

WINTER CARE OF LAMBS.—From autumn to winter, from grass to hay (which probably the young animal has never seen before), the transition must be somewhat shaded off. I think it advisable to remove lambs from the pasture early enough (depending on the season) to leave some green feed in the field for them to be returned to a few hours a day for a week or a fortnight. It is far better to take them up in this way than to wait until a snow-storm has covered the grass beyond reach, for then the commencement of housing will be so abrupt as to be likely to produce colic or stretches. Turn them out in the morning, for a few minutes' airing, and sprinkle in their racks a little of the greenest, most aromatic hay at command. I like it as green as English breakfast tea for lambs. When turned back, they will eat the greater portion of it before noon, and then they may be driven afield for a few hours.

Many writers argue that Indian corn is too heating for sheep, and especially for lambs, asserting that it causes loss of wool, "pot disease," etc. It is undoubtedly too oily and heating a grain to be given in unlimited quantities to young sheep for months together. In regard to corn as a feed for mature fattening flocks, I shall have more to say elsewhere ; in this place I shall only give my experience with lambs. Until about January first I feed bran, oats and corn—two parts bran, one of oats, one of corn—all they will eat. Oats are a very unsatisfactory crop on our river bottoms, and about the time above mentioned we generally use up our small harvest of them. I soon take out the bran, also, and for the remainder of the season carry the flock through on corn alone—about three or four gallons a day to one hundred lambs. I do this, first, because corn is our one great staple, and, second, because, after many experiments, I have satisfied myself that it is a thoroughly good feed for lambs. I do not wish to be understood as asserting that corn is better than other grains, or so good as oats. What I would say is that, where the farmer can grow corn to better advantage than oats, and cannot exchange it conveniently, he can safely give it to lambs in about the quantities above indicated, without fearing any evil results at all, if he will observe the following precautions: Use the white corn (the yellow is better for hogs, being more oily), give the lambs constant access to salt and all they will drink of temperate water, and let them have two or three hours' exercise daily.

GRAIN-FEED AT NIGHT.—It is not a good practice to give sheep grain early in the morning, unless they sleep out of doors and have an opportunity to get up and stir around briskly awhile before feeding. In a flock of sheep there will always be some that resemble certain persons—destitute of appetite in the morning. If the grain ration is given out then they will not come at all, or so listlessly that they will not get a fair proportion, and they will lose condition. I have found that in a flock of one hundred and fifty lambs, ten or twelve would scarcely touch grain in the morning, but at night not one would stand back.

WATERING SHEEP IN WINTER.—I can hardly lay too much stress on the importance of looking well to the matter of watering sheep in winter. "You can lead a horse to water, but you cannot make him drink." This adage would hardly be true of the sheep. It will drink after awhile. When a sheep comes out of the stable a trifle chilly, with its blood stagnant after twenty-four hours, quiet, it feels touchy, and it will sniff and sample here and there in a way which is aggravating to the shepherd who is waiting on its motions. It may be fifteen minutes before it can suit itself. It may utterly refuse to drink, whereas, if it could go off and take a run of an hour or so, it would return and drink a surprising quantity. If that sheep had been hastily shut up by an impatient shepherd, it would have suffered before twenty-four hours elapsed, and would not have eaten as freely as it ought, and consequently would have begun to lose condition. Hence, the belief of so many flockmasters, that sheep "do not want water only about every other day." Chilly and slow-blooded as they are, from inaction, they cannot force themselves to swallow the ice-cold water oftener than that; but if it were temperate they would gladly drink every day. Sheep fed freely on roots do not require so much water.

FEED-TROUGHS.—The old-fashioned V-shaped grain-troughs are objectionable because they allow the stronger sheep to thrust the grain along with their noses into heaps, so that they get more than the weaker ones, which need it most. All troughs ought to be flat-bottomed. I find the following dimensions very good: Sixteen feet long, six inches wide, four inches deep. They must stand on blocks or supports about a foot high. I generally set them radiating from the door, like the spokes of a wheel, so that when the lambs run out they do not have to leap over troughs—an operation to which they are very prone at any rate, and one which fouls the troughs in muddy weather. I frequently litter the yard to prevent the same thing from occurring. The troughs may be set around the sides of the yard, but this reduces their capacity nearly one half.

SORTING FOR WINTER.—Sheep ought always to be divided into flocks, according to age, strength, sex, etc. Ewes should be by themselves, also the lambs, then the dry flocks may be parcelled out as yearlings, two-year-olds, etc., though the weaker ones in each lot should drop back one year. A weakish yearling is as difficult to winter, requires as much care as a lamb, and should be thrown into the lamb flock. Last of all is what may be denominated the poor-house—a flock consisting of toothless old crones, inferior lambs, yearlings and others, which the flockmaster had neglected to dispose of in the fall, and he must punish himself for this omission by nursing and coddling the flock of inferiors more carefully than the others. An inferior sheep in a large flock has a poor chance, indeed. It ought to receive more than the average ration, whereas it receives less. A sheep is a timid and defenseless animal at best, and when cowed by a few hard knocks from the masters of the flock, it presently stands back and goes off into a corner to die.

TEMPORARY SHELTERS.—A straw-shed, well built, is a good protection; but poorly built, it is an utter nuisance. In a long rain the water percolates down through it and falls in drops on the backs of the sheep, staining the fleeces a patchy, clouded straw-color. It will continue to drip twenty-four hours, or longer, after the rain has ceased, and here and there a sheep will get wetter than it would have done in the storm itself. The bedding or bottom also becomes saturated, ferments and gives off ammonia, poisoning the air, and the wretched sheep, with stained icicles hanging from its wool, reeks with steam when it rises in the morning or perhaps tears out a lock of wool which was frozen to the ground, while the baneful ammoniacal exhalation is laying the foundation for disease and a cotted fleece in the spring. If a straw roof is built not over eight or ten feet wide and so high that the straw will be ten or twelve feet deep above it when fully settled, it will afford passable protection. A back or wind-break may be made by stacking the straw partly on the ground (though this is apt to settle unevenly and lean), or by constructing a barricade of rails and stakes with straw stuffed between.

A temporary shed-roof may be built of boards, with a straw barricade or bundles of fodder standing on end for siding; almost any shelter which will exclude the snow will answer in the dry cold of winter; but when a long rain comes on, or the frost is coming out of the ground in the spring (the time of year when the system of the sheep is most likely to break down under the debilitating approach of warm weather, and when it most needs a dry bottom to sleep on and a wholesome atmosphere), these cheap roofs are apt to prove a failure, and leave the sheep in a miserable mud-hole. I speak from experience. The sheep had better sleep on a dry sodded mound without a straw overhead than to find themselves, at the break-up in March, left in a slum of manure and water. During the spring thaws there are days when not even the sight of growing grass would tempt the well-fed sheep, chewing its cud in a well-littered house, high and dry on an artificial mound, with an atmosphere clean and sweet, to step out into the bottomless mud. It is in the saving of sheep in March that the shepherd reaps his reward for the building of the more expensive permanent structure.

THE GAIN OF HOUSING.—It is one of the most prevalent and persistent errors of the farmer, that sheep need housing less than any other domestic animal because they have a better natural covering. We are told by these disbelievers that sheep will stand quietly for hours in a rain when by moving ten feet they could get under cover. There are generally two reasons for this fact. First: The house is so foul with ammonia (though the flock-master, whose nostrils are several feet above the floor, may not perceive it) that they will suffer before they will enter it. Second: Unless the rain is violent, it takes it some time to penetrate to the skin of the animals and cause them inconvenience. An animal bearing a pelage of short thin hairs, though it experiences discomfort from the falling drops sooner than one which has a dense coat, is really better prepared to resist the hardships of outdoor life than the other, for the reason that the water dries off sooner. In large cities the best horsemen have little machies for clipping horses; and in the fall when their coat has grown thick and furry, they shear it off close to the hide. If a horse is driven hard and has a thick mass of hair on him to become saturated with perspiration, he is much more likely to take cold when put in the stall than if he had a shorter coat, which would dry out sooner.

While the sheep is not so hardy as it was in its primeval state, it is compelled, if allowed to remain out during the storms, to carry a burden of wet wool, five times as heavy as it would have had to carry when wild, and which is five times as long in drying out. The cow or the horse, though degenerated from its ancestors in point of hardiness, has no greater coat of hair to carry about wet than they had. Therefore, I argue, the sheep needs shelter more than any other of the domesticated animals, and that for the very reason which some urge in excuse of their negligence in providing shelter—because it has a heavier coat to carry. A fleece weighing five pounds will, when on the sheep's back, probably hold ten pounds of water without dripping perceptibly. A man with a heavy ulster overcoat on might for the first half hour be almost oblivious to the fact that rain was falling on him; but after he was wet through to the skin, if he was obliged to stand still, it is quite possible that he would, for the next twelve hours, rather have the overcoat off than on. The more a sheep becomes loaded down with water, the less it is inclined to stir about and take the exercise which is needed to dry its coat and warm its blood. In our capricious American climate a soaking rain is generally followed soon after by brisk winds and colder temperature. Every tyro in chemistry knows that the act of evaporation withdraws latent heat; thrust the wet hand out of the window and it will grow cold faster and freeze much quicker than it would if dry.

Cold is an enemy of life, and chills are always a loss. As Colonel F. D. Curtis forcibly says (in a communication to the *Country Gentleman*): "It costs blood to fight chills, and it takes food to make the blood, which is the current of life and bears with it heat, action and growth." External chills drive the blood in upon the viscera and produce congestion in greater or less degree, pneumonia, fever, colds in the head, etc. The farmer who suffers his sheep to get a wetting every few days through the winter, wonders why they are snuffling so much, with their nostrils constantly plugged up with disgusting accumulations of dried mucus. He smears tar over their noses; he holds them between his knees, pulls their tongues well out and thrusts tar far back into their mouths to make sure of their swallowing it!

What they need is not tar on the roots of their tongues, but tar on the roof, dry footing and dry, wholesome atmosphere. They want plenty of warm red blood instead of tar.

CORN-FODDER FOR SHEEP.—If I were feeding cattle and sheep, and were limited to clear fodder and clear timothy hay, I should give the fodder to the sheep and the timothy to the cattle. That is, if the fodder had to be given out without cutting; and I do not believe it pays to cut the coarse cornstalks of our Western river bottoms, after the first of January, at any rate.

There is no operation about the farm in winter which I perform with more satisfaction than that of giving fodder to my sheep. I have it in ricks about seventy-five feet long, disposed conveniently on two or three sides of the yard, so that it can be thrown over from the rick. After a week or two of practice, a flock of sheep, even yearlings, will pick the coarsest fodder very clean, if it is bright—cleaner than any other stock will. They consume, not only the husks, but the "thimbles" or sheaths, the tassels and a foot or two of the top of the stalk, especially if the weather is a little damp. That is to say, they leave nothing which would really pay for the labor of cutting. I have known a snug, tidy farmer winter a small flock of sheep entirely on the leavings which they could gather from the fodder after his cattle were done with it, supplemented by a small ration of grain.

I confine my flocks in yards the greater portion of the day, and in a few days the stalks accumulate so as to form a good feeding-bed for cold, dry weather; though I find it pays well to throw fodder, as well as hay, into slatted racks, in the best of weather. It is necessary to look sharply after the manner in which the feeding is conducted. The feeder should be required either to move the racks every few days, or, better, to clean out the canes which have been picked over, every morning before a fresh ration is given. To enable the sheep to pick fodder clean, only a thin layer should be thrown in at one time, just about enough to fill the rack up level with the bottom board; then they will not pull it out and waste it. After a few hours another thin layer may be given.

When wheat straw is given in conjunction with fodder (and I consider a ration of bright fodder with straw, cut before it is too ripe, decidedly preferable to timothy for sheep), the straw orts form a good packing material for the canes. The sheep should not be compelled to eat more than half the bulk of straw given to them; the remainder, when thrown out of the racks, is speedily "fulled up" with the canes by the constant trampling of the flock, and assists greatly in retaining the liquid manure, which would escape if only the corn-stalks were beneath. When the sheep are turned out in the spring, a heavy coating of straw is thrown over the yard; this retains moisture, prevents leaching, and insures the rotting of all the canes, so that they can be hauled out in the fall.

In the Atlantic States, where the stalks are smaller than on the rich bottom-lands of the West, the best farmers now generally cut them into lengths of an inch or less, and often steam them and mix with mill-feed. I shall have more to say of this in the chapter on "Feeding for Mutton."

CLEANING OUT THE STABLES.—During the dry, cold weather of winter, a considerable body of manure may accumulate without detriment; but the risk in this is, that when the thaw and break-up come, which will compel the doing of the work speedily, the mud is so deep that it is a great abuse, both to team and land, to haul out manure. The flocks will either have to swelter and sicken in the ammonia, or the team will have to be strained to do the work in half a foot or more of mud. Hence, it is best to make a general clearing out just before the winter breaks up, while the ground is frozen or there is snow.

The reader will bear with my repeated recurrence to the necessity of the shepherd's knowing with absolute certainty whether there is a hurtful generation of ammonia going on or not. He should not allow a week to pass at any time through the winter without making actual test by the nostrils, at the elevation where the sheep are obliged to carry theirs, as to the condition of the atmosphere which they are compelled to inhale the greater part of their time. After the manure has been removed it is well to sprinkle the ground with lime, also with several inches of bedding as an absorbent of liquid manure and to prevent the manure from adhering to the ground.

MAKING AND SAVING MANURE.—I have my straw stack placed every season as close as possible to my main cluster of sheep yards, generally so close that the straw can be pitched directly from the stack into one of the yards. It is not desirable to let sheep run to the stack. The amount of chaff which lodges in their wool is no serious objection, for it is mostly expelled from the fleece before shearing time ; but sheep will bore up into the stack at an angle of about forty-five degrees, and they will soon get wholly atop of it, wasting the greater portion.

A cluster of fodder ricks is placed on another side or other sides of the yards and given in conjunction with the straw to the dry flocks, for I wish every flock, except the lambs, to receive at least one ration of fodder daily. We make a practice of shutting up our breeding ewes in the house every night, unless the weather is unusually warm, so they have a ration of hay to lie to. But I never wish to feed fodder or straw inside of a house. In the first place, the straw orts thrown out of the racks serve admirably to pack closer the loose-lying cornstalks and make of them a better bed for the retention of the liquid manure; and no farmer would think of attempting to manufacture cornstalks into manure under a roof. In the second place, pure sheep manure, deposited in winter under cover and daily trodden firm and solid, does not ferment; or, if the lower strata do, the upper are so dense that no ammonia can rise through them ; and the sheep can and will sleep on such a bed, which is dry and almost dusty on the surface, all winter, without injury. But in the house where the hay orts are thrown out and mingle daily with the manure, the latter has to be removed two or three times during the winter on account of the fermentation and the escaping ammonia. For these reasons the dry flocks receive all their feed out-of-doors, and at night are left at liberty to sleep in or out of the house, as they choose. In the yard where fodder alone is given, straw is occasionally thrown to compact the bed against the wastage of liquid manure. I would not tolerate hogs in a sheep yard, nor is it necessary. It is remarkable how the constant trampling of even the light-footed sheep will full together and press down a bed of cornstalks. They crack the flinty outer covering sufficiently to allow the urine to penetrate and be absorbed by the spongy pith. The addition of corn (about a bushel to the hundred head), or of oil-cake meal, or shorts to the ration, imparts richness to a manure already richer than anything else on the farm except the droppings in the hen house. From the time the flocks are turned away to grass until October or November these beds are left covered with straw to rot in the rain, during which process they will sink down six inches or more. True, there is a small amount leached out of the beds by the rain, particularly toward fall when they begin to lose their sponginess; and a water-tight manure-pit placed to receive this drainage would doubtless be a good investment. But, as it is, from the yards and houses together we secure about three hundred two-horse wagon loads of valuable manure per year, or somewhere near half a load to the sheep.

The manure coming from the houses is hard and tough as old cavendish tobacco, and has to be grubbed up with a mattock. Plowed under eight or ten inches deep, this is a powerful stimulant to corn, which will show its effects for years afterward. It makes an excellent top-dressing for weak places in the meadows, but it has to be scattered on in winter and exposed to the frosts and rains two or three months, after which a man with a stout dung-fork can fine it without much difficulty.

SHEEP LOSING WOOL.—There will often be noticed a sheep whose wool is ragged along the sides, with little locks pulled out and hanging; sometimes long seams showing in the fleece where the wool has wholly parted from the skin on the surface of wrinkles and fallen off. In searching for the causes of this loss of wool, the shepherd must first assure himself that there are no sharp edges, points, pins or nails about the racks or sides of the stable. Then let him watch the ragged-looking sheep and see if it is not addicted to the vice of "wool-biting." It is thought by many shepherds that this is caused by an eruption and itching of the skin, produced by ammoniacal vapors and the heat of fermentation in the manure. The following facts may be set down as established, respecting the habit of wool-biting:—

1. Young sheep are seldom addicted to it.
2. Sheep on grass never pull out their wool.
3. Sheep fed in winter on laxative feeds, as fodder, roots, bran, etc., are less inclined to the habit than those kept exclusively on hay and corn. Sulphur in the salt mitigates, to some extent, its manifestations.

Nevertheless, there are some sheep which, whether it is an idiosyncrasy with them, the result of a thin and sensitive skin, or a vice, are so addicted to wool-biting every winter that they ought to be dismissed from the farm.

Where wool is seen to peel off from the outer surface of wrinkles, it may be accepted as evidence of chilling having taken place in those wrinkles, almost to the point of freezing. Wrinkles are little else than simple folds or reduplications of the skin; they are ill supplied with the blood and warmth of the body, and if upon these conditions there supervenes a loss of condition in the autumn preceding, caused by excessive rains, slushy herbage, or short, frost-bitten grass, it is not surprising that the temperature falls so low in these remote extremities as to destroy the life of the fibers. These, then, are cast off and leave the surface exposed. This is one of the evils attending wrinkly sheep. It can be prevented only by housing and blanketing, which, of course, would not be practicable with a large flock.

CLOUDED FLEECES.—It is hardly necessary to say that if an attempt is made to house sheep it ought to be carried out consistently, for an animal housed awhile and then turned into the weather will presently look worse than the out-door flock. Most shepherds have probably noticed sheep, the fleeces of which were white on the neck, perhaps, while on the back they were yellowish, nankeen or saffron, and pasty-looking.

Sheep which are more or less deprived of exercise, even though their quarters may be kept clean and the sheep themselves in good health, are liable to have this spotted appearance. It is caused by a lack of vigor in the circulation of the blood, which latter is necessary to cause the proper liquefaction and equal diffusion of the yolky secretion throughout the fleece and to the extremities of the fibers. If the flock is exposed to the rain at all, and the fleeces are somewhat open and loose they are apt to part along the back, allowing the water to reach the spinal region sooner than it does any other portion of the frame. The wool fibers on the back are perpendicular and tend to conduct the moisture inward to the skin, while on the rest of the body they are more or less sloping and convey it away.

NECESSITY FOR GRAIN.—Some excellent flock-masters keep their sheep, even their breeding ewes, all winter without grain. Others, equally as good, do not begin to give grain until February or March. Unless the hay is exceptionally bright and fine, it is better to give a little grain all winter, though less is required in midwinter than at the breaking-up, when the sheep's appetite is rendered capricious by the increasing warmth. A little grain throughout the winter gives the sheep heart and thrift; it will consume its rough feed with less waste. A sheep that has fallen off through the winter, and is suddenly put on a ration of corn in March, is liable to lose its fleece, or a part of it. With the sheep, above all other domestic animals, it is necessary that the farmer should bear a steady hand all the year round.

SNOW-EATERS.—There will nearly always be some few sheep, especially in a flock of lambs, that have a depraved appetite for snow, and will not drink water if snow is obtainable. It is a habit as harmful to the victim of it as that of "wind-sucking" in a horse. I have thought, sometimes, that lambs acquired it from their dread of touching water which was ice-cold, when they were doubled up and shivering with cold themselves. With some sheep it becomes a confirmed habit, continued from year to year, and the observing flock-master will notice that the snow-eaters are the poorest of the flock. The remedy is obvious: Provide an **abundance of** temperate water and allow ample time for every sheep to drink, warming them up beforehand with exercise, if necessary; and either keep them out of the snow or drive them over it until it is all trodden down and dirty.

CHAPTER XIX.

FEEDING FOR MUTTON.

Merino Mutton.—We are indebted to our mother-country, England, for a great many moss-grown ideas and prepossessions, and not the least among these is the belief that the coarser and lighter the fleece on the sheep the better the mutton—this, of course, within reasonable limits. This belief has made a lodgment in our great Eastern cities, and from them it is passed on, at second or third hand, to our wool-growers in the West. The Merino has never been fairly tested by the mutton-eaters of the world, because it is, in its paramount function, a wool-bearing animal, and is not usually slaughtered for mutton until it has passed its prime. The only fair test would be one instituted between a Merino lamb and a Southdown lamb.

For all-winter feeding, Merinos are best; and wethers better than ewes, as there is a large discount on the latter. For an early winter market, probably heavy coarse-wooled sheep are preferable. The superiority of the fine-wools as feeding sheep in general, consists in this, that if the market for mutton is not brisk during the winter and spring they can be carried over, shorn early and sold as clipped sheep, bringing almost as much, shorn, as they would have commanded in the winter, wool and all.

Merinos as Feeding Sheep.—In the letter referred to below,

Mr. Isaac H. Frank, of Lake, Stark County, Ohio, who feeds several hundred sheep every winter, says: "In the first place, what kind of sheep are most profitable for fattening? Certainly those which bring the highest price in the New York market. Lambs sell always for more than sheep, and prime wethers sell better then ewes. *Blood don't make much difference if the animal is good size, fat, smooth, desirable wool and trim,* but the Southdown stands at the head."

I have italicized the words bearing particularly on the subject of breed.

Next, I will present an extract from the "Report of the Ontario (Canada) Agricultural College," on a series of feeding experiments conducted during the years 1882-3: "There is a remarkable uniformity in the annual value of wool and mutton from the grades of Cotswold, Leicester, Merino, Oxforddown and Southdown, resulting from differences in weight and value of both products."

In a conversation I had, in August, 1884, with Mr. W. M. Conner, yard-master for seven years of the sheep department of the Union Stock Yards of Cincinnati, I asked him: "What is the best mutton-sheep brought to Cincinnati?" To this he replied: "The Southdown." In reply to the question as to what held the second rank, he said: "The Merino." He continued: "I mean mature mutton. For early lambs, of course, the Merino ranks below the Down and the Cotswold. This is not because the mutton is inferior in itself; Merino mutton, when equally fat, is as good as any in the world—indeed, I am not certain but it is finer-grained than any other—but the point is to get your mutton fat."

"You never have Merino lambs brought to market, I presume?"

"Oh, yes, we have, sometimes; not often. They sell a little under the coarse-wool lambs—not, as I said before, because the mutton is inferior, but because the pelt is smaller and the butcher does not realize as much from the wool."

"Then I am to understand you as meaning that the main point of the English breeds is their precocity; that is, they put so much more flesh and fat on the carcass, and wool on the pelt, at an extreme early age? Is that it?"

"That is the point. They do their best work the first year of their lives."

"But for mature mutton you admit that the Merino is equal to them?"

"Not equal to the Southdown, but better than anything else, as I said before."

"What do you find to be the best feeding sheep?"

"There is nothing better than a bunch of nice Merino wethers for winter feeding. They herd better, in larger flocks; they hold fat better in the spring. If it were not for the Ohio Merinos we would have no mutton at all in the spring in Cincinnati. They come on in the nick of time all along in late winter and early spring, before the Kentucky early lambs begin to come to market."

Mr. W. D. Crout, of Wauseon, Fulton County, Ohio, who feeds for market from fifty to one hundred sheep every year, in a letter to the *Ohio Farmer*, says of Merinos: "I feed different classes of sheep almost every winter, and find that no other kind take to feed so kindly and fatten so rapidly, and have habits of quietude equal to them. Neither can I obtain so ready a market the last half of the winter, or any time much past the holidays. If I have long-wooled sheep to feed, I invariably turn them off early in the winter, but I believe I have never been fortunate enough to escape having some culls from coarse sheep. Do not understand me that Merinos are entirely free from this, but I do claim that they are less liable."

In conversations with me on this subject, Messrs. Miles Stacy, Jacob Dearth and Elvin Miller, all of Washington County, Ohio, and all of them experienced feeders, have repeatedly stated that for the Baltimore market, for winter corn-fed mutton, they prefer good straight Merino grade wethers to those of any other breed.

The views of some of the above quoted witnesses may be considered slightly open to criticism as being influenced by local fashions and predilections; but the testimony of the Canada Agricultural College and the Cincinnati yard-master is entitled to be accepted as entirely impartial.

WHEN TO FEED.—An all-winter cramming on grain is unprofitable with any kind of stock, especially with sheep. Equally true is it that to allow any fattening animal (or store animal either, for that matter), to get on the down grade for a single day is a double loss. I have had some experience in feeding Merino wethers for the shambles, and I find that the most profitable method to pursue is, when practicable, to keep the flock running on a stiff old sod or meadow rowen (when on rich river bottoms), until well along in February, making, of course, proper provision of housing in inclement weather, with enough, grain—say, a bushel of shelled corn per hundred a day—to make up for any deficiency in the frozen grass, and keep the flock gaining a little. This plan of preparation operates very much on the same principle that a clover field does on a bunch of hogs through the summer, keeping them loose in the bowels, growing in flesh and fitting them for the six weeks or two months cramming with grain in autumn.

Most farmers who carry, through the winter, a bunch of feeding sheep, do so with the expectation of selling the wool before grass comes. Hence, I have found that there is an interval between the strictly grain fed and the purely grass fattened flocks, coming in the month of April or May, when sheep will generally sell to best advantage. The wool market has been opened by that time, and yet has not been subjected to the "bear" influences of the regular spring clip coming later. Local manufacturers are about that time beginning to look about briskly for small stocks to start on, not having the capital to hold over a supply of wool through the winter, and not wishing to wait for the regular clip. There is also about this time a sort of interregnum in the beef market.

MANNER AND MATERIAL.—As to the manner of feeding and the material given, there are three points of great importance.

1. Sheep should be fed with the utmost regularity.

2. Though fond of variety, and requiring it for an attainment of the best results, feeding sheep resent a sudden change to an unaccustomed feed stuff.

3. Hence, combination of feeds is better than change.

Supposing the flock to have been on the range until the 1st or 15th of February, on a ration of a bushel of shelled corn per day, we would now yard them, and set about conducting them up to the regimen on which they are to finish off the fattening process. If accustomed to it, they may be put on corn fodder once a day for a month with great advantage, but after the middle of March, fodder begins to be distasteful, and is not so well relished by any stock. I give one hundred mature sheep twelve to eighteen bundles of fodder during the forenoon, generally in two feeds in slatted racks. A sprinkle of bright wheat straw or chaff may be given at noon; at night, all they will eat clean during the night of clover hay, Hungarian, June grass or Timothy (I name them here in the order of my preference).

If it has not been found convenient to let them run on the

range, the best substitute for it will be fodder and straw, or clover hay and straw ; though it is well to reserve enough clover to give, toward the end of the yarding season, in conjunction with the heavy grain feed, as the best coarse, cooling distender for the heated stomach.

A good grain feed for fattening sheep is shelled corn, one-half ; barley or rye, one-quarter ; oats, one-quarter ; but to the majority of farmers perhaps corn is the most available feed. I do not think, after many trials, that it is profitable to crush grain of any kind for sheep, much less the cob with the corn; the cob being, in my opinion, not only useless as a feed, but a positive damage.

It is wasteful to throw out corn unhusked, as some Western feeders are accustomed to do. There is too much of the grain to the amount of leaves, and, besides that, I never succeeded in feeding unhusked corn to sheep in any way in which they would not, before they managed to get the corn stripped and shelled, waste a good deal of the foliage. Some sheep are a great deal more expert and vigorous than others in husking and shelling the ears and get more than their share. In short, there is every reason for husking corn before it is given to sheep, and none (of any considerable value), in favor of giving it out unhusked.

As to oil-cake meal, or cotton-seed, most sheep are not accustomed to either before they are penned up to fatten, and they must be broken to them with caution. Sheep are fond of variety, but they want that variety to consist of articles to which they are accustomed. If the flock-master's sheep are used to oil meal, by all means let him give them some, perhaps a daily ration of it, during the fattening process. But if not, he must proceed with caution in breaking them to it. Let it be given in very small quantities at first, not over a tablespoonful per head, mixed with four or five times its bulk of wheat bran or some other coarse ground feed to which they are accustomed ; and then let the proportion of oil meal be increased until it forms one-half, three-fourths, or even the entire feed, if they are found to relish the article, which they undoubtedly will. But the limit with this rich feed stuff is easily reached ; it will not do to go beyond a few ounces per head, according to the size of the sheep.

If shelled corn alone is given, it can be so dispensed as not to injure the sheep at all, though it requires great watchfulness and good judgment to give fattening sheep all the corn they will eat without doing them serious mischief. If, by a trifling negligence on the part of the feeder, they get a little "off their feed," one or more of them will vomit up corn about the shed.

From the time the flock is put into the yard to begin the fattening process, it should be nearly or quite a month before the ration of shelled corn is increased up to their full capacity to consume. An increase of two quarts a day will carry the feed in that time from one bushel up to three per day; and that is about as much as one hundred Merino wethers can be induced to eat, with an abundance of clover hay. It is best to divide this amount into three feeds, and every feed should be given under the eye of the master himself. If, on account of warm, muggy weather or other reason, the most of the sheep run away from the grain-troughs before the corn is all eaten up, the remainder ought to be at once chased out of the yard and the residue of corn removed, else a few will linger and eat too much.

The yard ought to be kept well littered; the heated condition of the sheep and the strong manure getting into the clefts of their feet induce "scald-foot." Once a week all limping ones should be caught, their feet examined, pared clean, and a little finely powdered blue vitriol sprinkled in the cleft. Of course, the judicious flock-master will supply plenty of water; and constant access to salt, in which one-tenth or one-twelfth of copperas has been mixed, is beneficial.

As soon as the grass is sufficiently grown to carry stock— from April 5th to 12th, according to latitude—the flock may be turned on it, after being tagged, and the grain ration reduced to a bushel per day. But they ought still to be yarded every night, and a little very tempting hay sprinkled in the racks and brined (all other salt being withheld). If they are not, by this means, or some other, induced to eat a little hay, the grass makes their teeth sore, and they will not eat the grain as they should.

METHODS OF A NOTED FEEDER.—Mr. J. H. Frank, already mentioned, feeds for market from five hundred to one thousand sheep yearly. His barn (fig. 22) is one hundred and forty by forty-five feet without the wing—has no floor except the tamped clay; the sides consist mostly of doors, so that it can be entered with teams at any place, for the storing of grain or hay, or the removal of manure. It is used in summer as a barrack for grain; this being threshed out early in the fall, leaves the space ready for the sheep when winter approaches. I copy (and partly condense) Mr. Frank's account in the *Ohio Farmer:*

"The pens are formed by the racks and a double line of fence (the latter making the feeding aisle); all these are removed in the spring when the manure is hauled out. They are stored in shelter during the summer until after threshing, when they are replaced ready for sheep. At the south end of the barn on either side of the aisle we place half-racks, D, D, figure 23, and ten feet farther up, a rack, B, fifteen feet long, with one end against the fence at the aisle, which leaves a passage for the sheep to pass over to the other side. Then ten feet farther up we place another rack, B, which extends from the side of the barn to the tank, making a pen twenty feet square. Then on each side of the aisle the same arrangement of racks is continued until we have the barn partitioned off into eight pens—four on each side. At the north end of the barn we put our

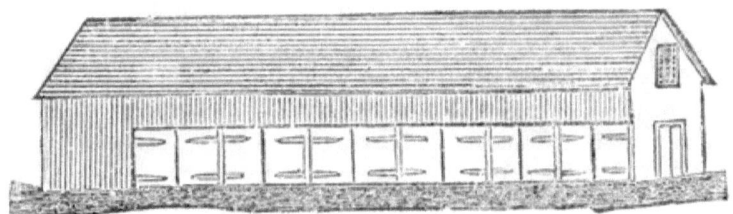

Fig. 22.—MR. FRANK'S SHEEP BARN.

hay, and as fast as it is fed we fill up the space with pens, so that by spring we have more pens. An empty space is left, however, between the hay and pens for throwing down hay, driving in sheep, etc. The racks are so made, that they are used both for feeding hay and grain, as shown in figure 24. H, shows trough for feeding grain, and K the hay-rack. It can be closed, as shown in figure 25, for sweeping and putting in grain, as the wings keep the sheep away until the grain is evenly scattered in the trough, when the wings are turned and all the sheep come up at once. These are by far the most convenient sheep racks that have come under our notice, and I doubt whether there is another rack near its equal for cleanliness, convenience and saving of both hay and grain.

"Now we do not wish to be understood as indicating that it requires a barn and racks just like ours to make a success of feeding, but we think ours are excellent, and if the genius of the feeder can find a better way or method, we would be pleased enough to follow his plan.

"We take it is an indisputable fact that whatever adds to the comfort of the sheep, will add to the profit of the feeder. The sheep must be kept quiet. No dogs, cattle or boys should be allowed to chase or worry them, so that they can eat and lie

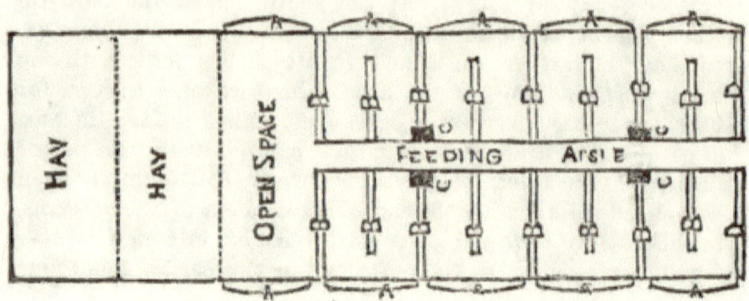

Fig. 23.—GROUND PLAN OF MR. FRANK'S BARN.

down unmolested until they again wish to eat or drink. Pure air is one of the essentials. A stable full of foul odors, damp and dirty, cannot be a place suitable for keeping an animal as cleanly as is the sheep. We secure fresh air by opening any or all of the double doors, *A*, *A*, in figure 23.

"Figure 24 represents the hay-rack and feed-trough combined, with the wings, *W*, *W*, turned in and buttoned fast, giving the sheep access to the feed-trough, which runs along at the bottom of each wing. There is a raised board walk along the middle, between the troughs, on which the feeder walks while pouring

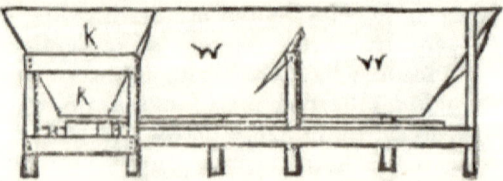

Fig. 24.—FEED-RACK, OPEN.

grain into the troughs. While the wings are turned in this way, they also constitute the sloping sides of the hay-rack, from which the sheep pull out the hay through a four-inch crack at the bottom. Figure 25 shows the wings turned perpendicularly and fastened, excluding the sheep from the troughs while the grain is being poured in. Under no circumstances must the sheep get wet, for the wool requires a long time for drying, and makes the animal cold and uncomfortable. Too much can not be said about cleanliness. Good bedding must be secured, and removed whenever the sheep have no clean, nice and dry place in which to lie down. The racks and troughs should always be swept perfectly clean before either hay or grain is put in them. Do this always. Another requisite is pure water. The water tank is liable to be fouled by droppings or particles of feed falling into the water. The tanks should be emptied often and rinsed out, so that the water is clear, sweet and clean.

"The water must be convenient so that the sheep need not go any distance to secure it. They will eat a few mouthfuls, then drink a little, go and eat, and so on until they are satisfied, when they lie down and chew their cuds. If the sheep must go into the storm or into mud or wet, or a little distance, they often do without water rather than go and get it. The water should be kept so warm that it does not freeze, as they will not

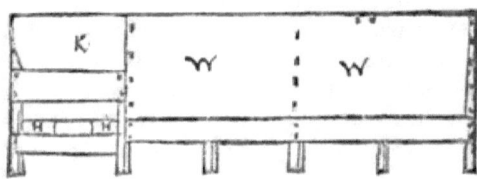

Fig. 25.—FEED-RACK, CLOSED.

drink enough to make them thrive well if too cold, and besides, it requires feed to warm the ice water. I have often seen lots of sheep which were well taken care of in all respects, but failed to do well because they were not well supplied with pure water.

"For the best results, sheep should not be kept in large lots, and those of a size should be kept together, as the smaller ones are crowded back by the larger ones, and knocked about so that they do not thrive as well. Hence, the small pens, twenty by twenty feet, as shown in figure 23. If I wished to do the very best with a sheep, I think I would put it into a stall by itself, as we feed cattle and horses. The nearer we approach this the better, and it would be more profitable to put sheep into smaller pens than ours, if convenient. Ours hold from forty-five to fifty-five, according to size, with comfort. Care must be taken not to crowd them too much. Our sheep never leave the pens save when carried to market, and we find they do much better than if they have the range of a barn-yard, or even a special yard made for them. We feed a variety of grains, and mix them. Corn ground in the ear, oats, sometimes wheat screenings, and always middlings and bran. We have fed some linseed meal; think it very good, but have not tested it so as to speak authoritatively on the matter. We grind our corn because it is easier digested, and it requires longer time for the sheep to eat it, and each has a better chance to get its share of the meal. Bran is of prime importance, we find." Mr. Frank gives the following as his rule for regulating the quantity of feed: "Begin in the fall with a small amount; increase gradually until the amount is reached which they eat up clean, and no more. They will eat a little more in cold than in warm weather.

"We feed clover hay exclusively, and find it far superior to any other. If timothy hay must be fed, let it be to the old sheep, for lambs will not do well on it. We feed all the hay that they will eat without wasting. Flocks are often fed so much that they waste enough to bed them, which is no advantage to the sheep, but wasteful and extravagant. Cut the clover before it is very ripe, as it is better relished and has more nutriment in it. We often cut clover on our wheat stubbles, and find that sheep like it better than any other.

"Regularity in feeding is very necessary. Sheep should be fed grain and hay twice each day, and at the same hour as nearly as possible. Salt is kept in a box in each pen, so that the sheep can get it whenever they want it. Now to recapitulate:

"1st. Select lambs or good wethers.
"2d. Confine in close quarters and small lots.
"3d. Cleanliness.
"4th. Keep sheep quiet.
"5th. Pure air.
"6th. Good shelter.
"7th. Pure water near at hand.
"8th. Variety of ground grains, with bran and salt.
"9th. Clover hay.
"10th. Regularity of feeding."

It will be seen from the above, that Mr. Frank's system contemplates an all-winter cramming on grain, and a rigid confinement of the flock to the pens throughout. My experience has been had on river bottoms; and on these the rich rowen of low meadows will, in an open winter, carry a flock of fat wethers far into the winter, with no more corn than above mentioned. But where the feeder has an abundance of clover hay, roots or corn-fodder, as a cooling laxative diet and a corrective to the grain, the feeding term may extend through the whole winter with profit, and the sheep may be closely housed.

IMPORTANCE OF QUIET.—Mr. Frank's remarks on this subject are just and deserving of special attention. All dogs should be kept out of sight and hearing. Not even shepherd dogs should be allowed about the pens where sheep are fattening. No sheep, unless mingling familiarly with a dog every day of their lives, will become so accustomed to him as not to be disturbed by his approach. The greater the quiet, the more rapid will be the gain in flesh. To this end there should not even be a change in pens or troughs or the feeders during the winter. The same person should take care of them in the same place throughout.

VARIETY OF FEED.—Many good feeders, including the celebrated John Johnston, give sheep no other rough feed than straw for considerable periods of time. I have myself done so through the whole period of feeding, except the last six weeks or such a matter. For a feeder who buys all his material this would not be advisable (though it would be good policy to invest a small amount of money in green, bright straw, oat or wheat, rather than to purchase hay exclusively); but it might be profitable for a farmer who has a large amount of straw on hand, and who also wishes to manufacture manure on a large scale, to give it to fattening sheep, early in the winter, very liberally. Sheep fed very highly on grain will consume with relish the coarse stuff, which a flock subsisting almost entirely on hay would reject. Hay mixed with bitter weeds or other trash may be given in occasional feeds to fattening sheep with evident advantage.

In warm, muggy weather, if the flock are rather mincing over their corn or corn-meal, it is well to mix a portion of oats with it, or give oats exclusively for a few days.

Timothy is too binding for an animal whose system tends so easily to fever and constipation as does that of the sheep. Almost any kind of straw, except buckwheat (which is apt to poison the lips), is better as a coarse feed for fattening sheep than clear timothy. In fact, there is no kind of hay, except clover, which is as good, unmixed, as the same would be with a judicious alternation with straw.

CLEANLINESS.—No other domestic animal is so easily disgusted with its feed by mustiness, dirt, foul odors, etc., as is the sheep. The breath of the animal itself soon renders its feed distasteful to it, and for this reason it might not inaptly be set down as a maxim that no feed should be placed before it which it will not consume within an hour. The least taint in the water-trough is offensive to this most cleanly and fastidious animal; it will go hours without water, to the point of actual suffering, rather than drink that which is polluted. Hence, the troughs and vessels must be kept clean, and the sheep which are observed to be dainty must have a fair allowance of time to find such place in the water-trough as shall suit their capricious appetites.

A Device to Keep Troughs Clean.—It is often urged as a sufficient argument against feed-troughs inside the stable, that it is impossible to have them in order without cleaning them out at every feeding-time. To obviate this troublesome task, the feeder may set the troughs around the sides of the building, from four to six inches too high for sheep to reach them; then put under them a piece of timber, or a bench, upon which they can step with their forefeet, but too narrow for them to stand on with all fours. This will keep all dung out of the troughs.

CHAPTER XX.

FROM HAY TO GRASS.

I have before made some remarks on the importance of turning the sheep afield frequently during the winter, at least when the ground is bare. But as soon as the grass begins to grow, even a little, upon the approach of spring, it will be necessary to exercise caution. The old grass which they get during the winter, the long tufts of rowen mixed about equally with dead grass and lurking under bushes or briers on some north hillside, which the sheep neglected during the summer, has a different effect on them from that of the young growth. The latter is washy, and scours them, and "takes away their appetites," as the old farmers say. During the winter I frequently let my sheep out twice a week, if the weather is favorable; and I find no injurious result from it, even though they remain on the grass all day and fill themselves. In fact, I generally give them only their grain rations on these days and no coarse feed at all. Nor do I find their appetites, the following morning, anywise dulled for their hay or fodder. But after the grass starts a little this will not answer. If they are left on it even half an hour, the next day they will mince over their dry feed and not consume a quarter of it. They are purged and they stand with hollow bellies and look through the gate all day long. They must now be restrained in their run on the grass; it must be greatly curtailed. The last two or three weeks, or month, before turning them out to pasture, I do not suffer them to run on grass more than a quarter as much as they do through the winter. Fifteen or twenty minutes a day, or say while you are putting a fresh feed into the racks, is long enough; and with this restriction, the privilege may be granted them every day. I do not lengthen the time at all up to the very last day of feeding them on dry feed. Then catch and tag the flock, shorten the long hoofs (it is well to attend to both these matters several days beforehand, on rainy days, or at any other convenient time), and let all go.

Hoove.—I find that suckling ewes are more liable to this trouble than any other class of sheep. The insatiable appetite created by the constant drain on the system during lactation, is apt to make them overeat. It is dangerous to turn ewes and lambs on a field of clover (white or red) until they have been long enough on grass to get their bowels toned up, their alimentary systems recovered from the winter torpor. Hoove is caused, primarily, by a lack of mucus, superinduced by the dry feed of winter. Mucus is needed to assist the peristaltic action of the stomach and bowels. This it is which makes necessary the gradual wonting of the sheep to the more succulent feed of spring. Sheep ought never to be turned upon clover when it is wet, and very carefully at all times, until it is in blossom, unless they have been thoroughly prepared for the great change by plenty of soft feed, roots, bran mashes, green rye, etc.

A suckling ewe will succumb under hoove more easily than a cow; her muscular and vascular system is frailer. When fermentation has already set in, and the paunch is distressfully distended with gas, a teaspoonful of turpentine may be administered in a little cold water. A two-ounce dose of Epsom salts, with ginger and gentian, should be given as a preventive of its recurrence.

Depasturing Wheat with Sheep.—It is when spring has fully set in that the uses of depasturing appear. The ground is then seamed with frost cracks, puffy; the wheat tufts are more or less thrown up; the earth needs to be compressed and packed about the roots. Cattle make deep foot tracks, with the wheat thrust down to the bottom of them; but sheep pack the surface gently an inch and a half or two inches deep with their innumerable tracks, covering it all over (I have found that, even when they run on naked ground during a winter thaw, they do not pack it over two inches).

Sheep are very peculiar and capricious in their way of grazing growing grain. They do not fancy it much at best, and they avoid all long stalks, seeking to bite close to the ground. I have seen them depasture a field of rye in a singular, patchy way—here a spot a rod or so square eaten to the very ground, while close at hand is another with the rye heading out three feet in air, never having been touched. This happens when they have too much, or are allowed to stay on it too long at a time.

If permitted to graze wheat in this fashion, it would work mischief. When it is seen that they are inclined to do so, they must be broken up, herded, pushed about, not allowed to settle down on their favorite spots and gnaw them down to the earth. The rank patches, which need depasturing most, they will graze least, if they are not somewhat controlled.

They ought not to be turned into a wheat field in the spring until it is dry and settled enough to be fit for harrowing—dry enough to be a little crumbly. If there comes a sudden March freeze, followed by a thaw, I do not allow them to return to it for a few days.

It is not a good practice to allow a large flock of sheep to run into and out of a wheat field a number of days at the same place. If a flock of young sheep can be turned on at the proper time and kept in the field day and night until their work is done, or turned in at a different place every day, better results will be accomplished. I turn them off before the wheat begins to joint.

The effects of depasturing wheat are: That the amount of foliage is reduced, the tufts are rendered lower and more stocky, the whole field has a cleaner, more open and more even appearance. There is a freer circulation of air through the growing grain, and a reduced tendency to rust. A judicious depasturing hardens and toughens it up. This is my experience in the latitude of Southern Ohio. Further north it would probably seldom be the case that sheep would be beneficial to green wheat.

CHAPTER XXI.

FODDERS FOR SHEEP.

RED CLOVER.—This is the best of all foliage plants for sheep-feed, if well cured; and the curing and saving of it are so apt to be ill-done (making it one of the poorest of fodders), that a few directions, founded on experience, will not come amiss to the young shepherd. If the soil is very rich, clover is apt to grow coarse and lodge. To render it fine enough for sheep it is best to sow it thick, say one and a half gallons to the acre (one gallon on thinner land). A gallon of timothy seed per acre, sown the preceding fall, is a good addition; the timothy will assist the clover to stand up and make it finer.

When the earliest clover blossoms have turned brown it is time to set about the cutting, though it may be well to delay a few days if the barometer does not indicate settled weather. If there is a fair promise of three or four days of clear weather, and help is abundant, five or six acres may safely be cut down at once; this should be done in the afternoon. A half day's steady sunshine will wilt it sufficiently, if the thick bunches at the corners of the lands are shaken out a little in the morning. The farmer should twist a handful of it to see how much moisture the stalks contain. If no sap can be wrung out of them, he may proceed to rake it into windrows, and leave these over night, unless rain is threatening. If so, it should be made up into cocks, about as high as a man's head, and rather slender. On the third day, as soon as the sun has thoroughly dried the hay, the cocks (or windrows) should be turned, bottom-side up, and, perhaps, the lower half of each cock (now the upper half) pulled aside, thus dividing it into two equal portions. At night cock up again, in larger cocks if desired. On the fourth day, as soon as the dew is off, haul the hay in without opening the cocks.

In catching weather, the farmer will have to vary this programme to suit circumstances; but on no account should clover hay be hauled in when very damp, especially if damp from extraneous moisture. Better let it stand several days, thrusting the arm into the cocks occasionally to see if they are heating. If they are getting a trifle warm at the bottom and there is not time between showers to dry them out and haul in, build the cocks over, putting the bottom on top, handling carefully, running them up high and sharp to shed rain.

Salt should never be sprinkled over hay of any kind when it is mowed away; it attracts moisture and discolors the hay. Air-slaked lime is good; it may be put on in almost any quantity without rendering the hay distasteful to sheep. If the clover is pretty heavy, and the farmer has straw convenient, it is well to put a layer of straw about six inches thick, alternately with a layer of clover, about a foot thick, and not allow it to be tramped much.

Where clover is sown on wheat, and comes on very rank after harvest, it is best to mow and cure it. This leaves the ground cleaner for next year's mowing, and the sheep will readily sort the clover from the wheat stubble. If the soil is very rich, and the season rainy, however, clover rowen is not safe fodder for sheep. I have had a few animals killed by it, and a large number in the flock were miserably "slobbered" and sickened.

It has also come within my experience that clover hay (first growth), cut very green and succulent on rich river bottoms, has caused pregnant ewes to "slink" their lambs. This is a very rare occurrence, however.

CORN FODDER.—To the casual observer it might seem quite a hopeless undertaking to winter an animal, which is so dainty, and which searches the ground over so carefully for the finest herbage as the sheep, on such coarse provender. But after sheep have once been trained to eat it, corn fodder is one of the very best feeds for them; superior to every other except clover hay.

For cattle, corn may be kept until it is yellow almost to the tassel, but for sheep, the best fodder will be secured by cutting as soon as the husk shows the color. When husked, it should be bound into bundles with tarred twine; this will prevent the rats and mice from gnawing the bands. With a knot in one end, slipping into a noose at the other, such a band can be easily unfastened in winter, slipped into the pocket and saved for another season.

All corn-fodder ought to be ricked near the feeding-yards. This may be done the last of November or first of December—if the fodder is not wet—without danger of molding.

My way of ricking fodder is as follows: I lay down a double row of bundles, top to top, lapping to the bands. To keep the middle full, I make every other course or layer a single one, consisting of bundles laid butt to tip alternately. I do not draw in any. At a suitable height I lay a stringer of bundles endwise on the rick, three or four to each length, which sharpens up a basis for the roof. The roof consists of a single course on each side, the bundles sloping up to a peak.

Colonel F. D. Curtis says "cornstalks are wasteful food for sheep," and he recommends that the leaves be stripped off when green, cured, and bound in bundles for suckling ewes and early lambs. The farmers of the Atlantic slope can probably best dispose of their small cornstalks by cutting them for cattle; but I doubt if it will pay to cut the large stalks of the West; and when given out uncut, sheep will pick them far cleaner than will cattle or horses.

I never wintered lambs on corn-fodder, but a neighbor of mine, Mr. W. S. Gray, a careful, practical shepherd, has done so several times with excellent results.

FODDER CORN.—In Vol. I., No., I. of *The Shepherd's National Journal*, Mr. Arvine C. Wales, of Stark County, Ohio, gives a very valuable account of his mode of growing this kind of forage for sheep. He states that he sows about two or two and a half bushels of common corn per acre, with a Buckeye wheat drill, in the first week of June. His only cultivation is to run the Thomas Smoothing Harrow once or twice over it when about three inches high. He harvests with a Champion, side-delivery, self-raking reaper, beginning about the first week in September, when the lower joint is turned to a bright yellow. I copy his own words: "Besides the driver of the machine, there are eight men, divided into four gangs, of two to a gang. The 'stations' are measured off and assigned as in reaping wheat. Each gang of two men is provided with a 'corn horse,' which is simply a light rail, with two legs at one end, and a loose four foot pin in the middle. Each gang is also provided with a quantity of wool twine, cut to a suitable length, and hung on a hook in the end of the 'horse.' When the men are in their places, and the machine starts, one of the men passes two of the gavels or sheaves, and sets up his 'horse.' He then goes back and picks up the two gavels, one at a time, and puts them into two of the angles formed by the 'horse' and and its loose pin; his comrade does the same with the two gavels in front of the 'horse.' Then one draws out the pin and moves the 'horse' on by two more while the other, with a piece of wool twine a yard long, binds the top of the shock. Here it stands for ten days or two weeks, till it is partly cured. Then the men break the shocks open, each shock generally separating into the four original gavels, and bind it into sheaves with the fodder itself, which by this time has become tough and withy. Twelve or more sheaves are then put into a great shock and the top of it bound by the wool twine used in the first place.

"I had almost forgotten to say that one is far less dependent on the weather in curing fodder corn than in making hay. Several years it has rained nearly every day while cutting, but I never lost a hundred pounds through wet weather, unless it had blown down and been allowed to lie on the ground. Here it should stand until wanted for feed. It is so full of sugar, and starch, and gum, that it cannot be safely stored in barns or stacks. It will heat and ferment. A near neighbor lost his entire crop last winter, although carefully stacked in long, low, narrow ricks. This is the greatest objection to fodder corn. It is hard getting it up when repeated freezings and thawings have glued it to the ground towards spring; and it is hard hauling when the wheels sink through the soft ground to the bottom of the furrow.

* * * * *

"I cut and steam all my fodder. It is cut on a cutter with a capacity of three or four tons per hour.

"The yield of dry fodder has been from five to seven tons per acre, and I carry as much stock and get as much and as good feed from seventy acres of fodder corn, as I used to get from one hundred and fifty to two hundred and twenty-five acres of meadow. Whoever sows corn for the first time will be astonished at the amount of feed he will have in September."

TIMOTHY.—Every farmer is presumed to know how to cut and cure timothy, but there are few who sow it thick enough and cut it in the right stage to make fine, palatable sheep-feed. On mellow, rich land I have found it advisable to sow two gallons of seed per acre. Coarse, ripe timothy is about the poorest sheep-fodder that can be imagined. It ought to be cut when in blossom; and if there is a large amount of it to be harvested, it is well to commence even earlier, reserving the finest and greenest for lambs, and the later cut for older sheep. Unmixed timothy is very objectionable for pregnant ewes: it is too constipating. They will eat off the heads and the leaves, avoiding the stalks to the last; and if these are over-ripe and woody they may become impacted in the stomach, causing heat and irritation. If the ewes are suckling lambs, the latter are apt to die of constipation at the age of one or two weeks. For fattening wethers it is somewhat better, but to any class of sheep it ought not to be given more than once a day, with an alternation of some more laxative fodder.

Objectionable as clear timothy is, I do not think it advisable (with the exception of a light admixture of timothy with clover, as above noted) to sow meadows with mixed seed. Pastures may well be composed of various grasses, thus affording a succession of feed; but a meadow has a set time for harvest, and the different grasses do not ripen simultaneously. It is best to grow and harvest each by itself; then, for variety, feed the flock from different mows. To this end there should be a succession of meadows, as, for instance, clover, orchard grass, timothy, red-top, Hungarian grass; then each can be harvested when it is at its best stage of growth and ripeness.

ORCHARD GRASS (*Dactylis glomerata*).—This, if allowed to become over-ripe, is even poorer for sheep than timothy, if this is possible. It ought to be mown as soon as the seed-stalks have attained their full height, before the pollen begins to fly about. Well secured in this stage, it is so thoroughly good that I have for years always had one, and sometimes two, of my meadows in orchard grass. It does not yield quite so much weight per acre as timothy, but it more than compensates for this by the dense and vigorous aftermath which it throws up, affording luxuriant pasturage for four or five months, while, if the autumn is dry, the timothy stubble will remain gray and parched. Most farmers make a failure with orchard grass because they do not sow it thick enough. Two or two and a half bushels of seed per acre are required to prevent it from growing in tussocks and to make it fine enough for sheep. It should be sown in March on a very fine, well-harrowed seed-bed.

HUNGARIAN GRASS (*Panicum Germanicum*).—This, too, should be sown very thick for sheep-feed, say a bushel per acre, on strong soils. Otherwise it produces heads so large as to be distasteful to sheep, and they will leave them lying in the rack after eating the stalks and leaves. This is especially liable to happen when the hay has not been cured enough, in which case the large, succulent heads will become moldy. To prevent this hay from molding is, in fact, the chief difficulty in its management. It ought to be exposed nearly, or quite, three days to the sunshine. At best it is suited to grown sheep, rather than to lambs, and it ranks high as a milk-producing feed for suckling ewes.

RED-TOP (*Agrostis vulgaris*).—This makes excellent hay for sheep, but, like timothy, it must not be allowed to stand too long, and become dry and woody. In Southern Ohio, on red and yellow clay uplands, it succeeds better than timothy, which it will eventually supplant; and it makes, also, better hay, because it is finer and more nutritious. Sheep fed on bright, red-top hay will winter as well as those fed on timothy with the addition of a half-bushel of shelled corn per hundred each day.

MEADOW GRASS.—Under the various names, June grass, Blue grass, etc., rather loosely and indiscriminately applied, most Western farmers are familiar with one or both of two species, *Poa pratensis* and *Poa compressa*. They are so nearly alike in feeding value and other respects, that most farmers recognize no difference between them. They are the richest of all grasses, native or cultivated, and are incomparable for pasture, but for meadows they are unsatisfactory to thrifty farmers, as they yield so light a weight of hay. I have, however, found it very advantageous to mow small areas of them, natural hillside meadows, strips of creek-bottoms, etc., as they furnish for lambs by far the finest and richest hay obtainable.

MISCELLANEOUS.—The vines of beans and peas are better relished by sheep than by other stock, and are excellent for a variety. Clover chaff, the refuse material left after the seed has been threshed out, if not too much bleached, will be eaten by sheep to some extent.

It is often the case that there are patches in the cattle pasture too rank to be eaten green; these ought to be mown and cured for the sheep. A certain portion of weeds and bitter stuff, ragweeds, morning-glory vines, docks, etc., will be more acceptable to sheep occasionally, than an unbroken regimen of the best of hay.

The orts in the racks ought to be thrown into a separate rack and brined; if there is still a remnant left, the horses will consume most of it. Cattle dislike the leavings of sheep.

CHAPTER XXII.

SYSTEMS OF SHEEP HUSBANDRY.

ON THE ATLANTIC SLOPE.—In the famous standard or stud-flocks of Vermont, the rule is said to be, twenty tons of hay for every hundred sheep. It is probable that the celebrated breeder, Edwin Hammond, of that State, did more to increase the wool-bearing capacity of the sheep of this country than any other dozen men; yet we have the testimony of Doctor Randall, that he fed his ewes nothing but hay.

But, on the Atlantic slope, the Merino is no longer preëminent, except in limited areas—of which the most notable is Vermont—for the production of wool and mutton is an entirely subordinate industry, owing to the fact that it is a region very much superior to others, devoted more largely to wool-growing in requisites necessary to success, in mixed farming. And this fact makes it creditable to sheep that they retain as permanent a foothold as they do there. The breeds of sheep are perhaps not as well defined or as highly improved as in the West; there lingers a greater proportion of the old native American stock, described by Youatt as being a sort of mongrel scrub Leicester, mixed with Southdown and Cotswold.

The limited product of grain and the great cheapness with which it can be produced in the West render it too high-priced to be given to sheep in any quantity. Eastern farmers endeavor to winter their stock or breeding flocks without grain—on clover hay, chaff, pea, bean, wheat and oat straw—thus making them serve as scavengers or consumers of refuse products. This for the reason that there is a cash market for nearly everything, even rye and wheat straw. A prominent object with them is the growing of early lambs for the market. They buy ewes shipped from the West, generally those which have passed their prime; rangy, good-sized, open-wooled grade Merinos; on which they cross a Southdown or Cotswold ram two years old or upwards. The earliest lambs are dropped from January 15th to February 15th; the ewes are well sheltered and fed to improve their condition, so that they generally yean fine, strong lambs. When the latter are a few weeks old they are allowed access to a separate apartment, and are fed bran, meal, and ground oats in troughs. They generally bring four dollars and fifty cents to five dollars per head when they will weigh thirty or forty pounds gross; sometimes as high as ten dollars! If not too aged, the ewes are retained for further service; if they are, they are fattened for the fall market. A Southdown ram generally costs from ten to twenty dollars. They are preferred to the Cotswold, Lincoln, or middle-wool rams, because their lambs, though smaller, fatten better, have better hams, and produce a marbled flesh.

The Merino's share in this, oftentimes very profitable, business is a somewhat humble one, yet it seems likely to be enduring, because, while the crossing with a Merino does not impair the quality of the mutton, the Merino ewe brought from the West offers the cheapest medium through which this mutton can be produced. A ewe too old to do further service as a breeder in the vast flocks of the plains is still, in most cases, capable of doing one or more year's excellent work in a small, well-fed flock; and she can be transported and sold to the New York or New Jersey farmer for less money than it would cost him to raise either a Down or a Merino on his own farm.

One of the curious by-products of the sheep that may be mentioned is the manure, which is sought for by the tobacco-growers of the Connecticut River valley. Mr. J. F. C. Allis, of East Whately, Massachusetts, in a letter to Hon. John L. Hayes, states that Merinos, crossed with long-wools, are the best for this purpose; they are better feeders and take on fat more easily than the long-wools. The feeders buy large wethers from Michigan, from three to five years old, and have them pastured till November. Then they are closely housed, forty or fifty in a pen, and well bedded; about December 1st they begin to feed grain lightly, gradually increasing, until they eat a quart apiece daily. They seldom eat more than that. Mr. Allis further says:

"The cause for feeding so many sheep for their mutton in this valley is the high value of sheep-manure for tobacco-growers, it having the effect on our light soil to produce a dark-colored silky leaf, of good burning quality, suitable for wrapping fine cigars; the tobacco burns white, and has a good, sweet flavor, perhaps owing to the potash it derives from the manure. So valuable do we consider this sheep-manure that we have shipped, since 1870, from West Albany, from fifty to one hundred and fifty cords, costing from eight to ten dollars a cord, every spring. On our light soils, called pine-lands, after raising crops of tobacco, two thousand pounds to the acre, we have sown wheat; yielding thirty bushels of a plump berry, and a heavy weight of straw, on land which without this manure is fit only for white beans. We, of late years, feed with our sweetest and finest hay, and mix with our corn, one-third cotton-seed meal; by so feeding our sheep fatten more easily, being more hardy and better conditioned, besides increasing the value of the manure and rendering it more full of plant food."

Wm. Ottman & Co., wholesale butchers in Fulton Market, New York, state that at the time of writing (March, 1885), well-fattened Merino wether carcasses are selling at one and a half to two cents per pound less than corresponding Southdown carcasses. Yet, so little do the latter excel the former in size that the butchers, to prevent their fastidious customers from imposition, are obliged to leave the dark skin on the legs.

Notwithstanding that the Merino has these odds to contend against, as a mutton-producer, on the Atlantic slope, still there are, undoubtedly, many localities in that section where, owing to the unfavorable conditions for turnip growing, the superiority of the Merino as a dry-feeder and also as a wool-producer, it will be advisable for the farmer to choose this breed. I have, in the preceding chapter, indicated, briefly, the best methods of growing and curing the various dry fodders for sheep; it remains now to consider the subject of growing those roots and green crops which are found to be best adapted to the wants of the sheep.

Of roots, the best is the sugar beet; then follow, in their order, mangels, ruta-bagas and turnips. Mustard, rape and rye are valuable for green-feeding or for that system, originated in England, which may be called open-air soiling. Rape may be sown as early as August, on a wheat or rye stubble, for fall and early winter pasturage; and again in September, or early in October for spring grazing. Mustard sown in the spring affords summer pasturage, and turnips may be sown so as to furnish feed in the field as early as September, or even earlier, while the beets, mangels and ruta-bagas will mature later, to be harvested for winter. Mr. Henry Stewart, in giving his experience, says: "One acre of either of these crops will feed fifty ewes from fourteen to twenty-one days, as the yield may be small or large; a fair yield upon good soil will last the longer period; but it is necessary in feeding these crops, to give the sheep only a narrow strip each day—thus, one acre being about two hundred and ten feet square, ten feet in width may be given to the sheep for their daily supply, which will give forty square feet for each sheep. Anyone who has grown mustard or rape, will see in a moment that the supply of food would be ample, and after one has had some practice in growing and feeding these crops the provision may be made to furnish a full supply to twice as many sheep as has been mentioned.

"It is an essential part of this business that the fields should be well arranged. The most convenient method is to have no larger fields than five acres for fifty sheep, and to have them long and narrow—that is, about two hundred feet wide and twelve hundred feet long; the fields being divided from each other by portable fences, so that they may be changed at will. A long field of this kind may be put into crops, sown successively one to follow the other, and at the above rate of feeding, five acres would feed fifty sheep for one hundred and twenty days before it was all gone over once; and by replowing and sowing, behind the flock, a new supply will be coming on to be used as soon as the end of the field is reached.

"This system is thus admirably adapted to mixed farming, in which a flock of sheep can be utilized with great economy and profit, as well as to a special sheep farm. It is perhaps most available for a mixed farm, because of the fine condition of the ground thus fed off, the soil being well and richly fertilized by the sheep, and the manure being distributed far more evenly than it could be done by hand. It is, in fact, a method of summer fallowing land without labor and with much greater advantage and effect than could be gained by the usual way of doing it, and at the same time making a considerable profit."

THE SUBMONTANE DISTRICT.—There is a large submontane region extending along both flanks of the Appalachian chain, from Lake Champlain to the Kanawha River, which may be considered the home and stronghold of the American Merino, where it will permanently resist the encroachments of the Cotswold and the Down. Here the mountaineer will for all time find these sheep the sheet-anchor of his humble system of husbandry, believing the old Virginian saying, that "they are an unhappy flock." By their fertilizing droppings, scattered on the summits of the knolls and hills where they delight to spend the night and the heat of the day, they will counteract the erosion by the rain and the frost and prevent that suicidal waste of soil from hillside plowing by which the farmer feeds the rivers from the heart of his pocket-book.

In this region the basis of sheep husbandry is Indian corn, hay, and fodder. The size of flocks increases as we go West.

FOR WOOL AND MUTTON. 213

Fig. 26.—SHEEP RANCH, ELLSWORTH CO., KANSAS.

In Western New York, Pennsylvania and Virginia, Ohio and Indiana, Merinos and their grades prevail, of established breeds, though in the southern half of this region there are still immense numbers of the old natives, or "mountain rangers," whose bald heads denote a mongrel Leicester blood coming from Virginia. The Pan-handle and adjacent regions still have some large flocks, yielding the superfine or electoral wools. Washington County, Pennsylvania, is the home of the Black-tops or Delaine Merinos. Western Pennsylvania, Virginia and Southern Ohio grow a plainer sheep and a longer staple than Vermont, Western New York, and Northern Ohio. Pennsylvania, West Virginia, and Ohio sheep are accounted the truest representatives of the American Merino, and their wool has long been quoted highest in the Eastern markets. But in Ohio, of late years, the breeding of very wrinkly and yolky sheep to cross on the coarse Mexicans of the West has somewhat debased the staple—as happened in Vermont from a similar cause—which, together with frauds and carelessness in the preparation of the clips for market, has hurt the good name of Ohio wool. Michigan and Wisconsin fleece, long holding the second rank, is now pressing for admission to the first.

In this region, wool holds precedence over mutton. Hay, principally timothy, some clover, red-top, blue-grass, with corn, oats, and bran constitute the staple feed. Some careful flockmasters grow turnips and fodder-corn for breeding ewes, but a vast majority depend on bran and clover-hay for a laxative. Shelled corn is the principal grain-feed for fattening wethers, while the favorite ration for lambs and tegs is corn, oats and bran, mixed in about equal proportions. Mutton wethers are shorn unwashed in March, April or May, sold at four dollars and twenty-five cents to three dollars and fifty cents a hundred, and shipped East. Many young ewes are sent West to found new flocks; oldish ones to the East, for the use above mentioned. The flocks are washed the latter part of May, shorn about two weeks later, and the wool sold to agents, who generally receive one cent a pound commission.

THE PRAIRIE REGION.—This cannot be termed a good section for the Merino; there are some fine flocks in Northern Missouri, Wisconsin, Indiana, and especially Kansas; but the English long-wools are less subject to that plague of the country—the foot-rot. Nor is the sheep generally well treated in this region. Almost as soon as one leaves Indiana, going west, he begins to see all kinds of stock in the same field, which is large, however, owing to the scarcity of fencing. In hard winters, thousands of sheep are driven east from the plains to the cheap corn of Missouri, Kansas, and Iowa, which is given to them in the ear, on the ground, in a wasteful manner; or they are allowed to enter the standing grain itself. In Minnesota they winter well on clover-hay, alone; or prairie hay and corn. In Nebraska the maximum cost of keeping a sheep a year is one dollar; from that down to sixty-five cents. Twelve tons of prairie hay, costing twenty-eight dollars and twenty cents, and two hundred bushels of corn, worth thirty to fifty dollars, will winter one hundred sheep. A shed and racks of pine for one thousand sheep will cost five hundred dollars; a "Kansas shed" of poles, hay, sorghum stalks, etc., costs only a trifle.

PRAIRIE WOOL.—The following schedule of prices will show about how the wools of the prairie region are valued (bright and dark):

BRIGHT WOOLS FROM WISCONSIN, MICHIGAN, ILLINOIS, MISSOURI, INDIANA AND EASTERN IOWA.

WASHED.

Fine	27 @ 29
Medium	29 @ 31
Quarter Blood	27 @ 28
Coarse	24 @ 25
Cotted and Rough	21 @ 22

UNWASHED.

Fine Light	18 @ 19
Fine Heavy	16 @ 17
Medium	22 @ 23
Quarter Blood	20 @ 21
Coarse	16 @ 18
Cotted and Rough	12 @ 13

DARK WOOLS FROM WESTERN IOWA, NEBRASKA, DAKOTA, MINNESOTA AND KANSAS.

WASHED.

Fine	25 @ 26
Medium	27 @ 28
Quarter Blood	25 @ 26
Coarse	22 @ 24
Cotted and Rough	19 @ 20

UNWASHED.

Fine Light	15 @ 16
Fine Heavy	13 @ 15
Medium	17 @ 20
Quarter Blood	16 @ 18
Coarse	16 @ 17
Cotted and Rough	11 @ 12

"STALK PASTURE."—Sometimes a field of corn is planted too late, or for some other reason does not fully mature, but makes what is popularly called "mutton corn." The sheep are turned into it when the grass pastures fail, in the fall, and are kept on it throughout the winter, or until it is consumed. Sometimes, in a good field of corn, the ears are "snapped" or pulled off the stalks and hauled out in wagons, the stalks being left standing. Sheep are turned into the field to harvest the imperfect ears and the foliage, and are herded on a limited area during the day (about an acre per day suffices for three hundred head, when the stalks have been gleaned ordinarily clean); so going over the field, after which they are allowed to run at will, and receive a stated ration of shelled or ear-corn and prairie hay until the stalks are pulled down and stripped clean, when the flock is removed to another field. Of course, it is only the stronger and hardier sheep that can "rough it" in this fashion; the weaklings should be removed and fed in the regular way. In South Kansas the cost of wintering a sheep this way is estimated at sixty to seventy-five cents.

OTHER FEEDS.—In the latitude of South Kansas it is estimated by an experienced shepherd that the natural grass will supply half the feed required by the sheep through the winter. It is very nutritious and fattens stock rapidly when it is young and tender, but it soon becomes tough and sheep do not relish it unless it is closely grazed; and at best it is pretty much done for by the frost as early as November 1st. There are very few kinds of hay that sheep will eat better than early-cut prairie hay; but it alone is too binding. One ton is allowed to fifteen sheep.

One of the best dry fodders is sorghum, of which sheep are very fond; besides which it yields more to the acre than any other forage plant. Sometimes it is cut and cured in shocks like corn, sometimes left standing in the field; in either form it is highly relished by the flocks. The seed is similar, according to analysis, to corn. Rice corn or Egyptian corn (another variety of sorghum) is considered second best. Millet ranks third. In Kansas are seen many large fields of broom-corn, the leaves and stalks of which are very fair feed in autumn. Both sorghum and rice corn endure the drought better than Indian corn, and are highly prized in the semi-arid regions for that reason. Half a bushel of rice corn, or Indian corn, or millet seed, per hundred head, with millet or sorghum, are considered a fair allowance for ordinary winters.

The Beard grasses or Broom grasses (*Andropogon furcatus* and *scoparius*) are estimated to furnish sixty per cent. of the grasses of the plains. The distinctive feature of prairie haying is the "buck-rake" or "go-devil." The teeth, like those of a large horse-rake, are about one foot apart. It is capable of hauling half a wagon-load on the ground. Some farmers drag a vast mass together, driving the horses over it as long as they can and dumping; then tear out around the bottom with pitchforks, top it out in some fashion, and call it a stack or rick.

GENERAL MANAGEMENT.—In the eastern half of the prairie section the agricultural system of the Eastern States prevails, but in the western half there is a gradual shading away to the free-ranging system of the Far West. Even in the agricultural section, a great deal of trouble is experienced by the flock-masters in providing water for their sheep, both summer and winter. They are, in a majority of cases, compelled to dig or bore for it, and then, perhaps, draw up the water by horse-power or erect a wind-mill; and the violent winds (erroneously called "cyclones") often blow these down or damage them, or the severe weather of winter freezes them up. It is not an uncommon occurrence to see the farmer in the dead of winter driving his flocks some miles to water, or, perhaps, hauling it for them with a wagon. One, for instance, in Southern Kansas discovered, by several weeks' observation, that his flock of six hundred head would drink about four barrels of river-water daily. But there is one compensation, and that is the facility with which immense stock-cisterns can be dug and plastered directly on the rich, black prairie mold or on the yellow underlying "loess." The scarcity of lumber also tempts the flock-master to attempt to winter his flock with too little protection, or underneath a wretched straw or sorghum shed. Though the winters in the western section are comparatively dry, yet there is an occasional flood of waters, and then the flocks, sometimes compelled to share the same enclosure with cattle, are frequently seen standing or wading about in woful fashion in deep mud and water, which is productive of foot-rot and a malady sometimes taken for foot-and-mouth disease.

In Minnesota, sheep are generally very healthy, with a slight exception of scab; in the rest of the prairie section the same general statement may be made, with the exception of foot-rot; but this exception is so important that it constitutes a serious drawback, almost an estopper, to the growing of the Merino. There are fine, rolling belts, as in North Missouri, Wisconsin, Southern and Central Illinois, where the better drainage, and the presence of sand in the soil, exempt the flocks, more or less, from this great pest; but wherever the black, waxy "gumbo" prevails, even if the surface is rolling, the foot-rot is so bad that half the flock will sometimes be seen limping, and a large percentage grazing around on their knees. Pellets of the "gumbo" soil harden between the segments of the hoof, rendering the sheep lame; and the shepherd has to catch them and remove the lumps. This trouble occurs in flocks as far west and north as Central Dakota.

In all this region the general preference is to have lambing come on grass. Even in Southern Kansas the most experienced shepherds do not care to have lambs before April 25th, though some begin as early as April 10th. When lambs are weaned they frequently receive oats at the rate of a bushel to two hundred and fifty head; when winter comes on a bushel is given to two hundred head, together with fine millet or prairie hay; or they are turned into the stalk pasture. In Iowa, and northward, blue grass is becoming the main dependence for pasture, while timothy is grown for hay much more than in Kansas.

The latter State has some choice flocks of Merinos. In Greenwood County, for instance, Mr. Robert Lay has a flock of over one thousand, which in 1884 yielded an average of eleven pounds of wool per head; and that of Mr. C. T. C. White, numbering over one thousand, of which ninety per cent. are ewes, yielded over ten pounds per head of white delaine wool.

THE SOUTHERN STATES.—In the greater part of the South, sheep husbandry is conducted strictly on the *laissez faire* principle—the sheep take care of themselves, except when wanted by their owners for the yearly "wool-gathering" and for marking. The fact that they continue to exist at all, and even to increase—despite the ravages of darkies, dogs and buffalo-gnats—is infinitely to their credit and to the credit of the natural resources of that sunny land where the snow spirit never comes, and where spring flings her flowers into the lap of winter.

"Colonel J. W. Watts, of Martin's Depot, Laurens County, South Carolina, a life-long breeder of sheep, after testing, in South Carolina, Georgia and Texas, six different breeds, settled down upon the Merino for wool and the African Broad-tail for mutton. He found the actual cost of keeping a sheep to be sixty cents per year; and, after balancing the lambs and the manure against the expense, he found the fleece to be clear profit. This, at seven pounds of unwashed wool (from full-bloods), selling at twenty-two cents (in 1877), would amount to one dollar and fifty-four cents per head. The average number of lambs raised from the Merinos he **placed at eighty per cent.** His pasture was broom-sedge and Japan clover (*Lespedeza striata*) until after harvest, then he gave them the run of the grain-fields. For winter pasturage he usually sowed rye for the ewes and lambs, and gave all the flocks the run of oats sown in August and September; also allowed them the range of the corn-fields and cotton-fields. As a mixed feed he found cotton-seed wholesome, economical and profitable. His sheep were very fond of it; after feeding on green barley all day they would eat it with great relish. Some feed was needed for three months, on account of the scarcity of cultivated grasses. Sheep were very healthy in his section.

"He housed the flocks in winter and littered the stalls frequently; the manure thus collected he sowed broadcast or in drills, in July or August, for ruta-bagas. In the summer he used the Ruchman portable fence, and kept at the rate of one thousand sheep to the acre a week. The value of the manure thus deposited he regarded equal to about four hundred pounds of guano the first year, and its effects were perceptible for several years afterwards.

"He found the sheep great helps to the farmer in **eradicating weeds**—as, for instance, the cockle-bur, and, in fact, nearly all useless plants."—[Letter to Hon. **John L. Hayes.**]

" Richard Peters, Esq., of Atlanta, Georgia, tested **nine different breeds** and crosses between many of them, and settled **down on thoroughbred Merinos and Cotswolds**, with crosses between **the two.** For a general purpose sheep he recommended, **most decidedly, a cross between the full-blooded Merino and the native.** Like Colonel Watts, he **found the fleece** clear profit, and he estimated it at the same weight.

"When the winter was mild he found the flocks needed feed about thirty days; if cold and wet, twice that time. In North Georgia the pasturage consisted of sedge, crab and other native grasses; of the cultivated grasses, orchard grass and red and white clover succeeded on uplands, and redtop on lowlands. Lucern and German millet were cut for hay; and for winter pasture, the red, rust-proof oats (sown in September), also barley, rye and wheat could be grazed during the winter and early spring and then yield a crop of grain.

" In North Georgia the system of sheep husbandry prevailing in Ohio would be applicable; in Middle Georgia, that of Kentucky; in South Georgia, that of Texas and California, with shepherd dogs, etc."—[Letter to Hon. John L. Hayes.]

In the South Atlantic and Gulf States lambing is expected in January, and the lambs coming thus early are usually more thrifty than those coming later. The farmer helps himself to the wethers at various ages, and sells the small surplus to local butchers or for shipment to Richmond, Washington and Baltimore, where they arrive before the Northern grain-fed mutton. The ewes are generally kept until they die of old age, disease or dogs.

Wool is generally shorn unwashed in April, and the most of it is sold to Jews; but of late years some shipments to Boston and Philadelphia have realized better profits and led the way for further ventures in this direction.

Our Northern flock-masters are accustomed to give out cotton-seed with timidity and caution, but in the South the planters who feed their sheep at all, not unfrequently pour it into the troughs *ad libitum*, and the sheep help themselves without stint and without injury. In Tennessee five bushels of cotton-seed to the head have been given, during the winter, to a flock of half-bloods (Merino and Southdown). In Navarro County, Texas, one hundred pounds of hay and a bushel of cotton-seed per head are provided as a winter store. In Duplin County, North Carolina, twenty sheep received, during January and February, a bushel of pea-hulls and two ears of corn per day. In Arkansas County, Arkansas, two pounds of cotton-seed per day have been given to breeding ewes.

In the piney woods, sheep do not subsist to any considerable extent on the coarse grasses, but on herbs, "mainly upon one small perennial herb, growing flat on the ground, with broad and rounded leaves, resembling very much the deer-tongue (vanilla)," (*Liatris odoratissima*). In Bradford County, Florida, as I have myself observed, they avoid the grasses of the "flat-woods," which are almost as coarse and jejune as the pine leaves overhead, and select the smut grass (*Manisuris granularis*). Bermuda grass, crow-foot and crab-grass, besides herbs. All these four last named follow cultivation. In Eastern Texas, Louisiana and Florida sheep are exceedingly fond of the seed of the beggar lice. Guinea grass (*Sorghum halapense*) has become acclimated; in winter it dies down, but sheep find, deep down under the débris, a sweet and tender bite, and they may be seen buried to the shoulders searching for it. In winter they will penetrate the recesses of the canebrake, and they often have to be confined to the fields to prevent the lambs from drowning in the low, flat woods.

But in winter they need some provision of cultivated grasses or rye, or oats. The wonderful Bermuda—the pest of the cotton-planter, who is all his life "fighting General Green"—is in reality one of the greatest economic blessings ever vouchsafed to the South. Dr. St. Julian Ravenel, of Charleston, South Carolina, regards it as superior in value to timothy; his analysis gives it fourteen per cent. of albuminoids. Dr. D. L. Phares has demonstrated that red clover will grow in Mississippi, and I have myself seen both red and white flourishing, self-seeded, in the orange groves of Bradford County, Florida. *Paspalum platycaule*, also called *P. compressum*, is another great favorite of the sheep; it will travel miles in search of it. The Japan clover or bush clover has been mentioned above. All these can be propagated by cultivation, and are excellent for sheep. According to Dr. Phares, Japan clover contains 15.11 per cent. of albuminoids and 56.79 per cent. of carbohydrates, which makes it about equal to timothy.

In 1879 the Department of Agriculture sent out to hundreds of correspondents in the South a series of questions directed to the following points:

1. Proportion (percentage) of surface, exclusive of area actually cultivated, yielding grasses suitable for pasturage for sheep.

2. Average number of sheep such pasturage is capable of sustaining during the summer months.

3. Average number one hundred acres would sustain in winter.

4. Number of months in winter in which some extra feed is required.

5. Average weight of fleece in annual shearing.

6. Average value of fleece per pound.

7. Average number of lambs from one hundred ewes.

8. Average percentage of lambs lost by disowning, exposure or other causes.

9. Percentage of sheep (exclusive of lambs) lost annually by disease, theft, dogs, wolves, or other causes.

10. Percentage of sheep destroyed by dogs alone.

These returns, carefully tabulated, after the correction of obvious errors and the elimination of estimates not bearing the impress of accuracy of judgment—inevitable blemishes of general returns upon industries that are either new or of minor magnitude—present the following average results in tabulation:

STATES.	1.	2.	3.	4.	5.	6.	7.	8.	9.	10.
Delaware	10	50	20	4	3.9	28	92	19	8	4
Maryland	25	47	19	4	3.7	28	95	20	10	7
Virginia	42	55	22	3.5	3.3	27	95	19	12	6.5
West Virginia	50	60	20	4	3.7	32	90	16	10	4.5
North Carolina	52	53	23	3	3	26	90	20	13	6
South Carolina	50	50	22	3	2.9	25	91	21	15	8
Georgia	55	53	25	3	2.9	27	93	20	14	8
Florida	60	50	22	2.5	2.7	23	89	22	18	8
Alabama	57	55	24	3	2.8	26	96	23	13	7
Mississippi	50	60	25	3	2.9	25	92	22	14	8
Louisiana	45	70	30	2.5	3.2	22	95	20	11	5
Texas	75	70	33	2.5	3.5	21	90	15	9	4
Arkansas	65	60	30	3.2	3	27	94	18	12	7
Tennessee	45	62	27	4	2.9	31	90	20	13	6
Kentucky	40	90	29	4.2	4	31	97	21	9	4
Missouri	42	80	28	4.2	3.5	28	95	23	11	6

Column 5 shows how the influence of the Merino constantly diminishes as we go South; and columns 8, 9 and 10 show the hopelessness of sheep husbandry in that section until better management and better dog laws prevail.

CHAPTER XXIII.

SYSTEMS OF SHEEP HUSBANDRY, *Continued.*

TEXAS—HISTORICAL.—The substratum of the sheep of Texas, to-day, is the Mexican native, which is descended from the *Chourro* of the Basque Provinces. The introducer of the Merinos was George W. Kendall, founder of the New Orleans *Picayune*, who established his celebrated farm in Comal County in 1852. Besides Mr. Kendall, may be mentioned Captain Allison Nelson, of Bosque County; Mr. W. R. Kellum, of McLennan County; Mr. F. W. Shaeffer, of Nueces County; Mr. H. J. Chamberlin, of Milam County; and others.

The *Chourros* were long, lank and light, producing only one and a half or two pounds per head of a coarse, dry, white, strong wool, suitable for carpets. Being neglected, they had not increased much up to the advent of Mr. Kendall, but their hard life had developed a toughness which it was Mr. Kendall's happy conception to take advantage of and engraft upon it the incomparable fleece of the American Merino. After that date, sheep increased with extraordinary rapidity, especially in Western and Southern Texas. It is useless to cite statistics, on account of the imperfect returns made of those nomadic flocks.

RANGE AND PASTURE.—In Eastern Texas north of the Nueces, embracing about one half of the available pasture area of the State, there is an agricultural system very much like that of the Gulf States generally, with the same native and cultivated grasses, supplemented more and more as we go west by the mesquite, the grama and other grasses of extreme sections.

South of the Nueces are the great wool counties, Webb, Duval, Nueces, Starr and Encinal. Everywhere on the alluvial soil is the mesquite; on the coast, the sage and salt grasses; with some grama in the west, increasing as we go to the northwest —all excellent for sheep; the grama easily first, because it resists the droughts so common in this region. Stock are often watered from wells, from fifteen to fifty feet deep.

Along the Pecos and west, is a vast desert where even the antelope is sometimes hard-pushed for water. Close along the Pecos and Rio Grande there are strips of good grass; also ten or fifteen miles back—but no surface water. The pods of the mesquite tree come in early autumn, as fattening as corn; then there are the grama, the mesquite and the buffalo grasses—all with different varieties—with the black grama prevailing on the Rio Grande. Water is in springs, ponds and holes, of which only a few last through hot weather.

Between the 100th meridian and the Pecos, besides the above, is the *juahia*, eagerly sought by sheep in the spring, when it furnishes a juice of the color and taste of milk; the *sotol*, like the Spanish bayonet, of which the shepherd cuts off the top of thorns with his *machete* (knife), allowing his flock to eat the juicy interior, which is very fattening; the *nopal* cactus, on which, with the *sotol*, sheep will go without water for many days; the *saladio*, the *baradulcia*, or greasewood, extremely palatable and nutritious to stock in winter; and many other valuable herbs and bushes.

In the Panhandle the pasture is mostly too coarse for sheep, besides which there is found the poisonous "loco," which produces insanity, strange, fantastic capers, and lingering death.

GENERAL MANAGEMENT.—The most progressive owners are fencing their ranges with wire, to prevent quarrels between their shepherds and neighboring cattle-men. Where herding is followed, the flock is generally reduced to about eleven hundred; smaller flocks would do better, but would increase the expense of herding. Ewes and lambs are kept by themselves, leaving barren ewes and wethers—locally called "muttons"—to be herded together in "dry flocks." The corral is a simple circle of brushwood or a wattled fence; hard by stands the hut of the shepherd; both being generally on the southern slope of some knoll or creek, or on the south side of a cedar-brake, for protection against the northers.

The shepherd must rise early to give his charge the benefit of all the daylight hours. After his breakfast of mutton (goats are kept with the flocks to furnish this, where the shepherd is a a Mexican), pancakes and coffee, he opens the corral. If it is hot weather the sheep saunter out leisurely, but if it is chilly and wet they move away more briskly, and then the shepherd frequently, instead of following after, goes before, circling to right and left, to restrain their movements.

It depends a good deal on the disposition of the man whether or not he is allowed to have the assistance of a shepherd dog. If he is lazy and dishonest he can make the dog huddle them, while he sleeps or dawdles away his time, and the sheep go hungry. Besides that, an ambitious dog is apt to "circle" them too much, of his own accord, thus curtailing their feed; and it is highly necessary that they should be allowed to "take a spread" in order to fill themselves. Range sheep should be kept fat by all means; a poor animal will go down in a storm and get up no more. A cur dog is sometimes employed; having been suckled by a goat it lingers with the flock and will frighten away wild animals.

The flocks occupy the winter range from December until shearing time, and the summer pasture the remainder of the year. The winter range is selected with reference to its resources for protection against storms.

Rams (generally full-blood or high-grade Merinos bought in the North) are kept by themselves; they are given extra feeds of oats, cotton-seed or corn for a few weeks before service begins; then about the middle of September they are turned, with the ewes—three to every hundred—in the corral during the night, and removed through the day, or *vice versa*. The coupling season lasts six or eight weeks; after it is closed the rams still receive daily feeds of grain for some weeks.

The sheep are now put on the winter ranges, which are generally near the ranch headquarters. Sometimes a shed is provided here for a small flock of crones and weaklings, and for the rams. The shortness of the days and the scarcity of herbage now compel the shepherd to rise very early and to keep his flock out as long as daylight lasts. About once a week they are salted —say five gallons of salt to the thousand head, perhaps with a few pounds of ashes and sulphur mixed. If the sheep have grub in the head some shepherds mix with the salt a few pints of soot. Salt is not required on the coast or with mesquite grass.

LAMBING.—This comes on about February 15th. Two or three extra men to each flock of ewes are hired to assist the regular shepherd. Some small brush-pens are built near the corral for ewes disowning their lambs. In the morning when the ewes are let out of the corral they are restrained near by until all the newly dropped lambs and their mothers can be discovered, collected and removed to a separate flock. During the day the men are busy working homeward the lambs dropped on the range, frequently carrying in each hand three or four by the forelegs, stopping occasionally to let the ewes come up and smell them.

Soon there are three flocks; the main one; a second, with lambs a week old and upward; a third, with the youngest lambs. As fast as the lambs are transferred from the youngest flock to the older one they are marked, docked and castrated.

SHEARING.—About April 15th or May 1st, in South Texas, shearing begins; the dry flocks are shorn first, the suckling ewes last, to avoid loss of increase likely to ensue if the ewes and lambs are disturbed and separated too early. The shearing is generally done by Mexicans, who receive three and a half to four cents a fleece. A covered platform is provided for the purpose, and on this the sheep are thrown down, tied (about ten at a time) and shorn, while the flock-master and his assistants are busy receiving, tallying, tying up and sacking the fleeces.

South of San Antonio semi-annual shearings generally prevail: in the spring, extending from February to May 1st; in the fall, in September and October. North of San Antonio annual shearings are the practice. The spring-cut fleeces are tied up; the fall-cut are bagged without tying, being light. This latter practice, of course, operates against the grower, since it causes to be mingled together all parts of the fleece, which are graded by the buyer about on a level with the lowest.

Semi-annual shearings have their disadvantages as well as advantages. They cut the wool shorter and therefore make it worth from three to five cents less per pound—since the wools of Texas, if suffered to grow a year, would often be long enough for combing purposes—and they double the expense. On the other hand, they are very beneficial to sheep, especially lambs, in that hot climate, promoting their health and condition; they afford the shepherd a better opportunity to hold in check and eradicate the scab. They also put money in the shepherd's pocket twice a year, which is an object in a State where the merchant is frequently asked to advance money on fleeces still on the sheeps' backs.

PASTURAGE.—A great point of superiority in the Texas grasses over those of California is, that the former are perennial, and therefore do not suffer particularly if their seeds are consumed. Though they may seem to be dead in a drought, a rain will soon freshen them up and make them green in the heart. While cattle will not readily graze after sheep, the latter, by sharp tramping, close feeding and the tearing-up of grasses in a light soil, destroy pasture that would support cattle a long time; but where the land is strong and deep, and cattle would injure it greatly by poaching it when muddy, sheep are a benefit. Here they do not pull up the grass or poach the mud, while their light treading buries the grass-seeds and assists them to germinate, and they manure the soil.

The best flock-masters inveigh strongly against the old, shiftless way of allowing stock to go the entire winter without artificial feed. Not only does the short grass—dead and almost rotten—produce intestinal worms and fever, it is claimed, but even that fails, sometimes. An abundant provision of water should be made by means of wells, wind-pumps and tanks, for if stock have to wait long for water the weakest, which can least afford it, lose most time in waiting.

There ought to be some hay, kept from year to year, if necessary, and a field of sorghum or guinea grass, late-sown, for winter forage. Cotton-seed is excellent. One feeder in Kendall County reports that he gave his "muttons" six ounces of shelled corn daily for three months, and was well repaid by the superior quality of the mutton.

MUTTON.—The mesquite grass mutton is asserted to be the best in the State, destitute of the objectionable "sheepy" taste, and improving (?) with age up to the limit of five or six years.

It is often a "burning question" with the Texas flock-master, whether to ship his mutton-wethers shorn or unshorn. The loss of the fleece destroys the plump appearance, hence the sheep needs to be very fat in order to endure this exposure and the severe ordeal of the railroad journey. The Texas *Live Stock Journal* argues in favor of shearing generally, and makes this statement: "We have never known a market butcher to pay what we consider the amount the fleece and carcass of a well fleeced sheep would bring if separated. In these times, when the low prices on both wool and mutton make it a fine calculation, any man is liable to make an error in judgment, but if the sheep are good producers of wool, it is a safe rule to get off and make sure of the fleece before trusting the carcass to the tender mercies of the * * * railroad."

Conservative flock-masters wish to retain about one-eighth of the Mexican blood, to secure hardiness and fecundity; but the more progressive ones go on crossing without fear until they have practically full-blood Merinos; and their success in breeding seems to sustain their position.

In Northern Texas, south of Red River, the average fleece weighs, for the New Mexican sheep, two and one-fifth pounds; for the Merino grade, four pounds. The New Mexican mutton sheep weighs seventy-five pounds, live weight; the Merino grade, ninety pounds.

A scab-law, with enforced State inspection of flocks, rigidly carried out, is much needed. Fencing affords partial protection against scab, but not complete.

THE TEXAS SHEEP IN GENERAL.—The Texas sheep is lighter than it should be—probably averages the lightest of all improved sheep in the United States. Not to compare it with Northern animals grown under careful farm management—which would be unjust—let us place it beside some others which are to be found on the great ranges of the West. The French Merino wether of California weighs one hundred and twenty pounds; the American, one hundred and four pounds; the Merino wether of New Mexico, one hundred and five pounds; of Nevada, one hundred pounds, etc. We have seen above that the Merino "mutton" of Texas averages only ninety pounds. Even the French Merino, when brought from California to Western Texas, falls off; the wether only attains a weight of ninety-five or one hundred pounds.

The cause of this is undoubtedly lack of feed. The native

grasses of Texas are, perhaps, the most nutritious in the country, yet the sheep feeding on them are the smallest and their fleeces the lightest. It is because of neglect on the part of the flock-masters; they leave them to gain a sustenance wholly by the process known in the expressive local vernacular as "rustling." They have to "rustle" through the summer's drought and the winter's rain. Even where the feed is abundant and good the flocks are frequently mismanaged so that they do not obtain the full benefit of it. The result of this neglect is that Texas mutton and wool suffer when brought in competition in open market with those products from other Western States and Territories.

In large flocks, the average increase is seventy-five per cent. of the breeding ewes; in smaller flocks, eighty-five per cent. In seventy-one flocks, aggregating 139,968 head, one hundred ewes dropped eighty-three lambs; of these, 63.71 survived to yearlings. Texas has some really fine flocks; for instance, that of Hon. H. J. Chamberlin, of Milam County, numbering twelve hundred head, yielded in 1884 ten and a half pounds of wool per head, with stock rams running from fifteen to thirty-three pounds; and all showing stout, compact carcasses. Rev. W. H. Parks, of Bosque County, has another choice flock, many of his wethers at maturity weighing one hundred and twenty to one hundred and thirty pounds; while, as to fleeces he sold, in 1884, to Denny, Rice & Co., sixty-nine that averaged seven pounds of scoured wool to the fleece.

But it is a fact that a vast majority of wool-growers in Texas are quite too negligent in this matter of feed and care in winter and during droughts. The experience of flock-masters in Crosby and adjoining counties in the cultivation of alfalfa—which has been found so valuable in California and Colorado—will be conducive to good results. While alfalfa, if injudiciously given, is sometimes productive of scours, there remains no doubt that it is an enormously prolific plant in warm climates and lowlands, and that, in the form of well-cured hay at any rate, it is acceptable to sheep and very fattening, producing fine-flavored mutton. The Kansas experiments with sorghum are also very suggestive to the Texans, showing that it is an excellent sheep-feed, yielding two cuttings a year which aggregate a greater total of feed than corn will produce.

The dead and half-rotten grass of winter, and the rank growth of wet spells, produce worminess in the sheep and a tender, brashy fleece. The same results were remarked in Queensland,

and the *Queenslander* reports an experiment which is instructive to the Texas shepherds: "One of two young wethers, suffering from worms and greatly emaciated, was liberally supplied with good fresh hay, with a little bran at first. The result was that the animal became perfectly healthy and fat enough to be killed for mutton. The experiment was tried on a larger scale during the hot weather of December and January. Sheep have been thriving and fattening on a patch of lucern beside a flock pastured on indigenous grasses that was being decimated by worms. The lucern was comparatively green and succulent. In the other the most nutritious of grasses had been eaten off close to the ground."

Sellman Bros., of San Saba County, state that, of twelve hundred lambs, they "lost about sixty head, or five per cent. This last loss we richly deserved, as I think that anyone who attempts to carry lambs through the first winter without feed deserves to lose. Had we given the money, that those sixty head were worth, to the flock in feed, I feel confident that we could have saved fifty of them; besides, the flock would have clipped wool enough extra to have paid for it. To verify this, we have lost but one out of two hundred and twenty-five buck lambs which we wintered on worse range than the other herd had, and gave less than one-fourth of a cent's worth of feed a day per head."

Mr. E. A. Louis, of Kendall County, fed his "muttons," the past winter, six ounces of shelled corn per head daily. In a letter to me, describing his methods, he says: "I select a smooth, hard, clear surface and place the corn in small piles over a large area, and they all, weak and strong, get their share, and without injury to the weaker ones. Before this I had troughs, but found out that the stronger ones crowded out the weaker ones and often seriously injured them." Corn has formerly been ninety cents in Kendall County (the present price is fifty); but even at the former price, Mr. Louis considers it to be very profitable to give to "muttons" intended for the early spring market.

There is one scouring-mill in the State—in San Antonio; and it would probably be well if there were more. This vast State should prepare well for the coming struggle with Australia, for the New England market, for fine clothing wools. The growers should develop the system so long advocated of skirting their fleeces when the sheep are sheared, and grading and baling their wool, if need be, in their own State, before shipment. By this system their best wools would realize a higher price and find their way into the finest fancy cassimere mills, where they are now unknown and condemned unseen. It remains entirely with the wool-growers of Texas to change this state of things. They can do it from choice now, but the time will come when necessity will compel them to do it. The best Texas ranchmen are taking much better and much more uniform care of their sheep, and are allowing their wool to grow one full year as is done in Australia.

A SAMPLE FLOCK.—Following is a statement of the actual expenses and receipts of a shepherd in San Saba County, who began by purchasing one thousand ewes, shearing four pounds per head, at three dollars a head:

DR.

October 1, 1877. Original investment in stock, camp outfit, wages of shepherd for one year, etc.	$3,565 25
March 1st. Wagon, $60; pair of ponies, $50	110 00
Harness, $4; medicine, $1 50	5 50
Wages of Mexican and wife from March 1st to October 1st, seven months, at $16	112 00
Board of same, seven months, at $10	70 00
Grain fed to rams while running with ewes	20 00
Shearing 1,720 sheep, at four cents	68 80
Hauling 5,875 pounds of wool to market	29 38
Public weigher, weighing twenty-four sacks, at ten cents	2 40
Cost of twenty-four sacks, at sixty cents	14 40
Ten pounds twine, at fifteen cents	1 50
Needle for sewing sacks	10
	3,999 33

CR.

May 1st. Sale of wool from old ewes, 4,000, at twenty-five cents	$1,000 00	
October 1st. Sale of wool from 750 six-month-old lambs, averaging two and a half pounds, 1,875 pounds, at twenty-five cents	468 75	
October 1st. Value of stock at expiration of first year:		
950 old ewes, at $3	2,850 00	
750 six-month-old lambs, at $3	2,250 00	
Twenty merino rams, at $15	300 00	
Value of outfit:		
Shot-gun	10 00	
Bedding, $4; axe, fifty cents; bell, seventy-five cents	5 25	
Wagon, $50; wagon-cover, $1.50	51 50	
Span of horses	50 00	
Harness	3 00	
Net profits first year to balance		2,989 17
	6,988 50	6,988 50

These figures pertain to an exceptional condition, where there is no crowding of the pasturage, and no particular casualty interferes with the best attainable results.

Among Texas sheep inflammatory diseases and typhus fever are unknown. The only diseases reported are scab, liver rot, three kinds of worms, grub in the head and hoove, which will be treated under their proper heads.

NEW MEXICO.—For convenience I group New Mexico and Arizona with Texas, though they received their Merino stock and their system of sheep husbandry largely from California. In Texas the best flock-masters seek to breed out the Mexican blood entirely, but in New Mexico they wish to retain an eighth or a fourth. The winter storms in the mountains are very sudden, cold and terrific; but the Texas northers are usually dry. Hence, if the New Mexican shepherd carries the grading-up beyond three-fourths or seven-eighths, what he gains in symmetry of form, weight of fleece and fineness of staple will be offset by loss of hardiness and fecundity.

In Texas it is estimated that one-sixth of the sheep are Mexican, five-eighths are half-Mexican, and five-twenty-fourths are from half-blood to pure Merino. But in New Mexico it is only in the north-east corner—Colfax, Mora and San Miguel counties —where Americans have settled, that there is any appreciable touch of Merino blood. It is found that the first cross with Merino doubles the Mexican fleece in weight. One more cross —or at most two—which will bring a fleece of about eight pounds of unwashed wool, tolerably fine, yolky and of a fair, medium staple, is about as far as they think they can proceed without detriment to a "rustling," hardy habit and fertility.

In American flocks the average annual loss, from birth to weaning, is from fifteen to twenty per cent.; above six months of age, ten per cent. In occasional snow-storms the losses are fearful. Foot-rot is unknown, but scab is common. The Mexicans do not dip their sheep; they do nothing for scab except to drive them through deep water, which does little good; hence, their flocks infect the Americans. A rigid scab-law is needed, rigidly enforced; also fencing, which is found so efficacious a preventive in Texas. The Americans employ a dip consisting of thirty pounds of tobacco, seven of sulphur, three of concentrated lye, dissolved in one hundred gallons of water, and employed at a temperature of one hundred and twenty degrees Fahrenheit. In ordinary seasons about sixty per cent. of the ewes raise their lambs, an increase of about thirty-eight per cent. of the flock. In the north the coupling season begins about the third week in November and lasts six weeks; in the south, about the first week. Lambing is in April; shearing, in May. Fall shearing increases the total yearly clip about twenty per cent.; it has been common all over the Territory, but is gradually being abandoned in the colder north. Flocks are rather larger than in Texas.

Wool is the primary, almost the only, object. The average shepherd, in the keen mountain air of this region, will consume twenty-five sheep per year. This makes so marked an inroad into the flocks that some owners prefer to purchase beef for them at four cents a pound.

The pasturage and forage plants of New Mexico are better adapted to sheep than to cattle, and the former have always predominated. The characteristic feature of the topography is the number of vast, sandy, elevated *mesas*—sparsely covered with low but nutritious grasses—stretching between broken ranges which are themselves often covered more or less with grass and herbage. The white grama abounds on the levels, while buffalo and black grama are the principal highland grasses. On the ridges and rocky *lomas* are several varieties of cactus, the thorns of which are easily broken off, and these are troublesome to herdsmen and stock. There are few unavailable heights or forests, but there is much troublesome brushwood in the *lomas* that tears out the sheep's wool.

The most noticeable effect produced by grazing in this country is the destruction of the grass on the *mesas* and of the shrubs and herbs along the streams, as a result of which the flow of rain-water from the sudden showers is less impeded than formerly, and vast gullies are chasmed in the arroyos and watercourses through this sandy soil, which often compel the traveler to make a wide detour.

SHEEP DRIVES.—One of the peculiar features of the business has been the vast drives between California and New Mexico—both ways. New Mexico was fully stocked from old Mexico as early as 1800; when gold was discovered in California, sheep were driven in from New Mexico; and when the Pacific State became overstocked, it, in turn, filled up New Mexico with Merinos. In some of these drives thirty-four per cent. perished on the sandy wastes. Mexican sheep will travel ten to twelve miles a day; Merinos, four to eight.

SHEEP TAKEN ON SHARES.—This is much practised, and is

conducted in three different methods. By the first, the lessee makes payment entirely in sheep; by the second, partly in sheep and partly in wool; by the third, wholly in wool. Contracts generally run five years, and always at the end the lessee returns the same number and class of sheep he received. The following table will illustrate:

METHODS.	LESSOR GIVES:	RECEIVES: 1st year.	2d year.	3d year.	4th year.	5th year.
First.	1000 ewes. 30 rams.	Nothing.	Nothing.	1000 sheep. 30 rams.	Nothing.	1000 ewes.
Second.	1000 ewes. 30 rams.	200 wethers. 500 fleeces.	200 wethers. 500 fleeces.	200 wethers. 500 fleeces.	200 wethers. 500 fleeces.	200 wethers. 500 fleeces. 1000 ewes. 30 rams.
Third.	1000 ewes. 30 rams.	2000 pounds wool.	2000 pounds wool.	2000 pounds wool.	2000 pounds wool.	2000 pounds wool. 1000 ewes. 30 rams.

ARIZONA.—There is a vast amount of territory in the south, west, and north which is almost worthless, being either sandy deserts, or elevated plateaus, where the only water runs a half-mile or a mile below the surface in steep-walled cañons. In the south the country is better for cattle than sheep; only the hardy, acclimated Mexicans can endure the great heat and live on the coarse herbage; but in the north there are extensive ranges where sheep do best, because they can go without water longer than cattle. The scarcity of watering-places limits the grazing capacity of the land; for sheep cannot graze out beyond three miles in a day, and back, without losing condition.

But in the north-west and in the east, along the border of New Mexico, there are some fine grazing lands for sheep; and here are found about all the Merinos and their grades which are in Arizona, mostly derived from California. Yavapai County, which contains about all the sheep of the north-west, was stocked with a very fair quality of Merinos—American and French—and, with the exception of some old breeding ewes, the Mexican blood has been mostly weeded out. Many proprietors produce "heavy, fine Merino," though the bulk of the clip grades as "heavy, medium Merino," and is good, though dirty from the prevalence of sand-storms on the *mesas*. Flocks of California origin average from six to six and a half pounds per fleece, and that, twice a year.

Southern California systems of management are found on most ranches, though in the north the Mormons cling to the old-fashioned ways, as, for instance, "handling" for the scab —*i. e.*, catching and smearing with ointment, instead of dipping (as in Texas) or swimming through tanks (as in California). The scab is kept tolerably well under control; not much wool is lost from its ravages.

Apache County contains more than three-fourths of the sheep of Arizona, but they are mostly Mexicans. The numbers of these render mutton so cheap that the breeder of Merinos finds it best to keep his wethers for wool-bearing as long as they will live.

CHAPTER XXIV.

SYSTEMS OF SHEEP HUSBANDRY, *Continued.*

CALIFORNIA—HISTORICAL.—W. W. Hollister went to California in 1852, and immediately discovered its adaptability to sheep. Returning to Ohio, he again set out for the Golden State, in 1853, with a flock of six thousand of the best sheep that his native State could raise. They were reduced by death, etc., to one-third of the original number before the border of that State was reached. But they were the progenitors of the bulk of the sheep of California. His flock soon reached one hundred and fifty thousand, while his average sales ran up to one hundred thousand dollars a year. Success like this could hardly fail to inspire imitators, and soon a great number became interested in sheep husbandry, among whom may be named H. Hollister, Mr. Dibbles, T. and B. Flint, Jotham Bixby, W. W. Cole and J. Moore. These were followed before 1858 by H. A. Rawson, Peters, Murray Bros., G. W. Grayson and others. S. W. Jewett, of Vermont, shipped hundreds of Merinos to California, by sea. In 1870, J. H. Strowbridge introduced a flock of pure Merinos from Addison County, Vermont.

WOOL PRODUCT.—In 1854 it was one hundred and seventy-five thousand pounds. Next year it doubled. The following year, or 1856, showed a duplication of the previous one, while 1857 yielded over one million pounds of wool from the rapidly increasing flocks of the State. Thence afterward the increase was less rapid, but 1859 showed a duplication; and in 1862, or three years thereafter, the clip had risen to almost six million pounds. Five years then elapsed and ten million pounds was reached. The reason of the slower progress of wool growing was due to the greater demand of the markets, which thinned off the flocks, and the extended area over which the business was carried on, exposing the sheep to greater dangers, and the young to greater risks from cold seasons, etc. In 1868 circumstances were favoring, and the product at a bound went up fifty per cent. In the year 1870, sixteen years after the first serious attempt at the successful pursuit of the business, twenty million pounds had been attained as California's contribution to the wool product of the world. Three years afterwards witnessed another great stride—in advance, as in 1873 over thirty-two million pounds were placed to California's credit in the record. California now rapidly approached her maximum in the production of wool. The next year saw nearly forty million pounds produced, while in two years thereafter, 1876, she attained to her greatest height in that respect, the clip in that year exceeding fifty-six and a half million pounds. It is now estimated that seventy-five per cent. of the sheep of California are full-blood or high-grade Merinos. Having been engrafted on the old Mission or Mexican stock, they are generally hardy and prolific. When the time arrived —and it arrived full quickly, under the enormous stimulus of gold-digging—when the State became overstocked, California was ready to colonize the adjacent States and Territories with a class of sheep which could not have been equalled elsewhere in the United States in adaptation to the special requirements of the newly opened regions. The large, rangy Merino ewes, from California and Oregon, supplied the chief contingent in the whole region west of the Rocky Mountains, and have continued to do so even since the completion of the Southern Pacific Railroad, in Western Texas.

CHARACTERISTICS OF CALIFORNIA SHEEP.—The rich and copious pasturage of this State at an early day, and the use of a great many French Merino rams in the southern section of it, have developed, in California, a Merino from ten to twenty per cent. larger than any other in the United States, except the Victor-Beall Delaines, of Washington County, Pennsylvania. The mature California wether often weighs one hundred and twenty pounds; the ewe, one hundred and ten. The long, midsummer drought of the Pacific coast compels the flocks to be driven into the high Sierras, or the Coast Range mountains, thus conforming the sheep husbandry of California somewhat to that of Spain; and a large and strong sheep is required to endure the long drives and the severe climbing.

The California flock-masters, impressed somewhat with the conservative views of their Mexican hirelings, have generally hesitated to build up full-blood Merino flocks, believing them to lack in vigor. But Messrs. Kirkpatrick and Whittaker, of Stanislaus County, have handled their thoroughbreds, as nearly as possible, in the same way that the common sheep of the State are managed, and thus their stock has acquired a vigor possessed probably by no other thoroughbred animals, as few would care to hazard valuable stock on an annual journey to the summit of the Sierra, with its attendant losses, to secure a summer range of brush and sparse pasture; but prefer rather to develop their stock in the luxuriant alfalfa fields or in the well-filled barns of the valleys, thus, to some extent, at least, unfitting them, or their progeny that inherit their disposition, for taking care of themselves on an average sheep range.

The weeding out of the weakest, the survival of the fittest, the habit of hunting for their own sustenance, acquired and inherited by stock handled in this way, compensate for all losses sustained on the trip to the mountains, and is of immeasurable value to the wool grower who secures his breeding stock from such a source.

The disposition and ability to "rustle" is transmitted to their progeny as much as any other quality, and with that trait thoroughly fixed, all objections to the use of thoroughbreds disappear.

In Southern California the strong contingent of French blood has given a sheep somewhat too leggy; with stout shanks; thin-shouldered; the quarters not well developed; body, long and lank; constitution, inferior to the American; skin, wrinkly; a heavy fleece of rather coarse, straight, gummy wool. The best flock-masters are breeding away from these points by a free use of the modern American Merino, which gives an animal with shorter legs, a more compact and well-rounded body, a fleece of finer and longer wool, though, perhaps, not so heavy. A plain animal is generally sought after for a range sheep; one with not above a single, heavy fold about the neck.

The California Merino ewe excels the average range Merino of the older States in fertility, and as a nurse, by five or ten per cent. Pacific coast flocks have long been the favorites in the interior as breeders, and for this purpose they have been transported even to Minnesota and Western Nebraska. In Tehama County, ewes formerly raised one hundred per cent. of lambs.

GENERAL MANAGEMENT.—Ranges are not so generally fenced with wire as in Texas; the flocks are larger, running from one thousand to three thousand. Wethers are separated from the ewes in lambing time, as they travel too much for them. Mutton is much more sought after than in Texas, consequently greater pains are taken to segregate the wethers intended for the shambles. During the winter the sheep are frequently not "banded" at all; they run at large about the range.

In very large flocks, in Southern California, the ewes are separated into bands according to age- yearlings, two-year-olds, etc.

In the Sacramento and San Joaquin valleys, and in Southern California there are distinctly defined ranges for winter and summer. The great chain of the Sierra Nevada is the mainstay, in summer, of the flocks on the plains, for a distance of four hundred miles north and south; when the pasturage withers on the plains and foot-hills, they begin to ascend its slopes, gradually mounting higher as the snow disappears, until they reach the rich, natural meadows lying deep in the double crest of the Sierra, where they spend the summer. One acre here will support a sheep during the limited season. The mountain ranges between Kern and Los Angeles Counties have long been the resort of the flocks during droughts in Los Angeles, Ventura and Santa Barbara Counties.

In the extreme northern and southern sections of the State there is less distinction between the winter and summer ranges, except as the sheep themselves naturally regulate their movements, coming lower down when the snow begins to fall on the summits. At the south, the shepherds aim to remove the flocks from the tenacious "adobe" soils in rainy weather; there is danger of their bogging-down. The foot-hills are everywhere the favorite natural range in winter, being of a firmer soil, with rounded and thinly wooded knolls and patches of *chaparral* affording browse and protection from storms.

Some of the flocks of the great central basin, instead of summering in the Sierra, are driven into the vast tule-swamps bordering the Sacramento and San Joaquin, which afford much coarse herbage. After the water retreats from these swamps the tules are sometimes burned to freshen the growth. Wheat sown in the ashes and trodden in by flocks of sheep driven to and fro has been known to produce seventy bushels per acre. Flocks wintered on the black, deep soil of these tules get their wool much discolored; it contrasts strangely with the white fleeces just brought down from the Sierra in autumn.

Only rams or small and choice flocks are fed or sheltered in winter, though in the northern, mountainous ranges, when a snowfall lies on the ground a week or more, barley is scattered for them on the snow, at the rate of a half-pound per head. On the great wheat farms of the central plains no care is taken of the straw, and before the rainy season sets in it is burned in vast quantities. Within sight of the dome of the State Capitol I once saw a farmer, whose sheep were dying by the hundreds for lack of a little grain and the straw he had burned, construct a furnace and boil up the carcasses for hogs!

BREEDING FLOCKS AND LAMBS.—Very much the same methods of management prevail as were described for Texas. The lambing season comes somewhat earlier than in that State, however; in Northern California it begins in February; in Southern California, in January. In both sections it continues six weeks or two months. Ewes which "miss" in the autumnal coupling are put with the rams again in the spring, to drop their lambs from October 15th to November 15th. The ewes are nearly always corraled at night in the lambing season, although sometimes, when the corral has been allowed to become very foul, and there is no imminent danger from coyotes or other wild animals, they are simply "camped" for the night near the headquarters of the range. When there is six inches of manure in the corral and it has been rendered soft and thin by the long winter rains, it may well be imagined that lambing in such a place would be a miserable and disgusting business. Lambs are castrated when four to six weeks old.

Lambs are weaned at the age of four or five months. They are either wholly separated from the ewes, or else they are "cross-weaned"—that is, the lambs of one flock are put with the mothers of another, etc. After a few weeks the flocks are corraled again, and lambs and ewes put by themselves. In the wooded and brushy regions of the Coast Range and the high Sierras, however, it is difficult to keep flocks segregated, and they frequently run in masses as best they can.

SHEARING.—Indians are largely employed in this branch of the business, both north and south, and for herders, also; they being more patient and gentle than Americans. An Indian shears about three sheep to the American's five. The price paid is five or six cents per fleece with board (seven cents without). Indians and Mexicans frequently go in a club or company, traveling from ranch to ranch, under the command of a captain, who makes all their contracts, receives and divides the money, settles all disputes, and subjects all his followers to his command. In the balmy climate of the Pacific coast, especially at the fall shearing, generally no shelter is needed except that afforded by a clump of spreading live-oaks. Under these, long platforms are erected, and the swarthy shearers, with bared heads and breasts, their skins beaded with perspiration, bending to their work in an aboriginal silence, keep the shears clicking in a not unmusical concert.

In the arid climate of this coast, with its all-pervading dust, its sand-storms, its myriads of detached grass-seeds, with chaff and powdered foliage under foot, the fleeces get very dirty during the summer. A man who shears sixty sheep in the spring, will shear only fifty in the fall, though the fall fleece will probably be only about half as heavy in actual wool.

When the rains are not too severe, the spring shearing is done in March, in Southern California, and again in August; in the north, the months are May and September. One farmer, in Mendocino County, tried the experiment of shearing his lambs about July 15th, to free them from the grass-seed, and the results were so good that he continued the practice. In Placer County I have seen sheep, that were shorn too early (the owners of large flocks have to hasten matters to finish lambing and shearing before the pasture dries up in the foot-hills, and the sheep get impatient to set out for the mountains), huddled close in squads of twelve to twenty in the little, pit-like depressions of the *mesas*, a day or more at a time, during the long, driving rains; and when the rains were over, so that the sheep could go out to grazing again, there would be from one to three or four dead sheep lying in each depression.

GRADES OF WOOL.—In California, with its hundreds of isolated valleys and its hundreds of resultant sharp climatic contrasts—as, for instance, in going from semi-tropical, almost frostless Vacaville, a short distance over a low range to the cold ocean fogs and blustering winds of Marin—we find, perhaps, the most striking differences in wool values within short distances. The prevalence of the burs of the yellow or bur clover in some localities, mostly lowlands, contributes to this difference. There are six groups of wool counties: Sacramento and San Joaquin rivers, northern and southern coast, middle or foot-hill, and two mountain counties—Humboldt and Mendocino. They run about as follows: San Joaquin, free, thirteen and sixteen cents; burry, eleven and thirteen cents; southern coast, twelve and sixteen cents; northern, defective, fifteen and seventeen cents; Sacramento, free, eighteen and twenty cents; Calaveras and middle counties, best, seventeen and twenty cents; Humboldt and Mendocino, twenty and twenty-two cents.

The spring shearing in the south is done very early to anticipate the ripening of the alfileria and bur-clover seeds, which would injure the wool very much. In the fall there is no haste, for they will be in the wool at any rate.

The northern wools are better grown than the southern; they are brighter and freer from seeds and burs. Brightness results from the washings on the sheep's backs by the heavy rains of winter and spring. The shrinkage of the northern wools in scouring is less by ten to twenty per cent. than that of the southern clip. Even at the above prices a manufacturer complains, in the *United States Economist*, that "it costs me ninety-six cents to scour California wools." The southern flock-masters, however, claim that they are compensated for this lower price by the greater weight of their fleeces. As a rule, the southern sheep are the better bred of the two. Against the favoring rains of the north, there is in the south a more abundant pasture, especially bur clover and alfileria. The quality of wool throughout the State has improved progressively as the grade has advanced from the original *Chourro* or Mexican.

There are seven establishments in San Francisco which scour wool before it is shipped East, though some of them, I regret to learn, are about to be abandoned. Several years ago it was found that scoured wool could be shipped to the great mills at Cohoes, New York, at a price which would, and did, supplant the wools of Australia, heretofore principally depended on for a supply.

The best California wool, greatly to the credit of the State, is manufactured at home. California blankets have a deserved and world-wide reputation.

The United States Government Report, of 1884, gives the estimated wool-clip of 1880 as follows: Rams, fourteen pounds; ewes, 6.33 pounds; wethers, 8.11 pounds; lambs, 5.40 pounds.

MUTTON SHEEP.—The demand for mutton is so considerable

that nearly every flock has its separate "mutton band." When the rams are put into the breeding flocks in the fall, the old and otherwise undesirable ewes are culled out and thrown into the wether or mutton flock. Wool being the primary object of sheep husbandry, the flock-master likes to cling to his largest, finest wethers as long as possible; hence, has arisen, as in Texas, the delusive maxim, "Old sheep for mutton."

Wethers are never grain-fed, but after harvest they are turned on the wheat-stubble, six weeks or two months. So many heads are left by the wasteful machines used in harvesting, and so sound do they remain in the rainless months, that the stubble is rich feed; and it is a common saying of the California farmer that his stubble must pay his taxes.

American Merinos make better mutton than the French; they stand herding better, are more compact and round-bodied, are better feeders, being not so dainty in their search for choice herbage. California breeders unanimously condemn the cross with the English sheep, except where mutton is decidedly more important than wool.

At the fall shearing, old wethers, intended for the winter market, are sometimes left unshorn, as the protecting fleeces will keep them in better condition during the long, cold rains. A majority of the ewes, and a great many wethers, are kept so long for shearing that they die of old age.

The estimated average live weight of mutton sheep is one hundred and four pounds; dressed, fifty-four pounds.

SHEEP ON WHEAT FARMS.—A representative wheat-grower, in Tehama County, had fifteen thousand acres of wheat, and sixteen thousand sheep. After gleaning the stubble and cleaning off the weeds in the fall, they spend the winter on some rough lava-beds or *mesas* at the foot of the Sierra Nevada; in the spring they are driven into the mountains to remain until after harvest. The yield of wheat on the virgin soil was forty bushels per acre, and after steady cropping for about fifteen years it still remained the same. Being asked how he kept up the fertility without manuring, he replied: "My sheep furnish manure yearly to the stubble-fields that fatten them before they are turned out for winter."

EFFECTS OF SHEEP ON PASTURE.—With the exception of a limited number of cattle-men, inimical to sheep, the testimony of California stock-men is, that sheep produce favorable effects on the wild grasses. Messrs. Miller and Lux, after twenty-five years' experience, furnished the following statement to the United States Census agent: "Ranges are benefited by sheep if the stock is judiciously grazed; then they are sure to increase the yield and improve the quality. They must not be kept on too long in winter, so as to cut up the low land and tread out the roots, nor too heavily in spring, so as to prevent the grass from bearing seed. California land needs the packing that sheep give, and their tramping, when not excessive, prepares the soil to retain the surface rains to nourish better and more varied grasses. We have, most carefully, noted results, and know that, in California, land used for sheep with judgment is always improved."

It is undeniable that sheep have not always been "used with judgment," and it is the result of my own observations, extended through a period of over five years, that, while they have increased the production of alfileria (locally known as "fileree"), and bur clover—two very valuable forage plants—they have contributed more than cattle, by their close cropping and by the consumption of seed, toward the destruction of wild oats and bunch-grass, which, in the Sierra foot-hills, are succeeded by the worthless "squirrel-grass."

On Pit River and Goose Lake, sheep have also very materially decreased the production of the bunch, red-top, rye, blue-joint and salt (alkali) grasses. Such I found to be the opinion of a majority of the ranchmen at the time of my visit to that region in 1872.

It is remarked by observant shepherds that their flocks drink less on wild grasses than on alfalfa, though there are certain salt or alkali grasses which increase their thirst. Alkali grasses destroy the fertility of some cows after six or eight years, and render bulls impotent after three or four; but no such effects on sheep have been recorded.

SHEEP ON ALFALFA.—Mr. J. T. McJunkin, of Hanford, a wool-grower of much experience, has kept as many as nine hundred to thirteen hundred head of sheep on one hundred acres of alfalfa the year round, except for about two months, when they were turned upon the wheat stubble. During those two months he cut his alfalfa once, and stacked the hay for the cold weather of winter, when the green feed would be short. In a very good season he has kept as high as fifteen sheep per acre, and thinks that even thirteen per acre are more profitable than wheat. He finds that there is great danger of sheep bloating if moved from short feed to rank alfalfa, and he avoids this by not allowing them to graze down one pasture too low before they are turned into another. He also keeps lumps of rock-salt in the pasture as a preventive. Sheep drink a great deal when fed on alfalfa hay. Owing to the great scarcity of pasturage in the winter of 1881-2, Mr. McJunkin's neighbors reported a falling off of twenty-five per cent. in their clips; but his remained at the customary figure. For these facts I am indebted to an article in the *Pacific Rural Press.*

IN THE SIERRA NEVADA.—The great annual migration of the flocks in the central basin up to the rich, natural meadows in the double crest of the Sierra, is the distinguishing characteristic of California sheep husbandry. In the spring, as soon as sheep begin to show decided indications of thirst—they drink little or nothing on the fresh feed of early spring—the shepherds consider that the time has arrived to start into the mountains. The sheep are made up into flocks of four thousand to eight thousand for the ascent, which occupies a month or six weeks. When arrived at the meadows they are divided into flocks of twenty-five hundred to three thousand, for the summer. It is calculated that an acre of meadow will support a sheep through the summer; and that the fall clip will be increased by a pound and a half over what it would have been if the sheep had remained on the plains. When the coupling season arrives, rams are driven up in little flocks from the home-range on the plains. As soon as the first snow gives intimation of the approach of the rainy season the flocks are headed for the plains, but the descent is leisurely made.

THE "SHEEP-HERDER."—From the great number of "Diggers" and "Greasers" employed in the work of herding sheep, this occupation became degraded and vulgarized; and no one in California speaks of the "shepherd"—it is only the prosaic "sheep-herder." The great sheep-runs of California, like those of Australia, seem to be a sort of mild form of Botany Bay for their respective mother countries. Old flock-masters will tell you, out of their long experience in either country, of dozens of men, college finished, perhaps, who themselves or their families banished from home—not, perhaps, like Barrington's patriots, "for their country's good"—but for the suppression and healing of scandal, and who are now harvesting their traditional and unhappy crop of wild oats, at the same time they watch the sheep upon the hills pick their's (*Avena fatua*)—" comrades of the wolf and owl." One of the great flock-masters on the Nascimiento told me that during one year he had employed on his ranch a bishop's son, a banker, an editor, a civil engineer, and a book-keeper—at least two of them being college graduates.

Sometimes the corrals are simply enclosed with rambling strings of brush fence (the knaggy clumps of *chapparal* are easily pulled up with a yoke of oxen, and, being thrown together, make a fence impassable except to a grizzly or a nimble coyote). Sometimes they are made of cotton-wood limbs set in a shallow trench around a square, and wattled with the smaller twigs or with willows. In these the manure accumulates a foot deep or more, and then the indolent fellow makes another corral, or else sets to work with a span of horses and a scraper and scrapes it out. After a rainless summer of six months, it is dry to the bottom and as loose as an ash-heap. The operation of scraping stirs up an odor which is as pungent as mustard and smells to heaven. The sun, which is never hidden by so much as a capful of cloud, riots in these exhalations, and the air is filled with a fertile dust. A huge hillock is heaped up as they heap up a hay-rick in Nebraska, as steep as it will lie and left to waste, becoming in years a guano-bed, a score of feet deep, or even deeper. They often scrape it out into a little ravine or gulch, and the winter rains flush them out, rolling down to flats already abundantly fertile (if they had water), or sheer into the channel of a stream immense volumes of this valuable manure. Wasteful Californians!

Once in a fortnight there comes to the shepherd from the great outside world a donkey-load of beans, coffee or tea, sugar, and flour—perhaps a newspaper or two. Besides this he sees no soul unless it may be a hunter, or a solitary cowboy looking for strayed stock. At night the supercilious coyote inspects and pollutes the corners of his habitation. The long and hungry scream of the California lion floats athwart his dreams, and perhaps he is awakened at midnight by the heavy lumbering crunch of the grizzly over his brushwood corral, and the piteous bleat of some sheep (sheep will bleat with pain sometimes), whose ham the monster is scooping out. In the morning he follows his gadding flock over the rounded wild-oat hills, dotted with live-oaks; along the borders of the bright evergreen *chamisal*—too dense for his sheep to penetrate, but the minute flowers of which furnish pasture for myriads of bees; and at evening on the shelves of plains and in the little valleys, among the moss-streamered oaks, and the whited, plumy tufts of the bunch grass. Long thoughts are his as he lounges "mony a canty day" over the ripe and yellow mountains, which are frosted over, like a cake, with a tender lilac haze. Or, from some "specular height" he looks down on the saturnine and awful desolation of late autumn; the far dun reaches of rolling-tables, thinly flecked with dwarfish oaks, and the sharp-cut, purple peaks. Or, perhaps you will find him squatted with his faithful dog between his knees, while in the vast mustard plain around you cannot see a sheep, and only hear the multitudinous crackling and surging in the dry mustard. You will not heed his tatters, for his vagabond flock have led him many a chase through the sage and the rosemary of the adjacent foothills. There is no picturesque heirloom crook in his hand, but, instead, a plug of navy tobacco. This dissipates the poetry of the situation. Probably there is moored, even now, at San Francisco or San Diego, the ship from which he deserted.

THE "DODGE GATE."—The sheep of different owners often get mixed, especially in Southern California, when a high wind blows off the long Spanish moss; they are very fond of this and scatter widely in search of it. To separate them, there is provided a large corral, with a very narrow and long lane leading from it to two smaller ones. In the narrowest place is the "dodge-gate," in the middle of the lane and parallel with it; the swinging end points toward the large corral. The sheep are driven into the large corral and are slowly forced, one at a time, through the lane, while a man standing at the gate moves it to this side or that, parting Smith's sheep from Jones'.

California needs fewer "dodge-gates," and more wire-fence and cultivated fields.

SHEEP IN VINEYARDS.— The following paragraph is from the *Fresno Republican:* "Vineyardists should not forget the advantage derived from pasturing sheep in the vineyard as soon as the grapes are harvested. Vineyards infested with insects that lay their eggs on the leaves or on the ground are easily exterminated in this way. The sheep eat the leaves if they are yet green, and thus destroy the eggs. By packing the ground, many insects and eggs are also destroyed there. This is the best way to destroy the leaf hoppers, which some years are so destructive in our vineyards. That this year there have been none of these hoppers to do any harm, we have principally to thank the sheep for. After the frost has killed the leaves and they become dried, the sheep will not eat them."

PREPARATION OF WOOL FOR SHIPPING.—Freights by railroad are one cent and a half per pound on wool costing twelve cents per pound or under; one cent and three-quarters per pound for that costing from twelve cents to eighteen cents per pound, and two cents per pound for that costing over eighteen cents per pound, but the quantity of the latter is not very large. By ship to New York, around Cape Horn, it is one cent per pound. The freight on scoured wool is three cents per pound.

The sand-storms, the dust of the long, rainless summer, the grease created by the great heat, and the abundant seeds and burs, make the unwashed California wools very dirty. It is claimed in the East that the spring clip shrinks, in scouring, sixty-seven to seventy per cent., and the fall clip, seventy-two to seventy-five per cent. California wool is packed in compressed bales in its uncleaned state, pressed as hard as a block of wood, and the bales bound with iron hoops. In this condition it is shipped East for sale. Ohio, Pennsylvania and other similar wool, when received for sale in the Eastern market, is opened out and graded; each quality being piled separately in the warehouse, so that buyers can easily examine it. California wool, on the contrary, is kept in the compressed bales, standing on end, with the burlap cut on top about a foot in length and width to expose the wool. It is not difficult to imagine a buyer turning from looking at the sightly piles of bright fleece-washed Ohio and Pennsylvania wool to examine California wool in the condition above mentioned; perhaps eighty per cent. of earth and burs, with twenty per cent. of wool. It is a well-known fact that the eye must be pleased in buying raw products, as well as in the purchase of manufactured goods.

To remedy this, seven scouring mills have been erected in San Francisco, which, it is claimed, effect a great saving to the wool-growers. To illustrate, we will take one hundred pounds of wool in San Francisco, costing fifteen cents per pound in its crude state. It will shrink sixty-five per cent. in scouring, leaving thirty-five pounds of clean wool, which would cost near forty-three cents per pound. The same wool shipped and scoured in the East, adding freight—two cents per pound—would cost seventeen cents per pound, or near forty-nine cents per pound, scoured. Add freight—two cents per pound—to the San Francisco scoured wool, and it is in the Eastern market at forty-five cents per pound, while the Eastern scoured costs forty-nine cents, a difference of four cents per pound in favor of San Francisco.

Wool-growers all over the coast could materially help the sale of their wool by being a little more careful in sacking all dirty tag locks, of which there are always more or less, especially in "year fleeces;" these should be taken off before the fleece is tied up, as the injury to the price is always more than the gain in weight. Marking with tar is also very objectionable, as the ordinary process of scouring will not take off the tar, and the locks of wool to which it is attached must be sorted very carefully from the fleece, and as they are almost worthless, manufacturers must so figure on them in making their purchases.

It is also a great mistake to shear and sack wool when it is damp with the spring rains; this often causes mold, and a loss of several cents in the price received per pound.

ITEMS.—The amount of land required for the grazing of a sheep is, variously estimated by different shepherds, from two and a half acres to four acres per head, even on the rich pastures of Southern California. But the reader should bear in mind that an immense quantity of grass is consumed by the ground-squirrels, the gophers and the agricultural ants, which are three of the worst pests of the State.

The losses from all causes—poisonous weeds, disease, winter storms, dogs and wild animals—are estimated at 7.8 per cent. yearly, which is probably an underestimate. One owner, with twelve thousand, one hundred and fifty sheep, computed his losses from coyotes at one thousand dollars a year. This, not altogether by direct slaughter, but also by the corralling which their presence compels, which causes foul wool (bringing a lower price, loss of condition, and the engendering of disease.)

Sheep husbandry, in this climate, has been subject to great vicissitudes. Though very healthy, sheep are occasionally laid waste by drought, and by the rapacity of man, which causes overstocking of the pastures. In the "Government Report," the losses in Southern California, in 1877, are placed at two million, five hundred thousand. In the winter of 1881-2, thousands of lambs fell beneath the hammer-stroke at birth, this resort being the only means of saving the mothers' lives. I have known forty per cent. of the lambs to perish in a cold rain in the Eel River Mountains.

A SAMPLE FLOCK.—We will take a flock in Tehama County, consisting of two hundred rams, six thousand ewes, seven thousand wethers, two thousand lambs; total, fifteen thousand, two hundred.

15,000 acres of land leased, at twenty-five cents per acre	$ 3,750
Equipment in vehicles, harness, tools, etc.............	300
Four horses, worth......................................	400
Six dogs, worth	100
Investment in plant.............................	4,550
Investment in flocks	31,000
Grand total investment	35,550

Six men employed throughout the year, at $100 per month; five, at $25 per month each, board included.		2,700
At shearing time, five extra men for fifteen days at one dollar per day (to serve at corrals, handle and prepare all for shearers)....................	$75	
Shearers, at five cents per fleece (13,000 fleeces at last shearing)............................	650	
		725
Total expenses for labor.......................		3,425
78,000 pounds of wool sold, at twenty-seven cents (1879)		$21,060
1,000 wethers, at $2.50		2,500
Total...		23,560
Yearly outlay (land leased and interest on flocks)......		$ 6,850

But the average profit is reckoned at sixty cents per head.

HAY FOR SHEEP.—The fine native grasses of North-eastern California are well suited for sheep. In hauling hay to the stack, or barn, some farmers use racks from sixteen to twenty feet long, eight to ten feet wide and four feet high. The sides are made of small rods, usually willow, placed fourteen to sixteen inches apart. They are made in this manner so as to haul short, fine hay and to be able to work in windy weather. The other style of hay-rack is made just the same, except the sides are dispensed with and posts are put up at each corner and the sides.

In Sierra, American, Clover and Indian valleys, the greater part of the hay is put into barns, some of which are large enough to hold from two hundred and fifty to three hundred tons. A great many of these barns are so arranged that the wagons can be unloaded in the top of the barns. To do this they make use of a drive-way, in the form of an inclined plane, by which the wagons are either drawn up with block and tackle or the team is driven in at one end of the barn and out at the other. The sides of the rack are frequently fitted with hinges so they can be dropped down and the hay pushed off. A great many farmers use the Jackson fork and Church hay-carrier, which is fixed to run at the peak of the roof. The wagon is driven across the end of the barn or through the middle, crosswise. The hay is carried up with the fork to the carrier and then run back into the barn and unloaded at any point wished. Excepting in those valleys named, the great bulk of the hay is stacked, in some places covered, but much the greater part is not; the average width of the stacks is twenty feet, the length, from one hundred to one hundred and seventy-five feet, the height, before it settles, is from twenty to thirty feet. By far the greater part of the stacking is now done by horse-power; either the Jackson fork is used with a derrick, or nets, either rolling or lifting; but the latter kind is most frequently used.

The Winters' patent net and derrick has been partially introduced in Lassen and Modoc Counties during the past two years. Its mode of operation is to use two half-nets in the wagon rack and lift out each one separately, swing around on the stack, and drop the load by opening the net in the center.

NEVADA.—Northern Nevada presented, originally, a very fair opening to the sheep-breeder and grazier; there were fine ranges of bunch-grass, sand-grass, white sage, and stretches of meadow furnishing a varied and profuse feed of wire, red-top and rye grasses, with tolerable supplies of water. In the south the deserts were skirted with sand, bunch and gietta grasses and stunted white sage; but water was scarce. When the vast overflow of stock from California occurred in 1870, in consequence of drought, all these ranges were greatly overstocked and the pasturage injured. In a single drive in the south, eight thousand sheep perished from thirst in five days. On many places the seed was eaten up and the grasses entirely disappeared, the greasewood and other shrubs were stunted, and on wide areas the valuable white sage was wholly extirpated.

Nevada, Utah and Idaho are alike in this, that their elevated summer ranges are capable of "carrying" more stock than can be supported by the natural herbage of such lowlands and valleys as are suitable for winter grazing. This compels winter feeding.

SYSTEMS OF MANAGEMENT.—These may be divided into the nomadic, the semi-nomadic and the fixed. In the nomadic system the shepherd slowly works the band in at sunset to some spring sufficing for himself and horse; and, if the weather is clear, he simply "rounds up" or assembles the sheep on a level stretch of ground, assisted by one or more mongrel dogs common in Nevada, called "shepherd dogs." The herder sleeps by the flock. Aroused by the continual barking of the watchful dogs, he may find the flock scattered by the approach of a coyote; or, chilled by a sudden storm of drifting snow, they may have moved off in a solid body. Without a corral he cannot hold the flock in a storm; they will "drift" until a lull occurs, or they reach some protecting depression, or huddle behind a knoll. In their blind efforts to escape the cold, they may crowd into a gulch or a "dry wash," and large numbers be suffocated. Even if the shepherd should happen to have a sage-brush corral, unless the direction of the wind was toward the opening, he could not force the sheep to enter.

LAMBING AND SHEARING.—These are over about the last of May. If some stationary sheep-man near by has a set of dipping-vats, they are hired for the use of the nomadic flock; or, more commonly, the nomadic flocks are simply anointed with grease and mercurial ointment; or, perhaps, not treated at all. All classes of the sheep, except the rams, are now moved off together to the summer ranges. About the 25th of September another shearing occurs; for, with few exceptions, semi-annual shearings prevail in Western Nevada.

MORE PERMANENT SYSTEMS.—Where some land is held in fee-simple, or by squatter's right, this forms a winter headquarters; and frequently there are fenced fields here for meadow and for winter ranging for ewes and lambs. The allowance of hay is about one pound and three-quarters a day per lamb. In 1879-80 feed had to be given about six months, on account of the unusual severity of the winter. Some winters none at all is required.

The movements to and from summer ranges are about the same as above detailed.

WOOL AND MUTTON.—The American Merino is deemed the best breed for the climate and pasture of Nevada, though the best flock-masters do not desire to go beyond a three-fourths or seven-eighths grade. Many flocks have reached this standard, and show good handling. The close fleece of the Merino is a protection against the storms, and is less liable to be pulled out by the knaggy shrubs of the mountains than is the long wool of the English breeds. Besides, the latter are not adapted to these arid wastes and the alkali dust; they become lank; they cannot endure the close herding ; their descendents steadily fall off both in size and weight of fleece. The American Merino does better than the French Merino of Southern California.

The oftener sheep are changed from one pasture to another, and the less frequently they are corraled or camped on the same ground—littered with their droppings—the better will be their condition. Even the Merinos are very susceptible to disease when herded in such large flocks as are necessary under the present systems.

The scab acarus, or insect, seems to lurk in tufts of wool, bushes, or sticks—even in the manure, where sheep infested with it have been assembled a number of times. An instance is mentioned where a corral was occupied by scabby sheep in June, then remained untenanted until October, when some rams in perfect health were kept in it for a few nights, and were observed to sleep frequently in a corner where the sweepings of the shearing-table had been thrown. They contracted the scab in such malignant form that they had to be dipped, though they had not come in contact with any affected sheep.

Where the sheep are well graded up and kept in even condition through the year, the wool is as good as that of California. The average of the State is given thus : Rams, thirteen pounds ; wethers, seven and three-quarter pounds ; ewes, six and a half pounds ; yearlings, five and a quarter pounds, when shorn only once a year. Mutton sheep average one hundred pounds, live weight ; fifty pounds, dressed.

CHAPTER XXV.

SYSTEMS OF SHEEP HUSBANDRY, *Continued.*

OREGON—INTRODUCTION OF THE MERINOS.—The Hon. John Minto, of Marion County, Oregon, furnished the United States Department of Agriculture a very complete history of the introduction of the various breeds of sheep into that State, from which I extract a few items. The first Merinos were brought to Oregon in 1848 by Mr. Joseph Watt, of Amity ; there were seven pure Saxons and six high-grade Americans. Others were brought, in 1851, by Hiram Smith ; in 1854, by Dr. Talmie ; in

1858, by Martin Jesse—the last named being McArthur Australian Merinos, imported into San Francisco by J. H. Williams, the United States consul at Sydney. In 1860, Rockwell and Jones imported some pure-blood Merinos from Vermont. In 1861, Donald McLeod brought one hundred and fifty thoroughbred Merinos from Vermont across the plains. After that date, large numbers of pure American Merinos and some very fine French Merinos were brought to Oregon by different parties.

California and Oregon, from their differences in climate, are, to a certain extent, supplementary to each other. In disastrous years of drought the sheep are driven out of California, and the following seasons they flow back. In 1850, 1851 and 1861, the movement was toward the Southern State, eighty thousand going over from Willamette valley alone. For several years after 1864, California sent sheep to Oregon.

CONDITIONS AND MODES OF SHEEP HUSBANDRY.—From a very able paper contributed by Hon. John Minto to the *Willamette Farmer*, I extract the following paragraphs:

"From east of the Cascades to Western Kansas, and from Middle Texas to Alaska, is all clothing wool country, for which the Improved American Merino is the best known breed. The portion of coast moistened by the winds of the Pacific, now occupied as wheat fields, needs, as I have indicated, something approaching English methods of husbandry, both as to wheat and sheep, to make it carry combing-wool sheep."

*   *   *   *   *

"Western Oregon can excel, both in long combing and in fine clothing wools; but our experience proves that combing wool sheep require constant care on the part of the owners, to keep them in proper condition. There are a few localities in Western Oregon of which this is not true. There are a few ranges, of limited extent, that are better adapted to long wooled sheep than to any other. There are also farmers who so keep their flock, under conditions generally not favorable, that they bring to market a very good article of combing wool. But such are exceptional men at present. The general condition of the climate of Western Oregon, and the pasturage furnished either naturally or by the help of the farmer, are such that there is a steady deterioration from an average standard of Cotswold, Leicester, or New Oxford sheep. The flock grows gradually more and more leggy in appearance; the wool becomes shorter, drier and less lustrous; and in many cases the sheep, while comparatively young, lose considerable of this wool before ordinary shearing time."

* * * * *

"It is found, in practice, that in a flock of mixed breeds the long-wooled keep on the outside of the others in search of feed. Observation proves that when the short-jointed, round-bodied Merino grade, weighing one hundred and thirty pounds live weight, has fed to its satisfaction and is ready to lie down, the long-wooled, weighing one hundred and eighty pounds, has not had feed according to the requirements of its nature and size, and, in consequence, is restless at camping time. During feeding hours, such sheep require the constant care of the herder to prevent them from leading the flock to daily travel faster and further than is good for it. Then, when the season renders it difficult for a medium sized sheep to get a fair living—a condition suitable to growing fine wool of the best quality—the combing-wool sheep is not getting the amount of feed necessary to keep its wool in healthy growth, so both wool and sheep are deteriorating. On fresh range this is not the case, and for awhile a very good staple of long-wool can be grown on such range, but the causes I have indicated very soon begin to operate, with results that fully justify the wool-growers for breeding more and more towards the clothing-wool sheep."

It is Mr. Minto's opinion that sheep husbandry in Oregon is not so well conducted as it was in the earlier years of that industry, when the pastures were fresh and were devoted entirely to the flocks and herds. Since wheat farming has assumed a commanding importance, many farmers keep their sheep chiefly as gleaners of the stubble and to rid the fallow of wild oats, sorrel and other weeds, where they frequently suffer for water; and in consequence of this, and short feed during the succeeding winter, there is a tendency toward deterioration and dryness of fleece.

EFFECT OF ALKALI ON THE FIBER.—In Eastern publications reference is frequently made to the assumedly established fact that a great deal of the wool grown west of the hundredth meridian is weakened by the alkali which prevails more or less in the soil over a great portion of the Far West. This may, or it may not, be "trade capital" with the Jewish and other wool-commission houses. Any dust in the fleece, alkaline or other, is injurious to it, from the dryness and friction, which causes the roughening and discoloration of the fiber. But that the presence of alkali in the wool produces any deleterious chemical effect is, at least, as a Scotch jury would say, "not proven."

To test this matter thoroughly, Mr. Thomas T. Lang, of Rockville, Oregon, instituted a careful and extensive series of experiments by rubbing alkaline soil upon the sheep and into the fleece, taking care, in the meantime, however, to keep the sheep in a thoroughly good and thriving condition. In a letter to the *Wool Manufacturers' Bulletin*, of Boston, after describing his experiments, he states that "the dirt was scoured out, leaving a sound, strong fiber." It is true, as he asserts in this communication, that "the character of wool is dependent upon the grazing facilities; its strength depending upon the continued progressive character of the economy of each sheep." In support of this proposition, which can be corroborated by hundreds of shepherds out of their own experience, he mentions that on February 10, 1879, there came eighteen inches of snow, which lay on the ground for fourteen days, more or less. Two flocks, that he had at a distance from home, went without feed the most of this time, and after the snow melted and they began to find fresh grass in abundance, the fleeces slipped off of them, a joint having been formed by their long fast. Of course, no alkali had been flying for some months, and no part of this result could be attributed to its effects.

But it is necessary to present some opposing testimony. From advance chapters of Wade's "Wool Facts," printed in the *American Sheep-breeder*, I take the following:

"Dead tip is prevalent in all Merino wools, if the sheep are kept in immense flocks, even if grown in countries where pasture is fairly good—that is, a continuous sod. The rains, to which large flocks are subject, will by long continuance destroy the yolk on the outer surface of the fleece, and the wear and tear by rubbing together will destroy the natural lubricant, and decay at once sets in. Dead tip is simply wool which is decayed, or rather from which the life has departed. While this wool, or rather the tip, will scour easy, it will not retain the dye which it readily takes, and it is a source of great unevenness in fine, solid-colored, face-finished goods, as the color, even of indigo, is removed by finishing, and the surface has a gray appearance when shaded across the face of the goods. The French and English manufacturers must have learned this fact, for we do not see this defect in their finest goods, which are very uniform.

"It is in our territorial wools that dead tip is most prevalent. It is most common in the regions where bunch grass is the native growth. Where there is bunch grass there is bare ground, and where there is bare ground there will be imperfect wool. The yolk is the natural protector of wool, and will turn rain if unmixed by other substances, but when dirt falls on the sheep and intermixes with the yolk in the wool it will absorb moisture—that is, the dust will absorb the moisture, and by remaining moist will destroy the yolk or animal grease to the extent that it penetrates. In some localities, where flocks are large and exposure great, this penetrates to the skin of the sheep and frowsy wool is the result. Frowsy wool is that which has lost its nature, thereby destroying both the luster and felting properties, the animal grease having been driven from it or consumed by the dirt with which it is loaded, leaving the wool tender and freed from all that gives it its strength, and that which gives it its value as a material for clothing, either for the sheep or the human family."

The only remedy for this would seem to be the cultivation of tame grasses and a close sod.

BEASTS OF PREY.—Old shepherds have acquired much skill in tracing and trapping the various beasts and birds which prey on their flocks. If an eagle has done the work, there will be no large bones broken in the carcass; the flesh will be torn off in a ragged way, and there will nearly always be a few large, downy feathers lying about. If the ground is soft, coyotes and wild cats will leave their tracks in the mud. Panthers and grizzly bears generally carry off the carcass, and cover up whatever may be left, with grass or leaves. Coyotes and wild cats catch lambs by the head or throat; after a coyote has visited the flock at night, ten or fifteen lambs will frequently be found dead: bitten through the top of the head, perhaps, or a hole cut through the side to the heart and entrails. The coyote is a most fastidious and gratuitous assassin; he either slaughters many more than he requires, or else he wishes only a few quaffs of blood warm from the heart of each victim.

After much experimenting with various traps and poisons, strychnine has been found to be the most efficacious for all classes of *carnivora*. The best way to handle it is to cut the meat into mouthfuls, about the size of a hen's egg, and put a grain or less of strychnine in a very small bit of tissue paper, which, when rolled up, is no larger than a grain of wheat. Then pour a little melted lard over them; cut a hole to the center of each piece of meat, and then push a paper of poison in deep and close up the orifice. In this way the meat will not be made bitter by the strychnine, and the animal will not be deterred from swallowing it. The shepherd prepares forty or fifty baits this way, and thrusts a sharp stick into each. Late in the afternoon he starts out, dragging with one hand a beef's or hog's "pluck" on the ground in a wide circle, and carrying the baits with the other. Every hundred yards or so he sets a stick in the ground, which holds the bait a few inches above the surface, where the beasts will find it readily. Even if no carcasses are found, the quiet which reigns in the flock for several weeks gives the shepherd satisfactory evidence that the poison has done good work.

GENERAL REMARKS.—Under very close pasturing with sheep, some kinds of wild grasses quickly disappear; but other plants—especially the alfileria, volunteer in place of them; and if nature is assisted a little with some seeds of white clover, narrow-leaved plantain (rib grass) or other good forage plants, a pasture will be created better than the original.

The average annual loss among adult sheep—from various causes, is placed at 11.08 per cent. The diseases which prevail in Oregon are scab, fluke disease (locally called "leeches," and treated with charcoal and salt), "blind staggers," "scours" (locally, alum and wheat bran are given). To prevent the attacks of the sheep gad-fly, the Oregon flock-master bores two-inch holes in a large log and throws salt into them, with tar smeared at the top; the sheep in licking the salt smear their noses with the tar.

One method of preparing wethers for market is to let them run on the green wheat—which is very rank in the fall and winter—until midwinter or thereabout; then finish them off with bright, green hay, cut in the blossom, and a daily feed of about a pint of oats and two ounces of wheat per head. They do not injure the wheat, which at harvest will look as well as that which was not depastured, and yield thirty bushels per acre.

Merinos from California and Australia are found to be more hardy and prolific than those from Vermont, doubtless because they are already acclimated. In Oregon—as nearly everywhere in the Far West—a sheep above a three-fourths or seven-eighths grade of Merino is not desired.

The average clip per head is about the same as in California. It is stated that sheep reared to maturity in Western Oregon and then removed to Eastern Oregon, will increase their yield of wool.

WASHINGTON TERRITORY.—The natural grasses and forage plants are rye grass, bunch grass, goose grass, four or five kinds of slough grass, blue-joint, cane grass, alfileria, white sage, willow, ross, greasewood and broom sage, besides rushes in wet land.

Sheep in Washington Territory have encountered bitter opposition from the cattle-men. "A band of scabby sheep at every watering-place"; "Cattle owners running from a pestilence—sheep"—are the phrases one encounters. The original stock were the so-called "native" sheep of Oregon, bred up from California sources to Merinos. Within the last twenty years many full-blood Merino rams have been brought in from California and Vermont.

Rams run in the flocks from November 1st to Christmas, in about the same proportion as everywhere under the careless husbandry of the Far West; from two to three rams to the hundred ewes. Lambing is in April and May; shearing, in May and June.

In winter the flocks are generally close-herded where there is both feed and shelter. Some hay is stacked for severe weather, and they are corraled at night. During summer—until the grass dries up on the plains—they range at large, a herder with each flock. When the shearing is finished and the prairie pasture exhausted, they are driven to the mountains.

Scab is very prevalent south of Snake River, and is the only troublesome disease. The losses are caused mainly by scarcity of feed, winter exposure and wild animals. The estimated average annual loss of adult sheep is twelve per cent.

SHEEP IN THE "CHINOOK."—Sometimes there falls a sudden and deep snow, attaining the depth of even two feet. Horses on the hills come down to the bunch grass and obtain feed in abundance by pawing; cattle browse on greasewood, willows and sage; but the sheep huddle helplessly under the junipers and refuse to venture forth. But when the storm is ended, a mass of white clouds will be seen flying overhead; on the western horizon is a bank of dark-blue clouds; and fitful gusts of wind—now warm, now cold (the welcome "chinook")—begin to blow from the west. The patient sheep smell the salt in the air; they look up at the junipers overhead, which now and then throw off a load of snow from their bending boughs upon their backs. In the afternoon they scatter over the hills, wherever the snow, being thinner, has already disappeared; soon every ravine, valley and cañon is roaring with snow-water; a Japanese spring is at the doors; and in a day or two the snow has wholly disappeared.

IDAHO.—The grasses and herbage in this Territory are about the same as those mentioned for Washington. The sheep have been brought in mostly from California and Oregon on the west, and Utah on the east and south. The American Merino blood has steadily gained on the British and Mexican, as it was found better adapted to conditions of soil and climate than any other, until now the majority of the flocks in Idaho are from three-fourths to seven-eighths Merino.

Flocks generally number about fifteen hundred; it is unprofitable to employ a man to care for a smaller number, while a larger flock would scatter and become more subject to the attacks of eagles, coyotes, mountain lions and wild cats. From May to July they are moving up into the hills, where the grass and water are in more abundant supply; from September to November they are working slowly down to the plains and meadows. They are not corraled except for shearing, dipping or counting. Blooded rams have sheds within a pasture, and for about six weeks before service begins they receive hay, oats and perhaps roots. They run with the ewes at night from November 15th to January 1st, to bring on lambing in April. The lambing season is earlier in Western Central Idaho, though attended with risks; the object being to secure early mutton lambs and strong growth for those destined to be wintered.

Hay is cut from fenced "claims" and wheat "in the dough," to cure for winter feeding; it is given mostly to ewes and lambs when snow lies long on the ground—an infrequent occurrence. Grown sheep receive three to three and a half pounds of hay per day; lambs, one and a half to two pounds. Some wheat straw is given to them, but it is found too constipating for continued use.

Severe storms early in April and late in October, followed by cold, while the fleeces are saturated with moisture, are a drawback to sheep husbandry in the western part of Idaho. In this region the annual loss among adult sheep is estimated at fifteen per cent.; and at five per cent. from storms, and ten per cent. from wild animals and dogs. For the whole Territory, the loss is estimated at ten per cent. The loss from disease is two per cent.; from storms, four per cent.; wild animals, three per cent.; poisonous weeds and snake-bites, one per cent. As in all these mining regions of the West, there is a small local demand for mutton for mining towns and camps and for military garrisons. It is said that the Indians on the reservations refuse mutton.

MONTANA.—This vast Territory has a climate which, though at times very severe on the elevated ranges, is generally favorable for sheep, being softened by the friendly "chinook" blowing over from the Japanese *Kuro Siwo*, or warm stream. There is a great extent of good average grazing, consisting principally of the bunch grasses, of which the most highly esteemed kinds are the *Boutelona oligostachya* and the *B. hirsuta*, the buffalo grass (*Buchloë dactyloides*), red-top, wild rye, blue-joint, and wild oats. Greasewood and white sage are hardly known north and west of Judith Valley. The grass does not grow—with the exception of that on moist or wet lands—more than about four months of the year. A peculiar feature of the bunch grass of Montana is that while it is apparently cured early in the season, from the latter part of July to the middle of August, when the range presents the brownish-gray appearance of dead grass, a close inspection shows about three or four inches of green and growing grass near the ground, which possesses surprising strength and nourishing qualities, while the top portion, having cured early and during the dry season, retains all its original strength. The grass remains in this condition until the frosts and snows of December appear.

Montana may be considered as preëminently the home of the bunch grass, as California is of the fileree, New Mexico of the grama, Arizona of the gietta, Texas of the mesquite, and the Missouri Valley of the buffalo grass.

It is only ten years since the first sheep were brought into Montana, and last year the wool clip was over three million pounds. The climate gives the finest fiber to the wool, and the sheep seem hardy and healthy. Last year the deaths were only two per cent. in the flocks. Many of the ewes have twins.

Some of the owners of the larger flocks may be mentioned: Geo. Myers & Bro., Perkins Russel, Poindexter & Orr, J. M. Sharpe, Crane, Headly & Co., McClintock & Dowd. The sheep owned by Paris Gibson & Son are of the Campbell stock, of Vermont, noted for their long or delaine wool, and their freedom from gum and black-top. Indeed, this is more or less characteristic of Montana flocks in general. And it has been found by experienced flock-masters that, to withstand most successfully the severe cold of that region, the sheep should be comparatively free from wrinkles and yolk.

It is the testimony of an experienced flock-master that tracts of land upon and around which his sheep had been corraled until they were so dry and bare that the dust raised by the sheep could be seen a mile distant, next spring, much to his surprise, were covered with bunch grass thicker and stronger than ever before, the bunches being closer together. Yet there is much variation in this respect in different parts of Montana; in some places the stock, particularly cattle, pull up much valuable grass by the roots.

MANAGEMENT OF FLOCKS.—This is much the same as that heretofore described for Oregon and California, with variations to adapt it to the colder climate and ruder civilization of this inland region. Medium, rather than very fine-wooled sheep, are considered most profitable. The wool is remarkably free from burs and dirt; and the sheep are very healthy, though the scab is prevalent and requires the same rigorous treatment for its eradication as elsewhere. Some hay and shelter are provided for winter. Fresh pasture is reserved for ewes in the lambing season, which comes the last of April and in May. Shearing is done without previous washing, and dipping follows shearing.

The sheep of Montana, being largely of California origin, are generally good; they will average above one-half Merino in grade; are valued at about three dollars a head as wool-producers; wethers, three dollars and twenty-five cents. At maturity the mutton sheep weighs one hundred and five pounds, live weight; fifty-two pounds, dressed.

Grub in the head and "dropsy" are sources of limited loss. W. H. Peck, of Fort Maginnis, in a recent letter, says: "About two weeks ago, I had about three hundred sheep poisoned by some poisonous weed. I succeeded in saving all but three by bleeding them, and giving each sheep two tablespoonfuls of vinegar. My experience may help some one else out." What the poison was, Mr. Peck does not state.

The *Montana Wool-Growers' Association Bulletin* gives the following advice as to the preparation of wool for market:

"Good, clean bags should be used; those known as "machine sewed" have closer seams and keep the wool cleanest. Plenty of good, strong twine should be ready at shearing time, thus

Fig. 27.—MONTANA SHEEP RANCH.

avoiding the necessity of using strands of rope, strips of bark, and such substitutes, 'because the twine gave out.' It would be well if growers would pack their buck, and also the very coarse fleeces, each by themselves; probably, however, the present practice of packing all together may give the grower some slight advantage over the buyer; but it is probable that separate packing will finally be generally adopted. When scab is all through a band, it is hardly of any use to try to separate the best from the worst; but where there is only a percentage of scabby fleeces they should be packed by themselves, as their presence among the sound wool will perhaps condemn the whole, in the eyes of the buyer. But do not let a grower be entirely discouraged if he is unlucky enough to have a scabby band. The scab can be cured before the next season, and if he will pay proper attention to doing up his wool as carefully as he would if sound, he will obtain better results than perhaps he expects. In these cases pack the loose wool by itself—do not attempt to tie up a lot of loose wool and odds and ends to look like a fleece. Scabby wool is often very light in condition, and if carefully handled will often sell quite well on account of its lightness. Never dip, in the fall, with those strong dips which turn the color of the wool. In the spring it is not of so much consequence, as the sheep have just been sheared and the dip does not find any wool to damage."

In order to make a success of the sheep business here, sheepmen have found that they must put up from twenty-five to forty tons of hay for every thousand head, besides building sheds in which the animals may seek shelter during excessive cold. The hay can be put up at from two dollars to two dollars and a half per ton, and is an absolute necessity to successful sheep husbandry here.

The average clip in this Territory is about six and one-half pounds per sheep, though isolated instances are reported where a clip of twenty-five pounds of wool has been sheared. The wools grown in this Territory are worth from twenty-two to twenty-eight cents per pound, and are counted among the very best of the mid-continent.

DAKOTA.—The Indians held possession of this Territory until the Black Hills gold excitement turned public attention thither in 1875; population then came in rapidly. In 1880 there were eighty-five thousand, two hundred and forty-four sheep in Dakota, mostly grade Merinos from Minnesota and Iowa. Both classes of stock-men pronounce the country excellent for their purposes. Though cold, the constant dry winds sweep away the snow; the foot-hills with timber afford shelter; and grasses, principally the buffalo, furnish tolerably good feed.

In 1880 there were sixty-three thousand, two hundred and six sheep shorn, yielding about five pounds of wool per fleece; and there were one thousand, seven hundred and sixty-five slaughtered for mutton, having an average live weight of eighty-eight pounds.

ALKALI.—When sheep are very thirsty and heated they will injure or kill themselves by drinking alkali water. But when they have constant access to it they will sometimes take it without injury, and will not require to be salted; though careful flock-masters consider this a slovenly practice. The following paragraph, clipped from the *Pacific Rural Press*, is full of suggestiveness for Western flock-masters:

"Alfalfa will grow in alkali soil unless it is very strong; but it seems that even the strongest will support Bermuda grass. Mr. J. A. Cary, of Tipton, has a spot of very strong white alkali ground by the side of his reservoir. He planted alfalfa upon it repeatedly, but it would not make a start even. A few months ago he procured some Bermuda grass seed and tried that, with astonishing success. Although the fowls from the barn-yard scratch in it incessantly, it grows rapidly and will soon make a perfect mat all over what was before an unsightly alkali spot. Let others try this grass upon their alkali wastes."

ITEMS ON MANAGEMENT.—Mr. W. B. Skipton, of Empire, Dakota, furnishes me the following facts: In the spring and summer, sheep feed principally on the buffalo and bunch grasses, but in the fall and winter they resort to blue-joint. The latter grows tall, and is not covered by snow. There is no rain during the fall and winter, and the buffalo and bunch grasses get so dry and harsh that the sheep do not relish them. There is a weed called the rosin-weed, which sheep are very fond of; it is very scarce, and when flocks are put on a new range they will travel a great deal in search of it, but after they have been held there a few days, and have picked these weeds clean, they become contented.

Good sheds are used in winter; sometimes they are covered with hay, but generally with boards. Sheep are shedded only during the storms; it matters not how cold it is, they are turned out to graze whenever it does not storm. If it storms all day, they receive some hay under shelter. The joint grass is cut for hay, because it grows heavier than other grasses, but any of them makes good hay. The hay is stacked, but not with much painstaking, as there is not much rain at haying time. Very little hay is used, as the sheep are turned out to "rustle" for their living nearly the year round.

There is a poisonous weed called the milkweed (a very small weed, and not like that of the Eastern States', so called, which is sure death to the sheep that eats it. There is no remedy used, for the sheep swells up and dies very quick. They do not eat it very often.

There is no foot-rot in Dakota, but sheep are troubled with the "gumbo" getting between their hoofs, which causes them to go lame. The shepherds catch them, and remove it. There are no maggots in summer; but sheep are troubled with horse-flies and bot-flies on the nose and legs, as they have not so much wool on these parts as have the Eastern sheep. They are rubbed on the noses and legs with kerosene and lard as a protection.

Fenced sheep ranges will carry two sheep where the open range carries one. This proportion changes as the range improves and becomes greater in favor of the pasture system.

CHAPTER XXVI.

SYSTEMS OF SHEEP HUSBANDRY, *Continued.*

NEBRASKA.—Among the foremost importers of Merinos into this State may be mentioned William Draper, H. H. Stoddard, J. M. Chadwick, Colonel J. H. Roe, R. F. James, Henry Goodyear and William Stong. In Central and Western Nebraska, which is the stock region proper, the large, strong, prolific Merino ewes of California and Oregon (all ultimately traceable to California), have always been the favorites. A three-quarter California ewe, crossed with a full-blood Vermont or Ohio ram, gives the most desirable result—a strong, rangy animal, capable of taking care of itself on the wind-swept plains, with a dense

FOR WOOL AND MUTTON. 265

Fig. 28.—NEBRASKA SHEEP RANCH.

fleece of deep-grown, nicely crimped wool, adapted to resist the storms and shed the rain from its thick, oily exterior. Such a sheep, after a storm of snow and sleet which pierces the thin-wooled British sheep to the marrow, will get up and shake itself, throwing off the snow, and move quietly off to the range. Such a sheep will shear eight pounds of wool of good quality, grading as medium clothing, and will develop at maturity a carcass of one hundred and ten to one hundred and twenty pounds (ten pounds less if a ewe), which, fattened one winter on the abundant corn of the Missouri Valley, or in Illinois or Indiana, will weigh one hundred and twenty-five to one hundred and forty pounds, live weight, and bring, in the Chicago market, five, five and a half or six dollars per hundred.

It is estimated, by a high authority in such matters, that a wether like that above described can be grown to the age of four years at a cost of one dollar and twenty-five cents; saying nothing about the wool return, which will be at least one dollar and fifty cents a year, at the prices prevailing prior to 1883. It will then be worth three dollars and fifty cents—the carcass alone.

The great demand for "feeders," to consume the superfluous corn of Iowa and the other rich prairie States, tributary to Chicago and St. Louis, has created in Nebraska a strong prepossession in favor of the mutton breeds, and the large-framed Merino of the Pacific. The Vermont Merino is sought after— to use the expressive phrase of a ranchman—simply to give "roofing" to the heavy-bodied Californian. In default of the latter, the Nebraska flock-master will resort to the Downs or to the Cotswold. Still, he always wishes a considerable infusion of Merino blood from some source.

A Merino ram from California, weighing one hundred and fifty to one hundred and seventy-five pounds, costs (1883) from one hundred to one hundred and twenty-five dollars; but this is not considered as against the compensating gain of a pound per fleece in each of his progeny, and fifty cents per hundred on the mutton. There are not lacking men in Central and Western Nebraska, who fearlessly assert that the Merino is every way the best sheep for that section, being not only hardy, but yielding more wool and mutton than any cross-bred sheep. This is doubtless true, with the qualification that the flocks shall be the large range-flocks, and the Merino of the large, hardy, acclimated California variety.

APPLIANCES.—Reaching Nebraska we have approached near enough to the great agricultural systems of the populous East to frequently find on the ranch the convenient farm appliances to which the Eastern shepherd is accustomed. Thus, on the great ranch of Colonel Roe, in Buffalo County, we find a barn capable of receiving two hundred and twenty-five tons of hay and fodder; five good corrals; a wind-mill, which distributes water to all the corrals; sheds, with grain-troughs; an ample corn-crib; a ram pasture of ten acres, etc.

WIND-BREAKS.—In Western Nebraska, as in Wyoming and the adjacent regions, resort is frequently had to snow-fences or breaks for the protection of sheep. These are constructed nearly in the same way as ordinary board fence, except that there are no permanent posts set in the ground. Each panel is made independent, with slats instead of the posts; and these panels are set in strings, from one hundred to five hundred yards long, inclined at a slight angle with the direction of the wind (which generally comes from the north-west), and supported by braces or brackets. These strings generally curve a little, with the convex side toward the wind; and, if the locality is one which is liable to very heavy snow-falls and strong winds, they are doubled, trebled, even increased to seven or eight in number, all running parallel and with a space of a rod or less between each two. Each successive string detains a part of the snow, so that, unless the storm is unusually protracted, the sheep hovering in the lee are sufficiently protected.

Belts and groves of trees are also found very useful for this purpose. White willow is preferred by most farmers; some plant cottonwood in regions where it will succeed. The white willow grows very rapidly and forms a dense screen. A snow-fence or a belt of this willow, advanced a little beyond a grove of the same, form an excellent protection; very little snow will penetrate a grove thus protected.

DANGERS FROM THE ELEMENTS.—There are two dangers, somewhat peculiar to the Platte ranges of Central Nebraska, and especially noticeable in the Republican Valley—southwest—miring and drowning. There are hollows and stream-beds of tenacious or shifting soil, that become dangerous traps after the spring and early rains, particularly for young stock. When the floods break into the narrow valley ravines, numerous in those regions, there are sometimes heavy losses from drowning, the animals being caught by the flood of sudden storms. Along the

Platte there are wide, treacherous quicksands which swallow up unwary animals. Other things greatly to be dreaded are the blinding snow-storm known as a "blizzard," coming suddenly when the sheep are on the open range ; a cold rain that reaches the skin, and sometimes, when freezing follows, encases the sheep in a coat of icy mail ; and deep snow covering the ground for several days. In November, 1879, when Indian summer weather and good pasturage were giving promise of safety to flocks that were yet at a distance from home, suddenly, without any recognized warning, a furious snow-storm broke upon the plains in the night ; in the morning the thermometer had fallen nearly to zero, the wind was blowing a gale, and nothing could be seen a horse's length. The hastily aroused shepherds and their brave dogs could not control the sheep, although one of the latter killed fourteen sheep in his vigorous efforts to check their course. The storm lasted three days and nights. Sheep drifted forty and fifty miles from their ranges. Some were buried in snow-drifts ; some died from exposure and want of feed. As soon as the violence and persistency of the gale were realized, relief parties were started out. One herder was brought in frozen and helpless : his dog had kept with the drifting sheep ; but he, overcome with cold and fatigue, had lain down to die. Another was discovered on his horse, but man, horse and dog formed a motionless group—the herder unconscious, the horse almost dead, the dog nearly frozen. The sheep, of which many were suffocated, were found hard by in a dry wash, where a snow-drift had covered them up. Man and dog had to be carried twenty miles to the nearest roof. He might have left his charge and gone in search of shelter while strength still remained to himself and horse, but he stood by his sheep to the last. It was said that the dog refused to eat or drink until his master, restored to consciousness, recognized him ; then, when fed, he endeavored to return to the sheep.

FEEDING IN WINTER.—A hundred tons of hay, costing one dollar and twenty-five cents a ton, will winter one thousand sheep, and three bushels of grain per head is allowed to mutton-sheep. Stock sheep are not fed regularly, but if a stress of weather arises, they are brought within the corrals and receive two pounds of hay and one-fourth pound of shelled corn per head daily, scattered on the snow. Shelled corn is the universal feed for fattening sheep, while ewes in lamb receive crushed corn or oats. If the corrals are protected by snow-fences or belts of willow, as already described, the sheep will generally do better than if housed. But sheds are needed for lambs, rams, "poor-house flocks," etc.

WATER IN WINTER.—When the shepherd sees his flock, of their own accord and with a tremendous rush, run to their accustomed watering-place in summer, he understands that they need water; but in winter he is very apt to neglect this matter. With snow evenly distributed over the range, sheep are apt to get all the water they want; but when there is no snow they may suffer severely. Lambs, particularly, have to be carefully looked after; they will sometimes become so stupid from cold that, even when driven to water, they do not appear to understand that it is water, but seem to think it is ice or snow. Many a lamb has been taken from a range where the snow supplied its wants daily, and, with one or two thousand others, been fed on hay and driven daily to a hole in the ice large enough for one hundred to drink from at once, and held there long enough for six or eight hundred to drink, then driven away (six or eight hundred others not having had any water), day after day without a drink, until it died, and the master did not know the reason. It requires the utmost patience, gentleness and tact to hold a flock of half-frozen lambs to a hole in the ice, and keep them moving about, so that the large and strong ones may drink and fall back, and the others may be brought around within view of the water and secure what they want.

THE DETAILS OF LAMBING, ETC., are about the same as heretofore described for the adjacent regions.

DISEASES.—Scab is the only disease of any importance; it is claimed by the residents that it would seldom occur if not freshly brought in by trail sheep. This dry, elevated region does not develop foot-rot; indeed, it is asserted that flocks suffering from it are cured by being driven hither.

Flock-masters, in Nebraska, generally do not care to breed their sheep beyond three-fourths Merino.

WYOMING.—Probably three-fourths of the sheep of this Territory are animals bred from original Mexican ewes, crossed with Merino and—to a less extent—Cotswold rams. From the "Government Report" I condense the following table:

ESTIMATED AVERAGE VALUE AND WEIGHT OF STOCK AND MUTTON-SHEEP AND THE ESTIMATED AVERAGE ANNUAL WOOL CLIP FOR 1880.

Kind of sheep.	MUTTON-SHEEP.		AVERAGE ANNUAL SHEAR.			
	Average live weight.	Average value as mutton.	Rams.	Ewes.	Wethers.	Lambs.
	Pounds.		Pounds.	Pounds.	Pounds.	Pounds.
Full Mexicans............	80	$2 20	3.5	2.5	3	2
Half-breed Mexican Merinos—*i. e.*, bred from Mexican ewes and crossed with Merino rams.......	90	2 75	....	4	4.5	3
Half-breed Mexican Cotswold, bred from Mexican ewes and crossed with Cotswold rams............	115	3 25	....	3.75	4	2.75
High-grade Merinos..	100	3 00	10	6	7	4.5
High-grade Cotswold	125	4 25	7	5	6	4

Laramie City is fast becoming a very important wool market, over two hundred thousand sheep already being grazed on the Laramie plains. Near the city, at a point which is termed Gloversville, Mr. S. H. Kennedy has erected dipping tanks and apparatus whereby sheep-growers are enabled to dip their flocks much cheaper than under the old arrangement of doing it at home. He has also, here, a large shearing pen with fifty stalls, with a capacity of several thousand sheep, a sacking-house, a large ware-house, boarding-house and engine, and pump-house. The sheep are driven to the works, shorn and dipped, and the wool is sacked ready for market. Scouring machines are talked of to prepare the wool for shipping direct to manufacturers. The wool product tributary to Laramie has reached nearly one million, five hundred thousand pounds per annum.

All these associated efforts toward a consolidation, a better classification and an improvement of the condition of the wools of the Far West, previous to their shipping, ought to be fostered by the growers. It is this policy which has, among other things, given Australian wools their advantage over domestic fleece in the New England markets. In the case of these wools the manufacturer can, as he does in England, send an order to the broker for the exact number of pounds of the exact grade of wool he wishes; and the professional stamp of the sorter, affixed "on honor" to a lot of wool, passes as unquestioned as a

Bank of England note. It is this cleansing and thorough classification of clips on a large scale which will enable the growers of the West to compensate themselves somewhat for the poorer grade of their wool, in competition with the small and scattered neighborhood clips of superior wool produced in the East.

There are some objections to this plan which ought to be candidly considered. One is, the strong opposition, amounting to a prejudice, of the manufacturers toward scoured wool. Perhaps the West is indebted to the Californians as much as to any other section for the creation of this prejudice. Some years ago it was made known that if the Calfornia wools were sorted and scoured they could be shipped to the great mills of Cohoes and Lawrence and compete successfully with the Australian wools. With their accustomed energy the Californians entered into this scheme; scouring mills were erected in San Francisco, and large shipments of "scoured" wool began to go East. But much of it, besides being originally defective, was so ill sorted, scoured and baled, that it gave a bad reputation to the whole enterprise. In the summer of 1885, unwashed wools sold more freely in Chicago than either scoured or washed wools. As the freight charges on dirt must ultimately be paid by the grower, the manufacturer is indifferent to a reform in this direction; and it is easy for the grower to see how much it is to his interest to reduce his clip as nearly as possible to the condition of pure wool, sorted to its last analysis, before it starts on its long journey across the continent.

Another objection is that every fleece must be sorted, so that each of the several qualities of fiber found in each fleece may be placed with fiber of like quality from other fleeces. This will make necessary the services of skilled sorters, and to such men high wages must be paid. But this sorting must be done at some time, and while wages would doubtless be higher in the West than in the East, the saving in charges for freight would probably pay a handsome profit over any difference there might be in cost of sorting and scouring.

COLORADO.—This State has attained a prominent rank as a producer of sheep and wool. East of the Rocky Mountains and north of the Arkansas River, which is the principal stock region, the number of sheep in 1870 was very small, and the feeling of stock-men against them was very bitter. In 1880 the sheep-men almost controlled this region by virtue of holding the lands with water on them. The sheep numbered one million, ninety-one thousand, four hundred and forty-three. Fourteen slaughtering establishments reported thirty-seven thousand, one hundred and sixty-six sheep butchered in 1879, having an average live weight of one hundred and four pounds. The wool clip averaged five to eight pounds per head.

Here, as in nearly all the lands west of the hundredth meridian, the pasture is much depreciated in value below its condition a quarter of a century ago, when it was overrun by millions of buffalo. In Colorado, perhaps more than in any other State, except California, irrigation and the cultivation of forage crops—chiefly alfalfa—have been prosecuted to supplement the failing natural resources. "Alfalfa mutton" has a reputation almost as distinctive as the turnip-fed chops of Dorset.

The seeds of the grasses and forage plants, it is true, are trampled in by stock, and especially by sheep, and in soils not too sandy they are thus defended from the drying and freezing they would otherwise suffer; and the finely distributed sheep manure enriches the land; but on the other hand, close herding and cropping do undoubtedly tear out very considerable amounts of certain grasses by the roots.

There are many dry regions on the great plains of the mid-continent, which do not absolutely forbid winter occupation, except in a few limited areas. The snow-drifts supply water under the warm breath of the stock, or are melted in holes in which the animals have trodden or wallowed. Stock can travel twice as far for water in winter as in summer, thus greatly increasing their available pasture area; besides which, there are tracts which are supplied with water in winter but not in summer.

I append the following paragraphs from the *Government Report* of 1884:

"We may consider the flock year to begin when lambing and shearing are done. The sheep go on summer range about July 1st. Changes from summer to winter grazing are generally but not always made. The main idea is to keep the flocks where there is grass, and in winter to have them sheltered, artificially or naturally.

"For summer management the stock is divided into flocks, here always called "bands";* the numbers vary according to kind from fifteen hundred to two thousand each, the ewes with

---

"* The word 'band' is used with very different meanings in different localities in the West; it is used for a flock, a herd, a drove of animals, a subtribe of Indians, etc. Among stock-men it is used as the common name for either flock, herd or drove."

lambs together, the wethers and dry ewes together, each flock under a special shepherd (here usually called 'herder'), who is accompanied by a dog. The flock is put in an inclosure every night, for better protection.

"The summer ranging continues until the last of November, when the sheep are moved to pasturage with sufficient cured grass, shelters from storm, and stored feed for emergencies. Before winter sets in, weaker animals are separated from the flock and made into an invalid flock, with which, perhaps, the rams are run during the day. This flock receives special feed —hay or grain—as occasion requires. As a rule, feeding is not necessitated oftener than, perhaps, four winters in ten; but a prudent administration will always provide feed other than pasture for lambs and weaklings, and for rams before and during service.

"Dry stock and wethers will stand almost any severity of weather on pasture alone; the use of hay on a sheep ranch is principally for the horses. About an ounce of corn per day to lambs, or two and a half ounces to other sheep, and twice the amount if oats are fed, is the usual ration for such as are fed. In 1880, corn delivered on the ranch in Central-eastern Colorado cost one dollar and sixty cents per hundred weight; in 1879 it was worth one dollar and thirty cents; in 1878, while the deep snow lay, it was had at ninety cents. * * *

"The rams are turned in with the ewes from about the 10th of December to the 20th of January. They are, as a rule, put in with the ewe flock at night and taken out in the morning.

* * * * *

"The more experienced flock-masters discard sheds except for lambing. High, tight corrals, with outlying snow and windbreaks, are preferred, as they afford sufficient shelter and protection, are more cleanly, less liable to induce disease, and, in storms, the sheep do not overcrowd and smother one another.

"Inasmuch as severe storms and exceptional years are such an important element in Colorado sheep-grazing, the following facts pertaining to the experience of previous years in this regard may be of value:

"In the winter of 1871-72 severe snow-storms caused great loss, and April 7th a terribly cold wind with fine snow (the 'blizzard' of the plains) was very destructive. Stock was then run without any artificial protection. The man who owned the largest flock in the State at that time, lost outright seventeen per cent. of his sheep. The years 1874 and 1875 are memorable for extreme cold weather. The late storms during and just after lambing and shearing were the most disastrous. About the middle of June, 1876, there was a two days' storm of wind, snow, and hail. In the spring of 1877 again a like disaster came upon the flocks. During six weeks of December, 1877, and January, 1878, heavy snows remained upon the ground, in many places covering the pasturage entirely. One ranch, eighteen miles east of Colorado Springs, lost five hundred head out of thirty-seven hundred, while that almost unexampled snow lay on the ground. The losses consequent were said to have averaged twenty per cent. of the sheep. One man who then had six thousand sheep without pasturage, but who had provided hay the summer before and bought Kansas corn at eighty-five cents per hundred weight, carried his sheep through with but little loss."

Sheep husbandry in Colorado, as in California, is subject to great vicissitudes, but, for the most part, from a different cause, to-wit: blizzards and deep snows. Fifty per cent. of flocks, for which no dry feed had been provided, have sometimes perished within a month. Yet the business has often been remarkably lucrative.

UTAH.—Two Merino rams are said to have been obtained from California-bound emigrant trains as early as 1858. In 1866, upon the establishment of a woolen mill in Utah, there was a demand for finer wool than had been grown before. Henry Bell traded to Brigham Young, for fat wethers, five thousand graded Merinos from California. Still, up to 1873, the quality of Utah wool remained poor, being little improved except by a few long-wool rams. In that year Daniel Davidson brought in four hundred high-grade Merino rams, an example followed by others; and in a few years a large number of Merinos had been introduced from Vermont, Kentucky and California, of both the American and French varieties.

In 1879 the commercial estimates placed the Utah clip at two million pounds, classing the product as medium to fine, an increase in six years of three hundred and forty-four per cent. in quantity with a great improvement in quality. A serious check was given to sheep husbandry by the severe winter of 1879–80, when, in the three Counties, Tooele, Millard and Juab, forty-three per cent. of the flocks perished. But the Mormon hierarchy has succeeded better than most of the Territorial authorities have done in reconciling or suppressing the conflicts between the cattle and sheep-owners, and the latter are steadily gaining on the former.

From the *Government Report* I take these two items:

"SOURCES OF LOSS AMONG SHEEP.—The migration of sheep flocks encourages the prevalence of scab. This disease was said by all flock-owners to exist almost universally throughout the Territory. Previous to 1876 no Mormon sheep-man practiced dipping his stock as an antidote to scab; but in 1879 many flock-owners dipped their sheep. "Handling" sheep for the disease was still extensively adhered to, which consisted in catching such sheep as were seen to be affected and rubbing grease mixed with mercurial preparations on the diseased parts. Close attention to the appearance of the disorder often kept it under control, never, however, fairly eradicating it. When flocks were intrusted to lazy, unreliable herders, who failed to apply the ointment frequently, the progress of the disease during the long season of absence on the deserts was often rapid, and resulted in great mortality from weakening animals, thus causing them to succumb to storms, while the wool product of the surviving flock would be much reduced. The custom of driving thousands of sheep each spring to the neighborhood of the river Jordan for shearing, convenient to the Salt Lake market, has also tended to spread the contagion. Flock-masters assert that the bed-grounds of infected sheep are a sure medium for disseminating scab to a healthy flock which may later occupy the same spot. Each season between one and two hundred thousand sheep approach this common rendezvous in shearing time. Of other troubles occurring among sheep, blind staggers was stated to be most common and fatal, though no great mortality resulted from it. Losses by alkali taken in too large quantities with feed or drink while heated with traveling, and from bears, mountain lions,* coyotes or wildcats on the upland summer feeding-grounds were occasional throughout the Territory, and in some localities of more frequent recurrence. A flock-master of White River reported the loss of thirty-two valuable blooded rams in one night by a mountain lion that entered their pen. This same owner attributed a five per cent. loss each year to wild animals, an uncommon complaint, however, in most sections of Utah.

"POISONOUS VEGETATION.—Among the plants eaten by cat-

" * Panthers are usually known, west of the Rocky Mountains, as lions, mountain lions, or California lions."

tle and sheep there are three which are commonly believed to be a source of slight annual loss among live-stock. The poisonous parsnip, growing in wet meadow lands, and fatal to cattle when its root is eaten, is indigenous to many localities, particularly in improved and irrigated pastures to which milch cows may have access. Another plant often fatal to young lambs is the poisonous "Sego" found along water-courses on the valley slopes and bench lands; the leaves of this plant are said to cause the difficulty, as its bulb, though known to be injurious, is too firmly planted to be torn up by an animal while grazing. In Juab Valley a larkspur (monk's-hood) was said by cattle-owners to be a frequent cause of death, in wet springs, among neat stock. Other localities of the western slope of the Wasatch Mountains were reported as nourishing this baneful growth, identical with the notorious 'poison weed' of Colorado and Wyoming."

AMOUNT OF STOCK PER ACRE.—From the *Government Report*, above referred to, I compile the table here given, showing the average density of stock (cattle and sheep) occupation. One head of neat stock is taken as the unit of stock, and five sheep are considered an equivalent to one "cow" in relation to the consumption of pasture:

	Acres to the Head.	No. of Sheep, 1880.
Texas	24.72	3,651,633
New Mexico	53.27	3,938,831
Indian Territory	62.85	55,000
Kansas	27.29	629,671
Colorado	42.15	1,091,443
Nebraska	40.42	247,453
Wyoming	69.90	450,225
Dakota	73.72	85,244
Montana	78.49	279,277
California	35.63	5,727,349
Arizona	122.24	466,524
Nevada	145.65	230,695
Utah	136.97	523,121
Oregon	51.63	1,368,162
Washington	91.68	388,883
Idaho	135.00	117,326

The reader will bear in mind that, not the total area of the States and Territories is given above, but the area of available pasturage. Also, that if sheep alone were considered, the allowance of grazing would be only one-fifth of that in the table. For instance, Idaho would have one sheep to twenty-seven acres, etc.

CHAPTER XXVII.

DISEASES OF THE MERINO.—"PAPERSKIN."

As a motto applicable to the management of sheep, I should be disposed to paraphrase the old saying, "An ounce of prevention is worth a pound of cure" to read thus: "Pounds of prevention will reduce the cure to ounces." A diseased sheep, generally speaking, makes accusation against the master. I should be profoundly distrustful of the fidelity, industry, and special fitness of that flock-master, who professed a wide and deep acquaintance with the maladies of sheep, their treatment, and the remedies applicable to each. That ceaseless and tireless application to duty, that ever-watchful care-taking, which is the main requisite to success in sheep husbandry, and which will always be the principal dependence of the experienced shepherd, will leave him small leisure, and less inclination, to "doctor" his flock. The "doctoring" of a sheep is one of the most miserably unsatisfactory, uncertain and unprofitable operations in the whole scheme of management.

One of the most successful shepherds that I ever knew adopted as his motto: "Love your sheep." To the reader who does me the honor to peruse these pages, I would say: Love a sound sheep; leave no stone unturned to keep it sound; but if, after you have done your best, it falls a prey to some of the, happily, very limited number of ailments to which the American Merino is liable, and the use of some simple remedies does not prove efficacious, it would be better—unless it is an exceptionally valuable animal—to dispatch it at once. At best the sheep is a frail animal; it goes off quickly under the assault of most of its diseases; and if the case is a lingering one, with any visceral disease, it is almost certain to result fatally.

Yet the sheep is one of the healthiest of animals if thoroughly well cared for. Good feeding and good care, are of transcendent importance; and I would have the barn so large, the atmosphere so pure, the hay so sweet and green, and the corn so sound, that there never would be room for a bottle of medicine in it.

GENERAL REMARKS ON DISEASE.—The respiratory system of the Merino is proportionately smaller and feebler than that of the steer; it loves and requires, above all other domestic animals, the pure air of high and dry lands for the maintenance of health. It is less tolerant of the vitiated atmosphere and noxious stenches of the stable. The large, round nostrils of the Cotswold, while offering a more ready asylum for the gad-fly in summer, on the other hand conduce to that fullness and rotundity of the lungs, which materially contribute to protect it from the diseases incident to the respiratory system. This, together with its complicated and retarding alimentary apparatus, with its four stomachs and many yards of entrails, render the digestive processes weak and easily disturbed.

The sheep is naturally a gormandizer; it consumes an amount of food disproportionately large for its size, and extracts a relatively small percentage of nutriment from it; hence, the richness of its manure. Hence, also, like all gormandizers with an overloaded stomach, it needs air, exercise, freedom, in order to work off this gorge without detriment. A Merino closely confined and fed sufficiently, leads a cold-blooded life; its ears and extremities are cold; frequently it has not enough animal heat to liquify the yolk and expel it to the extremities of the fibers. Hence, the latter become clotted and pasty with yolk of a greenish tinge or nankeen-colored.

The sluggishness of the sheep's vital processes renders it a small and infrequent consumer of water; and by the same sign it ought to have all it will drink, and be encouraged, by frequent exercise, to drink more. There are more flock-masters who err in regard to water and exercise than in feeding; the flocks are oftener stinted in the former than the latter.

In proportion to its size, the sheep has also a smaller brain and nervous system than any other domesticated animal. On this account it is not capable of very severe or long continued muscular exertion, and is not very subject to violent inflammatory diseases; but, on the other hand, it has little power to resist disease, or to recover from it after the force of that is broken.

For these reasons, the bleeding and purging recommended by English veterinarians for their high-fed, full-blooded sheep, are seldom called for with the feebler and more sluggish Merino. On the other hand, its overloaded stomach often requires purgatives; but these, on account of the weak nervous and muscular system, should always be accompanied with tonics, such as ginger, gentian, oil of peppermint, etc. In other words, while the stomach needs depletion, the general system needs at the same time protection against the drastic effects of the purgatives and it needs toning-up as a curative agency.

PURGATIVES.—The best purgatives for nearly all occasions are Eposm salts and raw linseed oil. Whatever may be the method employed with other classes of medicine, purgatives should always be given in a liquid form, to secure more prompt and thorough action ; as, generally (though not always), a medicine given with the feed has to go through the slow process of regurgitation and remastication before it can pass through the four stomachs and exert any effect. A long-necked wine bottle, or a cow's horn prepared and bottomed for the purpose, is the best implement for drenching. It is best to let the sheep stand naturally, with its head held between the shepherd's legs ; and the tongue should not be drawn out, but the bottle may be thrust well down between the back teeth, thus keeping the mouth open as long as desired.

BLEEDING.—The best place for bleeding a sheep is in the facial vein below the eye, or on the inside of the forearm. If it is desirable to draw a considerable quantity of blood, it should be taken from the jugular vein in the neck ; let a little wool be snipped off, the finger pressed on the vein below the cut, and an incision made lengthwise, not crosswise. Half a pint to half a teacupful are the limits within which the amount of blood abstracted should range.

PARASITIC DISEASES.—Not only do the malodorous secretions and exhalations of the greasy, gormandizing Merino attract many parasites (*epizoa*) to the exterior of its body in summer but the closeness with which it crops herbage to the ground, and its omnivorous habits, expose it to the assaults of many internal parasites (*entozoa*). Cobbold states, in his valuable work, "The Internal Parasites of our Domesticated Animals," that the sheep is infested at times by at least eight *nematode* (round or thread-like), parasites, of which seven are strongles (*Strongyli*), while the eighth is the common whip-worm (*Tricocephalus affinis*), of the ruminants. Besides these, the American Merino is subject (though less than the sheep of England) to the ravages of one of the *trematode* (or flat) worms, commonly called a fluke (*Fasciola hepatica* of Linnæus, *Distoma hepáticum* of Rudolphi), and less frequently—perhaps not at all fatally— of two others (*Distoma lanceolatum*, and *Amphistoma conicum*). Then, too, the sheep is the unfortunate host or bearer of one of the *tænioid* (or jointed) worms, popularly called tape-worms (*Tænia expansa*); of the brain hydatid (*Cœnurus cerebralis*), which causes the blind staggers ; of the "bat" or larva of the gad-fly (*Oestrus ovis*), in the frontal or nasal sinuses; of an *arachnidan* parasite (*Pentastoma tænioides*), also a tenant of the nasal cavities; and, also, of a species of *Ascaris*, although there is little positively known concerning the last named. Then there is an *entozoön*, termed by Dr. Cobbold the "mutton measle" (*Cysticercas ovis*), which is very rarely found in the flesh of the sheep in England, but has been discovered in African mutton; and three species of the *Echinococcus*, chief of which is the *E. veterinorum*, encysted in the lungs and liver. Lastly, there is the so-called *lumbriz* (presumably Spanish, from the Latin root *lumbricus*, and which I find spelled in six different ways), probably a species of *Strongylus*; and the "screw-worm" of the same State, a product of an oviparous species of the *Oestrus*, a worm more destructive than the common sheep maggot.

THE LUNG PARASITES.—I shall treat of these first, because they are of preëminent importance. Without doubt, they cause, in all the region east of the Mississippi, not only greater annual loss of sheep than any other one disease, but a greater loss than all others combined. Probably four-fifths of all the sheep destroyed by disease in the region mentioned are lambs under the age of one year, carried off by what may well be termed ovine consumption, the insidious ravages of the lung parasite variously known as *Strongylus filaria*, *Strongylus bronchialis*, *Filaria bronchialis*, etc. When mature, the male worms are about two inches in length, needle-shaped, yellowish, and smaller than the females, which are three or four inches long, and white. They are found in the *trachea*, *bronchi*, and universally diffused throughout the lungs. In the spring, the females are found full of thin-shelled ova or eggs, and free embryos or worms already hatched, though the worms yet in the eggs appear sufficiently developed to be able to maintain an independent existence.

How the germs of the *strongylus* reach the lungs is not clearly understood. It would appear that there are two stages of the disease induced by them; the first stage being while the germs are still encysted in the lung tissues, producing what is mistakenly called *tuberculosis*, from the number of small, hard nodules arising from the irritation of the foreign bodies in the lungs and the consequent exudations around them, together with the cretification of the parasitic deposit itself. Each nodule is small, often not exceeding the size of a pin's head, and if it is squeezed, there is felt between the fingers the hard body referred to, which seems to be designed for the protection of the egg or eggs within. In due time the latter hatch; the worm is coiled upon itself in its narrow prison, within which it begins to move about until it effects its escape into the open passage of the lungs.

Now supervenes the secondary stage of the disease, that of the mature worms, which have by this time taken up their abode in the *bronchi* and *trachea*, where they often become knotted into loose masses or balls invested with mucus. This stage of the disease, unless relief is afforded, speedily has a fatal termination.

As already said, we have nothing but conjecture to offer as to the manner in which these parasites gain access to the lung cavities. Dr. Crisp's view is, that the parasites are first ingested by the animal as it feeds upon the grass, where they are lurking; then, after being warmed and nourished for a short time in the stomach, they are carried to the mouth again with the regurgitating cud, and, by some hook or crook (for here his theory is defective) manage to descend the *trachea*. Dr. G. Stuart mentions that some French writers hold that the living parasites, which are swallowed by the sheep with mouthfuls of grass, find their way at once into the windpipe and lungs, without the intermediate journey to the stomach, then back to the mouth, etc. Dr. Cobbold wisely refrains from offering any speculations on the subject. He confines himself to the simple statement of that which is known concerning them—that the worms, when mature, take up their residence in the *bronchi*, chiefly in those of a medium size, where they produce either a simple catarrh or an inflammation which diffuses itself over a great part of the lungs, ultimately causing death.

Dr. Stuart states that these parasites are so tenacious of life that even after being dried for a month they show signs of life when moistened; and that they survive an immersion in spirits of wine.

After the above description of this most pernicious and tenacious pest with which the Merino lamb has to contend for life, it will afford the shepherd some slight relief to learn the partially compensating fact that the disease created by it is not hereditary, contagious or infectious. Lambs born of infested parents always have healthy lungs. They never contract the disease except by swallowing the parasite with grass or other herbage on which, after being coughed up by other sheep, it has taken refuge.

SYMPTOMS—"PAPERSKIN."—Some writers contend that the parasite, having been swallowed, bores through the tissues from the stomach to the lungs; and cite as a proof of this the fact that it is only lambs which are attacked, the tissues of older sheep being too tough to be bored through! It is only during the first fifteen or eighteen months of the sheep's existence that it is exposed, to any important extent, to the invasions of these parasites. After the teg has passed the month of May the second time in its life, all danger, both from the *tenioid* and the *nematode* parasites is practically over.

The symptoms of the disease are, in general, an anæmic condition or bloodlessness, indicated in part by a waxy pallor of the skin, which, in popular usage, by a substitution of effect for cause, furnishes a common designation of the ailment—"paperskin." The blood becomes resolved more or less into its elements, in some cases the fibrin being apparently consumed by the parasites, while the serum collects in a watery excrescence under the chin. In this case the disease has advanced to an incurable stage; I never knew an animal in this condition to recover. There is a very perceptible pallor about the nostrils, languor in the motions, generally much thirst, hardly any emaciation noticeable, but an indisposition to travel which becomes plainly manifest when the flock is driven a few moments. The affected animals, as if aware of their inability to keep up, retire to one side, and desire to be left alone. Frequently there is a deep, exhausting, but almost noiseless cough. In an advanced stage of the disease, the wool parts readily from the skin, the fibers having become so attenuated as to have no strength to speak of.

PREVENTION AND TREATMENT.—Concerning the source or origin of these parasites, the most important practical fact to be borne in mind is that they are most numerous, or at least are found in sheep in the greatest numbers, during wet seasons, or among flocks which pasture most on lowlands subject to fogs, and more or less overgrown with aquatic (not salt) vegetation. Hence, the necessity, as a prophylactic measure, of keeping young sheep off from such pastures as much as possible, and of not allowing them to leave the stable in the morning until the dew is dried off. These are preventive measures which would suggest themselves to everybody. Others will be mentioned a little further along.

The proper treatment for sheep suffering from this affection should have regard to two points: First, to support the strength of the sheep; second, to expel the parasites. To sustain the strength and vitality of a sheep already affected, though it is very important, is exceedingly difficult, because the appetite is feeble and capricious. The lamb can seldom be induced to eat enough, even of the most nutritious feed, to make any considerable impression on it in the way of betterment; and the danger in giving it, by force, stimulating gruels, etc., is that, owing to its bloodless condition, the process of digestion will be so illy performed that the feed will do it more harm than good by causing scours. High feeding is of transcendent importance as a preventive measure; but when the lamb has reached such a pass that vermifuges have to be employed, it is necessary to proceed with great caution in giving rich feed.

To expel the worms, some shepherds resort to fumigation. The lamb is made to inhale the fumes of burning sulphur, tobacco or chlorine gas in a tight room, as long as the operator himself can stand it in the lower strata of the atmosphere.

After trying tar, copperas, soot, sulphur and other anthelmintics, I find that the best is a mixture of turpentine and linseed oil in equal portions. The operator must be very careful in administering it, or he will strangle the lamb, already enfeebled by the treacherous disease. After much experimenting, I find that the safest method is to let the lamb stand naturally on the ground, between the operator's legs. With the thumb and fingers of the left hand inserted in the mouth, hold the jaws apart, and the head a little lifted up—a very little, only just enough to cause the liquid to run down the throat. If the head is held back too much, the animal is very apt to be strangled. Have about a tablespoonful of the mixture in a longish-necked vial, stout enough not to be easily crushed between the teeth, and pour about half of the amount at once, right down beside the tongue. Do not attempt to hold the tongue. If the animal chokes and coughs, let it have its head until it recovers; then complete the dose. This ought to be administered every day, for two or three weeks, on a stomach emptied by a twelve hours' fast. This small quantity of the spirits of turpentine is better than a large one, as a large dose will be eliminated from the system through the bowels and kidneys, while a small one will be removed through the lungs; and it is the vapor of it passing along the *trachea* and air-passages which are offensive to the worms.

But, unless the operator is much more careful than most farmers are, there is a good deal of danger in this method of treatment. It is an extremely easy matter to kill a lamb with a teaspoonful of turpentine, as I have found out more than once. And, at the best, it is a miserably unsatisfactory and disheartening labor to medicate a flock of paperskin lambs. Prevention is far better. And, indeed—unless it may be an exceptionally bad, wet year—it argues ill for the watchfulness and diligence of the flock-master to have paperskin get any considerable foothold among his lambs. True, it is a most treacherous and insidious disease—equally so with its congener, consumption, in the human family—and a flock of lambs may seem to be healthy and growing, looking plump, and continuing to eat about as usual, when, if one of them is chased smartly five hundred yards, it will fall all in a heap, and probably expire in ten minutes. And it is astonishing how much indifference farmers manifest concerning it—principally, I believe, because it is not a demonstrative disease—makes its inroads silently, stealthily, and carries off the youngest, and therefore least valuable members of the flock. A single case of grub in the head of a mature sheep, with its violent capers, its agony, its tragic death, will cause more excitement and remark on the farm than the loss of a dozen once promising lambs by this obscure disease.

The preventive measure of transcendent importance is high feeding. A thoroughly vigorous, well nourished lamb seldom falls a prey to these parasites. If the farmer has found by previous disastrous experience that the soil of his farm predisposes sheep to this pest (some soils, notably limestone, seem to escape it), he ought to adopt some plan of feeding his lambs all summer, or at least from the time when wet weather sets in. Better get them accustomed to wheat bran; then wean them a month earlier than usual, if necessary, in order that they may be fed liberally. If the flock is large, the smaller and weaker ones ought to be put in a flock by themselves, and receive all the wheat bran they will eat up clean, twice a day. If a little oil-cake meal is sprinkled on it in the troughs, better still. It goes without saying, that they ought to be housed from drenching rains, kept on highland pastures, limestone if possible, and not be allowed to go afield in the morning until the dew is well off the grass.

USE OF COPPERAS.—The summer and fall of 1882 will long be remembered by the flock-masters of Southern Ohio on account of the great mortality among lambs from paperskin. I addressed a large number of inquiries, orally and by letter, to the best shepherds of my acquaintance, and received many replies, from which I will condense the following general statement:

The great majority of them depended for protection on finely powdered copperas, kept in the salt, in varying proportions, from one-twenty-fifth up to one-fourth, according to the wetness of the season, the dampness of the ground and the liability of the flocks to the disease consequent on these conditions. *They aimed to keep it constantly in the salt for the first eighteen months of the sheep's life, or at least to keep them on salt and copperas two or three weeks, alternating with an equal period on clear salt, for the above named length of time.*

I am specially indebted for facts to Messrs. C. C. Smith, G. B. Quinn, W. F. Quinn, R. and A. T. Breckenridge, W. R. Stacy, J. Chadwick, W. M. Buchanan, L. W. Skipton, L. Leget, T. Fleming, C. S. Pugh.

My own experience with copperas has been highly favorable, since I have learned to *keep it constantly before the lambs until they have passed* their second summer.

Since the above date, Messrs. G. B. and Wm. F. Quinn have decided against the *constant* use of copperas, on the ground that it not only blackens and destroys the teeth, but creates a depraved appetite in the sheep which demands its continuance through life. The last named gentleman has abandoned the use of it, except as a temporary tonic and purgative. He gives it finely powdered, a pinch as large as a grain of corn, in a piece of apple; two or three doses like this; one on every other day. He claims that it will produce, in this manner, whatever good effects it is capable of producing.

Mr. W. B. Shaw, of Beverly, and Mr. E. J. Hiatt, of Chester Hill, use and recommend pumpkin seeds for paperskin. If pumpkins are ripe, they are split in halves and laid, flesh side up, in a feed-trough divided into compartments, each large enough to receive a pumpkin. If they are not ripe, a decoction is made by boiling the seeds. For tape-worm or other stomach or intestinal parasites, pumpkin seeds are undoubtedly efficacious; but it may be doubted if they would avail against the *strongylus.*

Dr. Cobbold attaches great importance to the inhalation of chlorine gas, in case the animal has not advanced so far in the disease as to be much weakened.

High feeding and elevated, dry pasture are of the first importance.

In conclusion, I will add that it is very important to break up the disease by expelling the parasites before winter sets in. The vitality and condition of the lamb ought to be fully restored before it is put on a regimen of dry feed, as then, even if the worms should be expelled, it is more difficult to build up the lamb than it would be on green feed. A paperskin lamb will frequently, if only slightly infested, linger through the winter in a feeble condition, and then perish in less than two weeks after it is turned to grass. The juices of the green grass, while weakening to the lamb, seem to impart to the lurking parasites within it new and greater vitality; a fresh installment of them are hatched out.

As a further preventive measure, no manure from a shed occupied by infested sheep should be scattered on a pasture occupied by lambs; nor should lambs be allowed to graze in a field lately containing others that were known to be the bearers of the *strongylus*.

"BUCK-FLY GRUB."—A correspondent of *The Shepherd's National Journal*, in Wayne County, Iowa, describes a "Buck-fly grub," which caused the loss of four valuable sheep of his flock. They had some of the symptoms of lung fever, such as quick breathing, frothing at the nostrils and grinding of the teeth. All had the same symptoms, and died within twelve hours. An examination revealed the lungs sound, but the windpipe was nearly full of a yellowish fluid and froth, in which were swimming eight or ten grubs, from one-third to two-thirds of an inch in length, and very much resembling the grub or larva of the *Œstrus ovis*. This was in the winter. An old hunter stated that he had found these grubs in the wind-pipe of the deer in winter, but never in the summer.

The same treatment would be indicated as for the *strongylus* or lung-worm.

CHAPTER XXVIII.

PARASITIC DISEASES, Continued.

LIVER-ROT.—The liver-rot, fluke, or "bane" has been known among sheep, and its ravages dreaded, from early times. There have been many theories as to the nature of the fatal disease; but of late years it has been shown that it is due to a parasitic animal which inhabits the bile ducts of the sheep's liver. For several years researches into the natural history of the pest have been carried on by Professor A. P. Thomas on behalf of the Royal Agricultural Society of England, and the completed investigations are given in part I. of the Journal of the Society for the year 1883. From this I condense the more important facts discovered and reproduce most of the engravings.

The fluke, figure 29, is a sucking worm somewhat resembling the common leech, of a flat, oval shape, pale brown or flesh-colored, and an inch to an inch and a third in length. Near the head is a sucker, y, by means of which the fluke attaches itself to the surface of the infested part. The mature fluke produces a large number of eggs, which are one-two-hundredths of an inch in length. In one case observed, seven million eggs were obtained from the gall-bladder of a single diseased sheep. Figure 30 shows the fluke's egg, much magnified, as it comes from the sheep's liver. It is an oval body, with a transparent shell, which allows the rounded masses of the contents to be seen. The eggs are carried with the bile into the intestine, and at length are voided with the droppings of the animal. If they fall upon wet ground, or are washed by rains into pools or streams, other changes occur. With the temperature at seventy-five to eighty degrees, an embryo forms in about two weeks. Figure 31 shows an egg with the embryo fully formed,

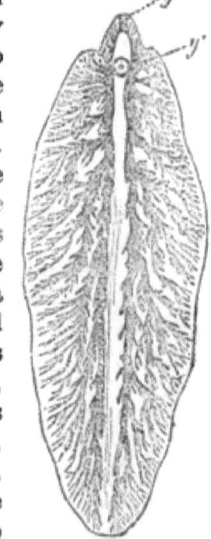

Fig. 29.
ADULT FLUKE.

and figure 32 represents the same when hatched, both highly magnified. The broader end is directed forward in swimming, and in its center is a peg-like projection which is used as a boring tool. When the embryo meets with any object, it feels about, and if not satisfied, darts off; but if the surface met is that of a certain kind of snail—*Limnœus truncatulus* (figure 33), it begins at once to bore into it. The young fluke spins around on itself like the handle of a cork-screw, by means of the many hair-like paddles covering its surface. The objective point is the snail's lung, in which the embryo fluke soon develops farther at the expense of the juices of the host. The form of the body of the embryo soon changes to an oval shape, shown in figure 39. This is distinguished as the first generation in the snail, and is termed the *Sporocyst*, which means a bag of germs.

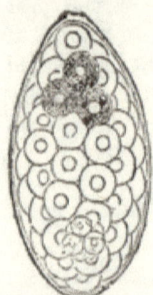    

Fig. 30.—EGG OF FLUKE.   Fig. 31.—EGG WITH EMBRYO.   Fig. 32. EMBRYO.   Fig. 33.—LIMNÆUS TRUNCATULUS.

These sporocysts grow rapidly, and develop offspring which are the second generation, and are called *Redia*. Figure 37 shows a magnified mature sporocyst, containing a number of redia. The largest one at the lower end is well developed, and will soon force its way through the walls of the parent—the wound healing up and the remaining germs continuing to grow. The redia are more active than the sporocysts and migrate from the lung to other organs of the snail. A full grown redia is shown in figure 36; it has a mouth and an intestine, and produces the third generation. The offspring of the redia are tad-pole shaped, and called the *Cercariæ*. It is this third generation of the snail parasite that is destined to enter the sheep and produce the liver fluke. The cercariæ, one of which is shown in figure 34, leave the snails, swim around for a time, and then become attached to and encysted upon grass stalks. These cysts remain dormant until picked up and swallowed by sheep feeding on the grass. The number of cercariæ descended from a single fluke egg is not less than two hundred, and under the most favorable conditions, over a thousand. A single live fluke may, through the medium of the alternation of generations above described, give rise to more than a hundred million descendants within a single season.

It is six weeks from the swallowing of the tadpole-like animal before the fluke becomes adult, and begins to produce eggs in the liver of the diseased sheep. It is seen that the fluke alternates between a particular snail and the sheep. The latter voids the eggs, and the developed embryos enter the snails, which, in turn, harbor them through three distinct forms, the last attaching itself to herbage, conveys the infection to the sheep.

Now that the life-history of the parasite is known, the conditions for its existence may be understood. Professor Thomas'

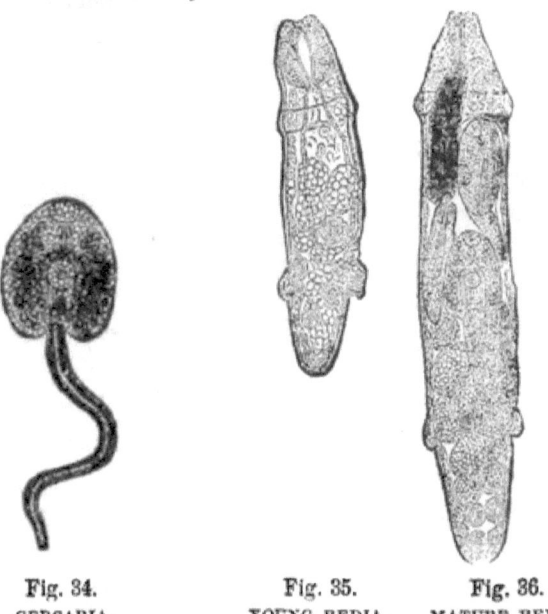

Fig. 34.   Fig. 35.   Fig. 36.
CERCARIA.   YOUNG REDIA.   MATURE REDIA.

summary is condensed as follows: For the production of liver-rot in sheep, there must be—1. Fluke eggs on the ground. 2. Wet ground or water during warm weather. 3. The snail, *Limnæus truncatulus*. 4. Sheep to feed on the ground infested by the fluke.

The eggs may be introduced in manure, in earth adhering to the feet of animals, or by running water, especially floods. Rabbits and hares are often infested with liver fluke, and may be a means of introducing it. The production of eggs is prevented by killing all sheep suffering from the disease. There is no cure for the fluke, as it inhabits an organ not easily reached by medicines.

SYMPTOMS OF ROT.— In the earlier stages of this terrible malady there is no way of discovering whether the parasite is present in the liver or not, unless one of them happens to be voided in the excrement. Indeed, in the beginning of its occupancy, it has a positively stimulating effect on the liver; it irritates and goads it into an increased activity, which causes the animal to improve in condition for a time. The shepherd often mistakes this improvement for a genuine gain in health and vigor.

Before death ensues, the animal falls into a deplorable condition. The belly becomes protuberant and pendulous, the spine

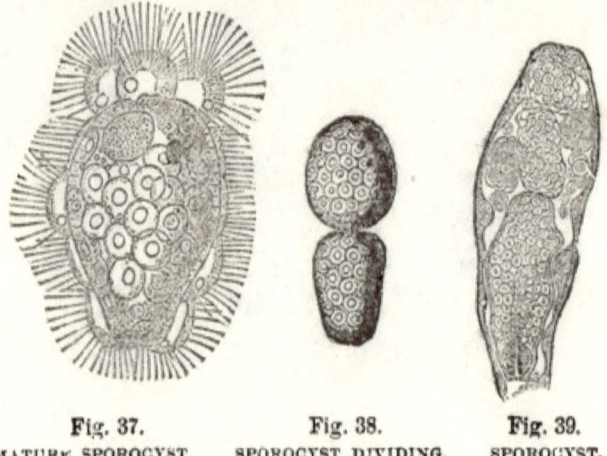

Fig. 37.   Fig. 38.   Fig. 39.
MATURE SPOROCYST.   SPOROCYST DIVIDING.   SPOROCYST.

is reduced to a condition which may be well termed "razor-back," the gait is feeble and tottering, the skin becomes dry and flaky and the wool begins to fall off, the subcutaneous tissues, especially under the chin, become swollen. Indeed, it is this dropsical excrescence under the lower jaw which has always, in my experience, constituted the one pathognomonic symptom of liver-rot as distinguished from the anæmia produced by the *Strongylus*. There is a rapidly increasing emaciation, frequently accompanied by scouring.

A *post-mortem* examination reveals a liver much swollen and discolored, and it is so rotten that the finger can be thrust into it anywhere, tearing it into pieces. The blood is nearly devoid of its coloring matter, and is reduced in volume; water abounds everywhere throughout the system, the flesh is pale and flabby, the skin waxen-white. If it is a ewe, the udder will become as white as paper. The heart is frequently surrounded by clear-looking water.

PREVENTION.—As with all parasitic disorders, prevention is highly important. Sulphur, eaten with the feed and passing thence into the blood, is distasteful to all parasites, internal and external. But common salt seems to be almost as valuable a preventive against the fluke as copperas is against the *Strongylus*. It is a fact which has been observed for more than a century, that sheep grazing on salt-marshes are exempt from the rot. Along the coast of the South Atlantic and Gulf States, otherwise so pestilential and malarious to men and animals, sheep running on the salt grasses are observed to be healthy.

If sheep are on a pasture of such a nature that the farmer is warranted in expecting rot, he ought to examine them carefully once a week. If any are found feverish, with the distended and rigidly rounded nostrils, indicative of hard breathing, eyelids and the sclerotic of the eyeballs yellow, breath hot to the hand, let him at once remove the flock to high pastures, give dry feed, and give each affected sheep two ounces of Epsom salts (one ounce to a lamb), and repeat, if necessary to physic the animal. Follow this with two grains of calomel mixed with one grain of opium daily, until the fever is wholly subdued. Plenty of common salt should be supplied all the while, and as soon as the use of the calomel is stopped, the salt should be given as a medicine—two ounces a day, with a dram each of ginger and ground gentian, mixed. This is the substance of Mr. Youatt's prescription, with the bleeding omitted, which is almost never advisable for an American Merino.

TAPE-WORMS IN SHEEP.—During the prevalence of the entozoic plague among the sheep of Southern Ohio in 1882, I received letters from a number of shepherds, in which occurred such expressions as these: "In one was a tape-worm five feet long, besides thirteen grubs in the head;" "I found in the intestines a flat worm about six feet long, jointed about as if you took a wheat straw, flattened it down, and marked it off with the thumb-nail every half inch," etc.

A sheep-owner in Hays County, Texas, states that during the summer of 1880, which was very wet, he lost forty-five per cent. of spring lambs from a cause not ascertained at the time, but which was attributed to wet. He writes, July, 1881, "that the summer has thus far been very dry," and that his lambs are dying a month earlier than in 1880. Upon examination he finds "in the small entrail, which is one hundred feet long, and runs from the bowels to the anus, one or more worms of great length, white, very soft, one-quarter of an inch wide and susceptible of parting at every eighth of an inch. I know no remedy, and the disease is very fatal. It is not identical with lombrez."

In the *American Sheep-Breeder*, May, 1885, M. C. Jackson, of Denver, Colorado, stated that he was losing about one hundred lambs a month from tape-worm. "Fat lambs go as quick as the poor ones. The first day they become stupid, second and third days they scour, and usually on the fourth day, die." He says a neighbor began the winter with fourteen hundred strong, healthy lambs, and had only three hundred and seventy left; all the rest died of tape-worm!

In Missouri it first appeared in 1875, when a correspondent of the *National Live Stock Journal* reported a falling off in his lambs, followed by mild diarrhœa. In less than two months from the time the disease made its first appearance, he lost half his flock, and the remainder were ailing. On examination after death the small intestines were found to be packed full of tape-worms.

In the spring of 1884 many flock-masters of Western New York and Pennsylvania reported to Mr. H. Stewart, the author of that excellent treatise, "The Shepherd's Manual," that there was a very prevalent and mysterious disease in their flocks, which some of them attributed to barley straw on which the sheep had been fed. As Mr. Stewart remarks, all bearded straw will sometimes produce disorder among ruminants, from the gathering of the sharp spines in the cellular coats of the paunch, yet this very seldom becomes a serious matter, and cannot possibly become a prevalent disease.

Cobbold states that only one species of *cestode* is known to be present in sheep in an adult condition; this is the *Tænia expansa*, or "long tape-worm." But there are two other tape-worms which occur in sheep in the larval or immature form (*Tænia marginata* and *Tænia echinococcus*). From this, it would appear that the worms seen by my correspondents must have been of still another species; for Cobbold states that the "long tape-worm" is thirty or forty feet long, and has been known to reach the enormous length of one hundred feet!

SYMPTOMS AND CURE.—Professor Stewart says, in an article in the *Country Gentleman*, "It is obvious that there is no cure by medicine, for no medicine could reach the creatures encysted in a diseased sheep." This is true only as it applies to those tape-worms which exist in the sheep in a larval stage, for the mature parasites, many yards in length, which have often been found in the entrails, are, to a large extent, susceptible to the action of what are denominated *tæniafuges* or tape-worm destroyers.

The symptoms usually developed by the presence of tape-worms are voracity of appetite, alternating with refusal of food, loss of condition, inclination to swallow earth, stone, sand or ashes; the passage of soft excrement, mixed with mucus, and evidence of internal pain. The only infallible indication of the presence of tape-worm is the occurrence of its sloughed-off joints or segments in the sheep's excrement. These, the flock-master, unless he made a careful examination, would probably mistake for the clots of mucus, such as are often voided by the sheep after it has contracted a cold. A close inspection reveals their annular structure.

It is usually lambs that have tape-worms; they pine away, though they may eat all the while; frequently have diarrhœa, or rather dysentery. After death, the white, many-jointed worms are found in the small intestines. As soon as the first death occurs, the shepherd should make thorough search and satisfy himself as to the cause. If there is tape-worm, the lambs should be fed liberally on mixed grains with salt, to keep up their strength. Some veterinarians recommend turpentine, mixed with linseed oil, as prescribed for paperskin in the foregoing chapter. But powdered areca-nut, or santonine, are considered better remedies for this parasite; a dram of either, given once a day, in a little milk or in grain feed every evening. After the flock is purified, two ounces of Glauber or Epsom salts may be given to each sheep (one ounce to a lamb) dry, or in water, if the sheep can be induced to drink it.

OTHER INTESTINAL WORMS.—Professor Simonds, of the Royal Agricultural Society of England, describes another kind of worm, which lays the foundation for diarrhœa, named the *Trichocephalus*, or hair-headed worm, which, though found in man, exists to a greater extent in sheep than in any other domesticated animal. These worms burrow their heads into the mucous membrane. The worm called *Sclerostoma* also exists in sheep in immense numbers, the two species being frequently found together. His method of getting rid of them is by the free use of common salt, and sulphate of iron. He gives the salt in quantities of a quarter to a half an ounce at a time; **and alternates it with sulphate of iron on every other day.** Half a dram of the latter is a full dose, even for a large sheep.

GID OR BLIND STAGGERS.—I have never seen a case of this, knowing it to be such, nor have I seen an American shepherd who has met with it. It was probably imported from England, and it seems to prevail chiefly in the Eastern States. Unwilling to believe, in my earlier experience, that a cause apparently so trivial as "grub in the head" could destroy an animal, I made many autopsies of sheep that I supposed had died of blind staggers, searching carefully through the entire encephalon for the bladder or cyst of this parasite; but I never found one. To-day I am satisfied that they all died from grub in the head, which I seldom failed to find, while the hydatid sought for could not be discovered.

Indeed, the only important difference between the outward symptoms of hydatid and grub is in the length of time which intervenes between the first manifestations and death. The grub does its fatal work much more quickly than the hydatid. The sheep infested with the latter ceases to eat, stands vacantly about with head held slightly up and frequently turned to one side; it seems to suffer from *hemiplegia* or paralysis of one side, as it is unable to ruminate or chew, and the mouthful of half-masticated grass lingers long in the mouth, while the green-colored saliva dribbles down the lower jaw. The animal begins to wander aimlessly about, seeking to rub alongside the fence or rack, as if for support; in the yard it has a tendency to go in a circle; blindness, total or partial, comes on; the wretched beast dashes in a sudden spurt against the fence; it goes down on its side in a tremor, rolling its eyes in agony; after recovering itself and falling a number of times it remains on the ground and dies in convulsions.

When the case is long drawn out, the bladder or tumor on the brain, by constant pressure on the skull, absorbs it to such a degree that a finger pressed on the spot discovers a soft place in the plate of the bone, or the latter even bulges out in a protuberance an inch or two deep and twice as wide. Twice I have seen this phenomenon in my own flocks, and, in rude fashion, lanced them, thereby saving the sheep.

LIFE HISTORY OF THE GID-HYDATID.—The veterinarians differ on some points, but they are agreed that the dog is responsible for the presence of this parasite in sheep, as it forms perhaps the best illustration of Steenstrup's scheme of alternate or dual generation. That is, two animals, the sheep and the dog are necessary for the complete development of this pest.

A dog infested with the gid **tape-worm** (*Tænia cœnurus*) passes through a field in which sheep are kept. His excrement will probably be **deposited near a tuft of grass**, or in the fence-corner; probably, also, it will contain one of the segments which the worm is continually throwing off, each of which is full of ripe eggs. Or, these eggs may have escaped from the segment within the dog's intestines, and, passing on out, may be **adhering** to the hairs about the anus. The dog enters a pond to cool **himself**, and some of these eggs are washed off, and the **sheep** when it goes to drink may swallow them, or may ingest them **as** it crops the grass upon or near which the dog's excrement **was** dropped, the eggs having been liberated by the rain and left adhering to the grass.

The eggs are hatched soon after they enter the sheep's stomach. From each egg there emerges a minute embryo, **oval or ovoid** in shape, having one end of its body armed with **sharp cutting** hooks. With these it bores its way into the **tissues of its host** in various directions. Some of them become **encysted in the** lungs, liver, mesentery, and even **in the uterus, each being en**closed in a bladder or tumor full of **yellowish** or greenish fluid. These are not further developed, but remain encysted, **perhaps** not materially injuring their bearer, unless their number is **very** great, in which case they sometimes produce death by the inflammation in the peritoneal membrane which they cause.

Others penetrate the spinal cord or the great nerve-trunks, where they produce obscure nervous troubles, or partial or total paralysis, and other incurable diseases. They either travel **along the** spinal cord, or enter the arteries and are borne along in the blood until they reach the brain. The substance of the brain seems to be specially adapted to them, and in it (there is generally only **one** in the brain at **a time, though there may be as many as four, on its upper surface, in some of its extremi**ties, or on its lower surface) they undergo a development different **from that** which they experience anywhere else in the body.

The embryo first surrounds itself with a wall formed of the brain tissue, **and in this it increases in size rapidly, casts its hooks and its outer wall, and develops into a rather thick skin.** There now appears in the center of the mass of cells which constituted the body of the embryo, a cavity which rapidly enlarges and becomes filled with liquid, until the organism consists of a simple spherical bag of about the size of a hazel-nut, filled with a milk-white fluid. The symptoms of gid disease—which first appear seventeen days after the hatching of the embryo—are now at their height. Upon the outer surface of the cyst, or "vesicular worm," there now appear at many points (from three hundred to one thousand) depressions, which gradually elongate out like the finger of a glove. At the bottom of these pockets hooklets now develop, and on their walls suckers, so that each depression becomes a perfect "head," *scolex*, of a future tape-worm. These "heads" are easily seen by the unaided eye. At this stage the sheep usually dies, and being of course useless, is generally thrown out where in all probability it will be devoured by dogs.—[R. W. SEISS, M. D.]

The immense number of these *scolices*, or heads, of the hydatid, on entering the system of the dog, would insure an equal number of gid tape-worms; but owing to the extreme difficulty of a dog's getting at and eating the brain, it is probable that only a very few enter his stomach. And, indeed, one is enough. Each *scolex* anchors itself in the stomach or intestines by its hooklets, and quickly develops into a tape-worm. The circuit is now complete. The tape-worm grows and forms hundreds of segments, each of which is capable, if necessary, of maintaining a separate existence; but they are usually cast off, one by one, into the outer world, full of eggs ready to do their deadly work in the sheep, as above described.

For some reason, not well explained, it is usually young sheep that are infested with these parasites, as with all others that take up their abode in the viscera.

THE OPERATION FOR STAGGERS.—I quote further from Dr. Seiss: "In the first stages of the disease the position of the tumor can only be decided upon by a thoroughly educated veterinary surgeon, but in the later stages, when the head bones have been softened and bulged forward, anyone accustomed to handling animals can locate the cyst by carefully feeling the head. In either case, the best mode of operating is as follows: With a sharp knife an incision, an inch in length, is made directly over the tumor down to the bone, which should be cleared just sufficiently to permit the passage of a small drill (one a fifth of an inch in diameter is quite large enough). The skull is carefully penetrated with the drill, great care being taken not to injure the membranes of the brain. The cyst is then punctured with the needle of a strong hypodermic syringe, its contents withdrawn, and the following fluid injected in its place:

Carbolic acid (pure crystals), forty grains; alcohol, one fluid ounce; mix, and inject half a teaspoonful; this is allowed to remain for a minute or two, and then in turn withdrawn. The wound should then be carefully washed, * * * left open, and daily cleansed with a gentle stream of pure, cold water, to each pint of which thirty grains of pure carbolic acid have been added, and kept covered with a compress, wet with the same fluid."

PREVENTION.—As with all these obscure parasitic diseases, prevention is a hundred times better than cure. All sheep which have died of staggers ought not to be thrown out for the dogs to devour, but should be buried deep, or burned. All strange dogs should be kept from running over the grazing grounds of the sheep, or bathing in the ponds where they drink; and if it is found necessary to keep shepherd dogs about, it is well to tie them up for a few days, occasionally, and free them from tape-worms by giving a few doses of some vermifuge. Professor H. Stewart recommends a half-teaspoonful of powdered areca-nut, mixed with butter, given three times a day, followed by a dose of castor oil, and continued for about four days.

"LUMBRIZ."—This is the name (probably of Spanish origin, and spelled in many ways), which, in Texas, is applied indiscriminately to the parasite and the disease produced by it. Mr. H. Chamberlain, of Nueces County, writes to the Department of Agriculture: "Lumbres, a complaint which up to 1868 had annually carried off many thousands of spring lambs, commencing in July or August, and operating upon them through the fall and winter, until the flock frequently became exhausted. This disease follows overflows, and a superabundance of rank grasses. It consists of something like a knot of long, small worms, resembling hair, in the stomach, the lungs invariably becoming infected; the outward symptoms resembling consumption in the human race."

Henry Bundy, writing from Atascosa County, Texas, to the *Shepherd's National Journal,* says: " Lombricze consists of little worms in the fourth stomach; they are about the color and size of a corn-silk and an inch to an inch and a half in length. They consume the chyle or nutritious parts of the food, so that the sheep or lamb simply starves to death, often living two or three months after becoming diseased. The symptoms are: A dropsical swelling under the chin, paleness of the eyes, lips, tongue, and, if a ram, pale around the roots of the horns; the entire skin is pale; the animal soon loses appetite, mopes around, is delicate about eating, but is always thirsty. The blood becomes pale, watery and of small quantity. It occurs principally in lambs, but sometimes in grown sheep, and the greater the proportion of Merino blood, the more susceptible are the sheep to the disease. There is no certain remedy known in this section."

A correspondent in Dewitt County, Texas, states, that by feeding a half bushel of corn per head through the winter, following a wet season, he checked the disease. The editor of the *Shepherd's National Journal* advises a trial of oil of sassafras, mixed with an equal amount of castor oil or of linseed oil.

"GRUB IN THE HEAD."—This does not require much notice in addition to what was given to it under the head of "General Management." When in summer, sheep are seen to stamp violently with the fore-feet, run with their heads held close to the ground, or huddled together with their heads under each other's bellies, they are trying to avoid the attacks of the sheep Gad-fly (*Œstrus ovis*). Many farmers entertain two erroneous notions concerning this pest, which seem to be very difficult to eradicate. The first is that the fly deposits eggs in the sheep's nostrils; the second is that the resultant worms or larvæ, after ascending the nostrils, penetrate the brain itself. The gad-fly is viviparous—that is, it deposits living worms in the nostrils. The farmer may convince himself of this by catching a fly and squeezing worms from the ovipositor. Dr. Randall's statement, repeated by Mr. Stewart, has perhaps done more than anything else to perpetuate the above mentioned error.

The larvæ, or worms, do not penetrate the brain itself, though many farmers would become angry if told so. The brain is separated from the nostrils and nasal sinuses, by a firm wall of bone, which these parasites never bore through. There are few shepherds who have not seen these "grubs;" they are occasionally found in the feed-trough, in the winter, having been expelled by the sneezing of the sheep when grain feed is given to it dry. The figures herewith presented are taken from Randall; they show the grub in its natural size. Figure 40 representing the half-grown worm, which is white, except two brown spots near the tail. Figure 41 represents it of full size, the rings being now dark brown. Each ring has darker spots, and below them are others. Figure 42 shows a full-grown larva on its back, the minute dots between the belly rings representing small red spines, the points of which turn backward.

The statements of Dr. Randall would leave the reader to infer—though he does not so state in terms—that the presence of the grub in the frontal sinuses, brings on a fatal crisis only late in the winter or early in the spring. This is true of sheep beyond the age of one year; but lambs, probably from the greater tenderness of their tissues, cannot carry these hostile guests so long. I have frequently known them to succumb as early as September, and that, too, with lambs yeaned in April! This is not to be wondered at, when the number of

Fig. 40.   Fig. 41.   Fig. 42.

THE "GRUB" OR LARVA OF THE GAD-FLY.

grubs in one lamb's head reaches as high as thirteen, as attested by my friend, Mr. J. Chadwick, a careful observer. We have only to consider how delicate and acutely sensitive is the membrane lining the nasal passages, purposely so made for the protection of the nose and the assistance of the sense of smell (to which end it consists largely of a ramification of the olfactory nerve), to understand the sudden and violent fatality following the apparently slight irritation produced by this parasite.

SYMPTOMS.—These have already been sufficiently detailed under the heading of "Staggers." Blood-streaked mucus is generally seen issuing from the nostrils when the grub is present, but not with hydatid on the brain. I may here repeat what was there stated, that a fatal termination may generally be expected in four or five days after the symptoms of grub begin to be manifest.

TREATMENT.—The natural and simple preventives are: plenty of dust for the sheep to lie in, and stamp in, long grass to graze, sheds for them to lie in during the middle of the day. Many American shepherds recommend tar smeared in the bottom of the salt-trough. Tar is alkaline, and is apt to take the hair off of the nose, besides rendering the hair follicles diseased and unproductive. The Scotch shepherds mix tar with whale oil; or, better still, apply the whale oil by itself, rubbing it over the nose frequently. They apply it toward evening, as the gad-fly is busiest just before sunset.

Another good preventive is an ointment made of: Beeswax, one pound; linseed oil, one pint; carbolic acid, four ounces; melt the wax and oil together, adding two ounces of common rosin to give body; then, as it is cooling, stir in the carbolic acid. This should be rubbed over the face and nose, once in two or three days, during July and August.

A canvass face-cover, smeared with this mixture or with one of asafœtida and tallow, may be hung in such fashion as not to interfere with the sight, or with grazing, and yet protect the lamb against the fly.

When the grubs have obtained a foot-hold, fumigation generally avails little. It is best to procure at the drug store (price about one dollar) an elastic bulb-syringe, with a small nozzle six inches long. Mix turpentine and linseed oil in equal parts. Accustom yourself to the action of the syringe so that you can gauge it accurately. Let the affected sheep stand before you in a natural position, and carefully probe the nostril with the nozzle until you find its bearing and depth (the nozzle will pass up a surprising distance—six inches in a grown sheep). Then charge the syringe, introduce it to the extremity of the nasal cavity, and with a quick pressure inject about a teaspoonful of the mixture. Withdraw at once and let the sheep recover somewhat from the effects of the shot, then treat the other nostril in the same way.

If the summer has been dry and hot, it is best to treat every lamb in this manner at weaning. Great care must be used not to strangle the lambs with the turpentine. Keep the mixture well shaken.

CHAPTER XXIX.

EXTERNAL PARASITES.

It is well known that the wool of sheep is inhabited by a considerable number of parasites peculiar to the animal, which seem to subsist on the yolk. M. Levoiturier, of Elbeuf, France, in a communication to the French Acclimation Society, stated the results of an examination as to the number of the *coleoptera* found in wools from different parts of the world. In Australian wool, he found forty-seven species of insects; that from the Cape of Good Hope, fifty-two; Buenos Ayres, thirty; Spain, sixteen; Russia, six. But these insects, while perhaps contributing something to what is known as "wool-sorter's disease," are not supposed to prey on the sheep itself.

THE SCAB INSECT (*Acarus scabici*).—This is the most universally distributed scourge of the sheep, though not the most fatal; perhaps nine-tenths of the area of the various countries principally occupied by the Merino are subject to its visitations. For, like the itch in man, the yellows in the peach, and many other diseases, the scab is caused by a parasite. That is, the actual scab, or crust, is formed from the exudation which comes from the puncture made by this insect. But there are, undoubtedly, predisposing conditions of the sheep itself, caused by climate, atmosphere, dust, heat, etc., which invite the attacks of the scab insect. Else, why does it prevail in one part of the United States and not in another?

The insect is hatched from an egg deposited by the female. One male insect suffices for the fecundation of several females; the latter are about twice as large as the former, but shorter-lived. The eggs are laid in the pores of the skin, or in the depressions of a scab already formed, vast numbers being laid by one female; the eggs hatch in three days, and the young insects at once bore into the skin, causing irritation and itching. Watery pustules arise where the insects entered; the sheep, in torment, scratches and bites itself, so that these pustules are broken, their contents mingle with blood and forms scabs or scales, which frequently cover large patches of the surface.

SYMPTOMS.—Sheep infested with the *Acarus*, first become restless; then, as the insects begin to bore, and the intolerable itching sets in, they rub against the sides of the shed, against posts, trees, etc.; they seize their wool with their teeth and tear it out in mouthfuls. The skin is whitened and thickened by the presence of the insects, and soon it becomes moistened by the yellow exudation from their punctures. The sheep continues to pull out its wool, making a desperate fight against its enemies if it is vigorous; presently a patch is stripped quite naked, while the infection continually spreads in all directions, as the parasites increase in numbers. Yellow and brown, or bloody scabs now cover the space they have devastated, while all around is an irregular ring where they are encroaching on the sound skin, and the wool is falling before the assaults of the sheep itself. There may be several of these centers of contagion and all increase at once, and if neglected they will ultimately meet and merge, covering the whole back of the sheep with a loathsome crust, if, indeed, the wretched creature has not already perished.

PATENT SHEEP-DIPS.—The live-stock journals of the West contain advertisements of a great number of proprietary compounds, recommended by their several owners, and by flockmasters who have experimented with them, as infallible cures for the scab. No doubt all of them have some virtue, and probably the majority of them, at the outset of their manufacture, were sufficiently efficacious; but soon cheap adulterations creep in, if not positively poisonous ingredients. Most of those who resort to them do so because, like the clothing sold on the ranches of the Far West, they are "ready-made," and therefore save some time and trouble, which would be required in compounding one's own prescriptions. But the shepherd who cares to do his work well had better stick to the home-made article composed of tobacco, or of sulphur and lime, as described below. Of course, this is more or less offensive to lambs whose mothers have been immersed in it; but the same objection lies equally against the patent dips, if they are strong enough to kill the *acarus*. If the novice has a friend who is a veteran shepherd, and who can unequivocally recommend some patent dip, from his own experience, that is well enough. If not, he had better use the home-made articles, even if they should be a little more expensive, rather than experiment with proprietary dips.

"HANDLING."—Mercurial ointment is sometimes applied by hand; a little wool is clipped off, and the ointment is well rubbed in around the scab indication. A pint of kerosene and a gallon of buttermilk, mixed and thoroughly rubbed in; or a quart of kerosene and a gallon of any cheap oil, say fish oil; or mercurial ointment, olive oil and a little turpentine; or sulphur, tar and lard, equal parts—all are effective if thoroughly worked into the wool. In California a "dope" of sulphur and linseed oil is applied with a swab or a "soap-root brush," and is found efficacious.

But with flocks of considerable size, or where the scab is making rapid headway, resort must be had to more wholesale treatment; either the "dipping" of Kansas or the "swimming" dip of California.

THE DIP.—Different formulas prevail in different States. In Texas and New Mexico the following is often used: Thirty pounds of tobacco, seven pounds of sulphur, three pounds of concentrated lye, dissolved in one hundred gallons of water.

In Nevada this is the formula: Sulphur, ten pounds; lime, twenty pounds; water, sixty gallons.

A California recipe is as follows: Sulphur, four pounds; lime, one pound; water, enough to make four gallons.

A recipe given in Kansas reads thus: Sulphur, twenty-two pounds; lime, seven pounds; water, one hundred gallons.

Sulphur and lime are probably the cheapest recipe, but the lime is apt to injure the staple; still, this recipe appears to prevail over all others in the scab-infested regions. The addition of arsenic is risky. The chief objections to tobacco are twofold; it is expensive, and if applied in the spring, when the wool is long, it stains the staple, and there is not time enough before shearing for the stain to wear off. Probably tobacco and sulphur form the best combination known for treatment of scab. To every hundred gallons of water there should be used thirty-five pounds of good, strong tobacco (if stems or other inferior parts are used, there should be more), and ten pounds of heavy sulphur. Flowers of sulphur should be used, and not ground brimstone, as the latter does not mix so well with the dip. This preparation, used at a temperature of one hundred and twenty degrees Fahrenheit, will kill all *acari*, ticks and lice, and leave the skin and wool in a healthy condition. But this, like all other dips, to insure thorough work, ought to be applied a second time, in ten days or two weeks, to destroy the *acari* that have been hatched out in the meantime.

DIPPING.—For flocks of a few thousand, and more particularly in the Rocky Mountain region and eastward, dipping is employed. I give a diagram and description of the appliances used by Mr. David Fox, of Wichita, Kansas:

APPLIANCES.—1, Figure 43, is the dipping-vat; 2, 2 are the boilers; 3, 3 is the dripper, divided into two compartments; one sheep-yard, with a small, three-corner pen, next to the dipping vat, which is of great convenience for catching sheep, all of which are shown in diagram. The vat is made of two-inch clear lumber, well braced and bolted together so that it is per-

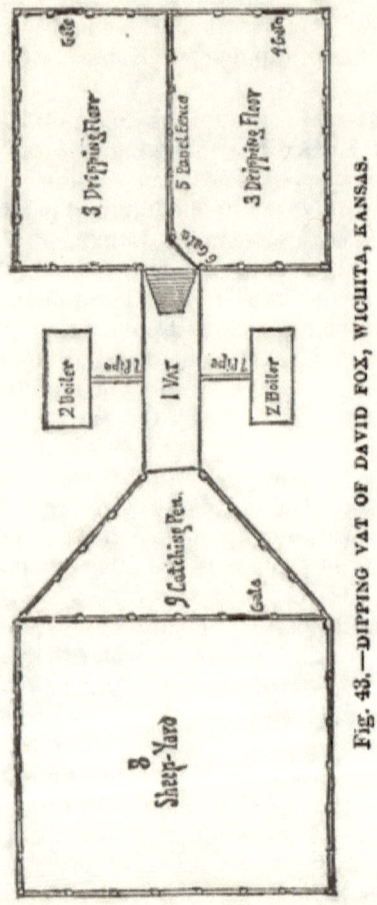

Fig. 43.—DIPPING VAT OF DAVID FOX, WICHITA, KANSAS.

fectly water-tight, sixteen feet long on top, twelve feet long at bottom, which gives four feet slope, with slats on the inside for the sheep to walk out of the vat to the dripping floor, six inches wide at the bottom on the inside, sixteen inches wide on top, four feet and a half in depth. Three and a half feet of dip is plenty to work with, but the vat should be deep enough to allow one foot above the dip to catch the splashing of the dip caused by the struggles of the sheep. There should be two bars across the vat, at the level of the dip, at equal distances, dividing it into three equal parts. At each division should be a good trusty man. His duties I give below. The vat is set into the ground two feet and a half, leaving two feet above ground. If the ground is conveniently located to lay an escape pipe to the bottom of the vat, it would be a great convenience for cleaning out the vat after dipping. This apparatus requires to be located where water is plenty. 2, 2 are the boilers, which are one on each side of the vat and about six feet from it. They are made of one and a half inch lumber for sides. I bought fourteen feet plank, sawed them in two in the middle; then cut a circle on each end for the ends of the boiler; then took sheet-iron thirty inches wide and eight feet long and nailed it solid to the plank, which makes the bottom of the boiler. Across the top nail three pieces of one by four lumber, at equal distances, to keep it from spreading. These boilers should be set on a furnace built up two feet from the ground with brick or stone. The space between the two sides of the furnace should be eight inches narrower than the boiler, giving four inches on each side for the boiler to rest on. The furnace should be open at each end, and a flue made of sheet-iron seven feet long, and one end made to fit either end of the furnace, so that it can be easily changed from one end to the other to correspond with the direction of the wind. This flue being seven feet high will conduct the smoke out of the way of men and sheep.

PENS FOR THE FLOCK.—3, 3 is the dipping floor, which is sixteen feet square, made of flooring well braced underneath with joist, and set up on a foundation high enough for the bottom of the dripper to rest on the vat. The foundation around the outside of the dripper should be built about three inches higher than the supports under the center, so as to spring the floor enough to make the dip run towards the center, with a strip across the two corners next to the vat, to conduct the drippings from the sheep into the vat. This dripping floor should be enclosed by fence. A panel fence (5) fourteen feet long is placed across the center of the dripping floor. Two of the bottom boards of the fence of the dripper on the side next to the vat should be cut out the width of the vat, and a small gate (6) fastened to the division panel, so that it can be swung to either side of the vat, that when one part of the dripper is filled with sheep this gate can be swung around, closing the pen that the sheep are in and leaving the other side open for the sheep to go into the other side. By the time this last half of the dripper is filled with sheep the first lot will be ready to go out, and continue in like manner until dipping is finished. 4, 4 are gates to let the sheep out of the dripper. 8 is the yard for the sheep before dipping is commenced. It should be built so as to make a small, three-cornered pen (9) next to the vat, large enough to hold fifty or seventy-five sheep, which would be handy to the vat and make it easy to catch the sheep.

NECESSITY OF THOROUGHNESS.—This yard should be made penitentiary tight and so strong that it is impossible for any sheep to escape undipped. Should a single sheep get out and mingle with those already dipped, unnoticed, that had a single living female *acarus* on it, it would in a short time infect the whole flock; hence, the importance of thoroughness from beginning to end.

THE PROCESS OF DIPPING.—Put the tobacco in gunny sacks and place them in the boiler, filled with cold water, and let them soak, at the least, twelve hours; then start the fire and bring the water to almost boiling. Then let it simmer for six hours so as to extract all the strength from the tobacco; this should be done the day before the dipping is commenced, as usually the boiling capacity is too small, and with the hurry to get done the strength does not get extracted from the tobacco. In moderate weather three-fourths of the dip required to commence work can be put in the dipping vat over night, and will still be very warm next morning. The boilers can be filled up again at the same time and a good fire left under them; then a good fire, started early next morning, will soon bring it to a boil. If run to the full capacity, which is about one thousand sheep per day, six good, strong men will be needed; one to attend fire and oversee the work and see that every man does his duty. This overseer should be the owner of the sheep or the one most interested. If the sheep are very scabby, two men should be stationed in the sheep pen with a curry-comb or stiff brush to thoroughly scratch and break up every scabby patch on the sheep, then put it into the vat head first. Now, the man that stands at the first division of the vat takes charge of the sheep and thoroughly rubs all the scabby spots, and moves it easily up and down in the dip in order that the dip can penetrate all wrinkles and folds. This man should occupy fully one minute with each sheep, then pass it under the cross-bar and on to the second man, who occupies an equal length of time in the same manner; then he pushes the sheep under the second bar and allows it to go out of the vat.

THE SWIMMING METHOD.—For very large flocks, and more particularly in California and other Pacific regions, this method is employed, requiring larger pens, vats and boilers. I give a diagram of those used by Mr. Charles Crane, of Millard County, Utah, with an elevation of the corral, figure 44.

A is the chute from a large corral; B is a sloping board over which the sheep in attempting to pass to decoy pen C, slide into tank D, which is generally twenty feet long, four feet six inches deep, and sixteen feet at the bottom, thus giving it a slope at the outlet of four feet, two feet wide at top and eight inches at bottom, thus compelling the sheep to swim in the middle of the tank. E is a board fastened in tank with cleats on it, to enable the sheep to obtain a foot-hold in walking out. F, F are draining pens (water tight), and sloping to sluice box in center which carries the dip again into tank D. C is a decoy pen containing a few sheep to entice the sheep in the chute. H is a pen into which to dodge sheep not required to be dipped. L is the dodge gate. This tank can be made of one inch pine boards, and lined with galvanized iron; No. 20 will do, which makes it water tight, and gives no footing to the sheep. K, K are pieces of two by six, twelve feet long, bolted lengthways of the tank and four feet from each end, and six inches from top of tank, leaving a twelve-inch space through which the sheep must put their heads, preventing those in the rear from riding those in front and thereby drowning them, at the same time keeping their backs under the dip. While in the tank, the scab can be broken up, teeth looked at, and, as the sheep pass out, branded. The draining pens, F, F, are regulated by a gate at E, and can be filled alternately, thus allowing one pen to drain while the other is being filled. An opening is made whereby the sheep escape from the draining floors, and can then be combed.

Many dispense with the sloping board, B, and use pen, H; particularly when the ewes are heavy with lamb, the sheep are dropped carefully into the tank, rump first. A boy is often placed behind the sloping board B, with a short stick, to push the sheep in as they pass over it. A man or two stands at the tank to regulate the passage of the sheep, examine teeth, break scab, brand, etc.; two thousand is a usual day's work, and that number can easily be passed through in ten hours.

308 THE AMERICAN MERINO

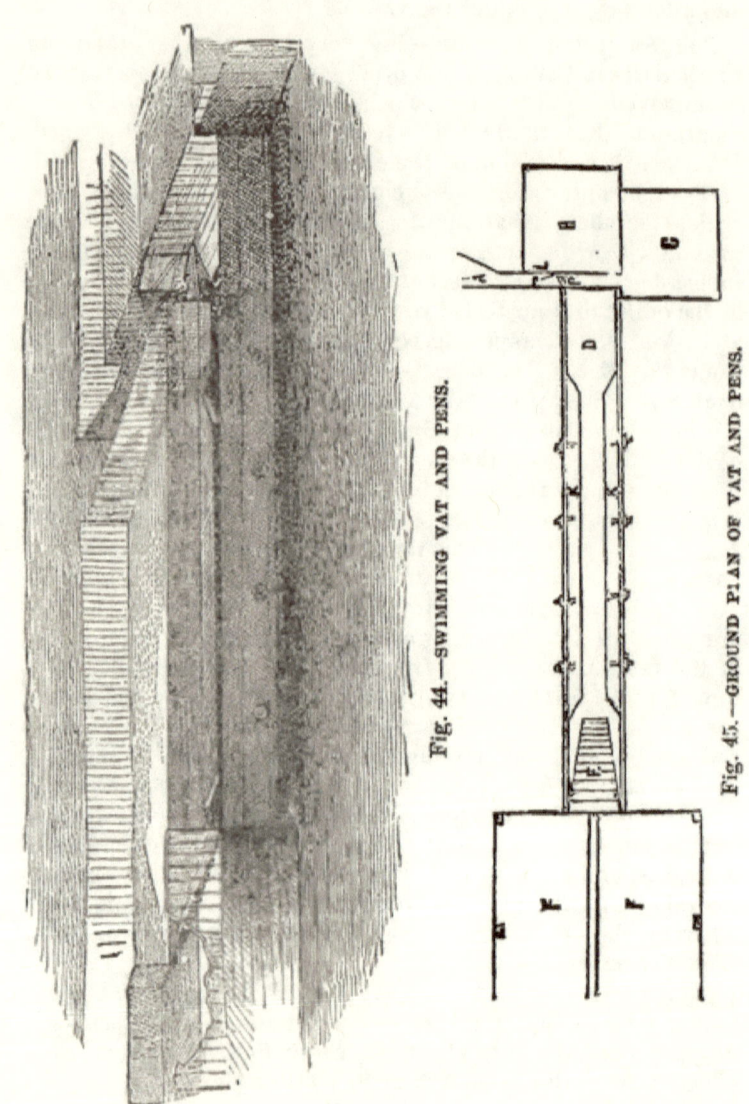

Fig. 44.—SWIMMING VAT AND PENS.

Fig. 45.—GROUND PLAN OF VAT AND PENS.

BOILERS AND VATS.—Two one-hundred gallon **boilers and** two four-hundred gallon store-vats (not shown in the plan) are **required** for a flock of two to four thousand **sheep when to-bacco is used.** Where the storage **capacity is limited** it is well to prepare beforehand a **very strong** infusion of tobacco, say one pound to the **gallon, and** dilute it when needed. The to-bacco should be **steeped in two waters to extract** all the strength.

In localities where convenient, and where **the** expense would be too great for one, several frequently join **and build a** corral and dipping-tank, each paying his proportion of the general ex-pense, according to the number of his sheep dipped. By a care-ful selection of ground expenses may be greatly **reduced.** The dipping station should be near a stream if possible. **A V-trough** carries the water to the boiler, **which, when** hot, is allowed **to** run into the steeping-tank through a hole in the bottom; **when steeped, another** plug is **drawn, and the** now **prepared dip runs into the tank.**

"SPOTTING."—The **whole flock ought to be** dipped **twice** within ten days—once before shearing **to remove the scab in-sects from** the wool, else the flock **will infect the shearing-floor, and** the contagion will be **perpetuated;** then immedi-ately **after shearing, when the skin is exposed, a second dipping** will ensure a **complete extirpation of the insects.** But if the flocks **are very badly infected, it is well to select** out the worst cases **and give them three or four** dippings, extra strong **and** hot. After the first dipping let the "spotted" flock be handled —that is, examined—and the scabs and scraps of wool removed, so that the second and subsequent baths will penetrate **to** every **lurking place of** the insects.

THE VAT.—This **should be at least sixteen** feet long, to keep the sheep in the water a minute or **so, and deep** enough to prevent it at all times from touching bottom. The vat should be tongued, grooved and pitched, and should be closely covered up when not in use, to prevent the sun from warping **and split-ting the boards.** If an inch gauge is placed on the **inside, and** the capacity of the vat ascertained **in gallons,** it may be known at any time at a glance how many gallons the vat contains.

DRAINING-YARDS.—These yards should hold **one** hundred sheep apiece, and be at least **two in number, to allow the** dipped sheep to drip thoroughly before they are turned out. Corru-gated iron, laid down in long sheets, or sheet-iron roofing, is the best material for the flooring of these yards. It should be laid down on sleepers close together, and the joints or seams protected from the trampling of the sheep by narrow strips of of wood nailed alongside. The same kind of strips nailed transversely on the incline will enable the sheep to come up out of the vat without slipping.

WHEN TO DIP.—In the spring the feed is better, the days are longer and warmer, the sheep are gaining in flesh, the sulphur is more abundant, the wool grows faster, and the sheep are thriving and lively; then is the time to annihilate the *acarus*. One gallon of dip applied then is worth two in the fall.

Never fail to dip the lambs; if the ewes have the scab the lambs will also, and if allowed to escape dipping will surely convey it to the flock; besides, it is extremely difficult to detect the scab in young lambs, particularly if short-wooled ones; dip previous to castration, and, if possible, ten days after, leaving no living *acarus* to propagate.

HOW OFTEN TO DIP.—The best flock-masters in the most scabby districts—which seem to be nearly coterminous with the alkali in the soil—practice dipping twice a year—once after each shearing. It not only cleans the sheep of the *acari*, but also of the ticks; and prevention, with both classes of parasites, is preeminently important.

FENCES.—In Texas great importance is attached by flock-masters to wire or other fences as a preventive of contagion. It is claimed that a three-wire fence will exclude all sheep from other ranges, and thus protect the shepherd who is disposed to keep his flocks clean from those who are careless in this regard. These fences would require extra strength on the side toward which sheep usually "drift" in northers or other severe winds, to resist the pressure of large flocks crowding against them. It may well be doubted, however, whether a single line of fence of any kind would wholly exclude the contagion, since it is known that the scab insect lingers for weeks on posts, stones or on the ground. Besides that, the well-known gregariousness of sheep would lead them to come into close juxtaposition on opposite sides of the fence, allowing the transit of the *acari* from one to another.

Sheep care less for the barbs than most other animals, and they will crowd through narrow spaces if they have to leave their wool on the barbs. The distance apart which will be necessary to exclude them securely will depend on the size of the sheep, the tension of the wires, and the attraction on the other side. Measure the size of the animals, and then place the wires decidedly nearer than these measurements would indicate. The accompanying engraving represents a successful fence employed on the farm of the late George Geddes. It was first made of three barb wires, placed at about the usual distances, the lower being about eight or ten inches from the earth bank, the next nearly a foot higher, and the third sixteen inches. The sheep crowded through the lower wires, and were not deterred by them. Two common, smooth and cheap wires were then added in the two lower spaces, and these, although insufficient in themselves, operated by crowding the heads of the animals against the neighboring barbs, and preventing further effort.

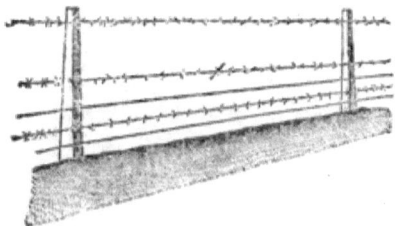

Fig. 46.—GEORGE GEDDES' SHEEP-FENCE.

The bank of earth, a foot high, made the fence a visible barrier, and this, with the shallow, open ditch on each side, deterred other animals from running or pressing against it. This fence may be modified by nailing a board near the bottom if the posts are near enough, but in the form shown in the figure the posts may be more remote and the cost less.

DESTROYING INFECTION.—When a sheep, from neglect, becomes wasted by its struggles to free itself, and by the fever and irritation produced by the scab insects, it is in a very poor condition for winter, and will probably succumb to the storms. In every State infested by scab there ought to be stringent laws for the inspection of sheep, by an official appointed for that purpose, for the protection of the community at large against the flocks of men too shiftless to cleanse them. All sheep which have died from scab ought to be buried at least two feet deep, or their graves covered with brush and stones, to keep away dogs; or their bodies should be burned or boiled down.

PREVENTION.—A liberal supply of sulphur in the salt in the summer (not in the winter—it is risky then), is efficacious in preventing scab. Flowing in the sheep's blood, sulphur is offensive to the *acarus*.

SCAB IN THE EAST.—Sometimes, though very rarely, a case of scab occurs in the Cismississippi, or other of the Southern States, where a resort to the wholesale methods of dipping practiced in the West would be impracticable. Where this is the case, the following remedy may be used: Take palm oil, one pound; lard, or beef suet, one pound; melt together and add soap, one pound; carbolic acid, eight ounces; potash, four ounces. Mix all together. Take a sufficient quantity that, added to one quart of boiling water, will make an ointment of the consistency of cream; part the wool of the sheep and rub the affected spots well over, twice a week.

TICKS AND LICE.—The sheep of the careless farmer is nearly always infested with the tick (*Melophagus ovis*), and, less frequently, with the louse (*Trichodectes ovis*). The tick is so well known as to need little detailed description. When filled plump with blood, its body is almost red; when depleted it is an evil-looking, grayish, round-bodied creature, about one-quarter of an inch long, and the body one-half that in width, covered with a very tough skin, which cracks audibly when squeezed to bursting. It is propagated by a "nit" (*puparium*), which is nearly round, brownish-red, and about as large as a radish seed. By means of a proboscis, as long as its head, it pierces the skin of the lamb and sucks the blood greedily, giving rise to the saying, "as full as a tick." When the number of them is large they make very serious inroads upon the vitality of the lamb, and it is a singular fact that they seem to thrive best on the blood of the lamb that is poorest.

Ticks seldom work the lambs any injury through the summer, or, at least, the abundant feed of that season enables lambs to resist their assaults, and keep in good condition; but as soon as they are confined to dry feed, and experience, as is generally the case, a slight falling-off, the ticks, in accordance with the almost universal law of parasitism, are stimulated into vigor, and begin to multiply rapidly. They do not cause so much restlessness as does the scab insect; probably their bite is not poisonous. But for this very reason, perhaps, they are even more to be dreaded than the irritating scab, *acari*, for the latter make the sheep demonstrative and it attracts attention to itself, while the abominable ticks, like leeches, pursue their silent, insidious work of sapping the sheep's life-current, without attracting the notice of the master until he is struck by the pallor, the debility, of the hapless creatures which are being literally eaten alive. The first thing the careless shepherd knows the lambs are so feeble they are not able to jump over the sill in the stable, and then, on catching them and parting their fleeces, he finds the inner ends of the fibers grown weak and spindling from deficient nutrition, and the whole interior defiled with the little black excrements of these disgusting vermin.

MEANS OF ERADICATION.—From this account it will be seen that it is very important to destroy the ticks before winter begins. The best time for this is soon after shearing; in a week or ten days the ticks on the older animals, which the shearing has enabled them to reach, will be so harried that they will either escape to the ground or take refuge on the lambs. And if these are now thoroughly dipped—though it is always best to repeat the operation in about ten days, to destroy any ticks which may remain or hatch out, for even the strongest infusion of tobacco will not kill them all—they should not have any further trouble from this source if properly cared for thereafter.

There are numerous remedies proposed for ticks, of which the favorite with the majority of shepherds has long been tobacco water.

When the weather is too cold for dipping, resort may be had to snuff or tobacco dust, which is cheaper; in fact, large dealers in tobacco always accumulate quantities of a gritty dust which is of no value to them, and which they will sell cheaply or give away. Ten pounds will suffice for one hundred sheep. Let an attendant lay a lamb on one side, on a box about two feet high, and hold it while the operator parts the fleece with his hands and sprinkles in the dust in four rows the entire length of the animal. Let one row be a few inches from the backbone, and the other about midway of the side; and on the other side of the sheep, two more in the same way. Let the dust be well worked down to reach the skin.

If tobacco water is preferred, refuse tobacco, or tobacco stems, may be bought cheap and will answer all purposes; four pounds of stems will make twenty gallons of dip. This amount of dip will suffice for fifteen lambs.

I have tried the arsenical dips and have been entirely satisfied with them. Dissolve three pounds of white arsenic in six or eight gallons of boiling water, and dilute with enough cold water to make about twenty-five gallons. Test the strength of it by immersing a few ticks in it; if strong enough, it will soon stiffen them. Two wash-tubs, or large iron kettles, will answer for a small flock; one to dip in, the other for the lamb to stand in while dripping. A suitable vat can be made by anyone with a modicum of ingenuity, the only point of importance being to provide a separate compartment or apron with a tight floor, inclined toward the vat, to carry the liquor running from the lamb's fleece back into the vat. This floor may be made of sheet-iron, painted with Venetian red; it may slope all one way, or be two-sided, sloping to the middle. Two men seize the lamb—one the forefeet, the other the hind feet—and lower it slowly into the dip, back downward, holding it in about a minute, until the wool is well saturated. It matters not if a little liquid enters the ears, but the lamb should be so handled as not to allow any to splash into the eyes; yet every lock of the fleece should be submerged. The operators must have whole hands, free from scratches; with this safeguard there is no danger in the arsenic water, if it is thrown into the fence corner after being used.

PREVENTION.—A judicious and timely use of that universal insecticide, sulphur, in the feed will save all this trouble. I have never had ticks on my lambs since I have employed the following preventive: If the lambs are "ticky" in the fall, mix sulphur in the salt, at the rate of three pounds of sulphur to five of salt, and give the lambs constant access to it, keeping them housed from the rains. It does not kill the ticks outright, but poisons and renders them nearly harmless, and if continued in the salt until shearing-time, the sulphur will drive them off without the dipping.

THE LOUSE.—The red sheep-louse is seldom found on the yolky Merino, preferring the dry-wooled breeds. Its head is red, the body pale yellow, marked with dark bands. It is found on the side of the sheep's neck and on the inside of the thighs and arms where the skin is bare. It requires the same treatment as the tick.

MAGGOTS.—These have been touched upon elsewhere, in so far as preventive measures are concerned. They are the larvæ of the blue-bottle or blow-fly (*Musca vomitoria*) and the flesh-fly (*Sarcophaga carnaria*). The Merino, on account of its peculiar qualities, the yolkiness and density of its fleece, which generate filth, and its wrinkles, which retain it, together with rain-water, urine, etc., creating sourness and stench between and under the folds, is more liable to the assaults of these **flies** and their progeny than the open-wooled breeds. Very wrinkly sheep are objectionable for this reason, among others; they require vigilant watching during the summer, especially if the weather is damp and muggy. But, wet or dry, very hot weather is propitious to the maggots with ewes, rams or ewe-lambs, which are more filthy than wethers. They ought to be caught and examined once or twice a week, and the folds about the vent drawn tight or doubled across the left hand, while the right, with a pair of shears, clips the wool with the greatest possible closeness to the skin beween the wrinkles. Then soft pine tar should be smeared very lightly over the surface wherever urine or excrement is likely to lodge. Though the daintiest of all our animals in its selection of feed, the sheep is the filthiest in its habits; its great fleece seems to render it indolent when once it has lain down, and its gormandizing propensities make it foul and stinking.

There is no part of the shepherd's work more odious than the fight with maggots; not one jot or tittle can he relax of his vigilance from the time warm weather sets in, until they are relieved of their fleeces. Maggots will sometimes make their appearance on the wethers, on the shoulders, almost anywhere on the fleece where a particle of filth has obtained lodgment in a wrinkle. If once they reach the skin, they begin to fret it away, a serous effusion begins, the wool adjacent soon becomes saturated and foul-smelling, and after that, even if the maggots are completely eradicated, it is almost impossible to save the sheep, as it is continually "fly-blown" afresh. There is no hope for it only in keeping it by day in a perfectly dark place, or enveloping it in a gunny-cloth blanket saturated with kerosene, benzine, or turpentine.

To remove or kill maggots there is probably no substance better than oil of sassafras mixed with alcohol—one part of the oil to four or five of alcohol. Turpentine is too severe on the sheep, applied to the raw, lacerated surfaces, already fevered by the worms.

THE "SCREW-WORM."—In a letter to myself, Dr. D. E. Salmon, Washington, D. C., says of this: "The fly of the screw-worm deposits eggs, but these hatch in a very short time—in fact, almost as soon as they are deposited; and it is thought by naturalists that they may at times be hatched at the instant they are deposited; but the fly is really classed as oviparous." It is the belief of Texan shepherds that the screw-worm is deposited alive, but it will be seen from the above that this view is not quite correct. At any rate it has the advantage of the common maggot, for the eggs which develop the latter, frequently dry up and do not hatch; but the screw-worm seldom fails to reach the sore or wound near which it is deposited. It is a great pest in hot summers, and is very persistent, sometimes being seen as late as December.

For the screw-worm, as for the common maggot, there is no better application than the oil of sassafras, diluted as above described.

SNAKE-BITES.—In Texas, California, and the Territories, a very considerable number of sheep are lost from the bites of poisonous snakes, especially the rattlesnake. It is related that a shepherd in Atascosa County, Texas, carried a long cane on which he cut a notch for every rattlesnake he killed; and that from January 1st to the end of the warm season he killed thirty-three. This is much better work than the poulticing or fomenting of a swollen leg. If the sheep recovers, the leg is likely to be permanently enlarged from the thickening of the areolar tissue. If taken in season, cupping will save the animal. The most effective treatment, however, is the prompt cutting out of the virus in a piece of flesh as large as a dime and half an inch thick.

CHAPTER XXX.

DISEASES OF THE FEET.

Thus far I have endeavored to present the diseases of the Merino, not with reference to the importance of the organs in which they occur, but approximately in the order of the frequency with which they attack the flocks.

Next after the parasitic affections which trouble the Merino, owing to its sweaty, foul pelage, its wrinkles, its gormandizing propensities, and its liability to the attacks of internal enemies, come diseases of the feet—not ranking high in fatality, but in the trouble which they cause the shepherd. The sub-tropical climate of Spain gave it its sweaty, greasy habit; and it was also the annual migration of eight hundred to one thousand miles from Andalusia to the north and return, kept up for one thousand, perhaps two thousand, years, which undoubtedly caused the hoofs of this breed of sheep to be so abnormally developed as to be the source of a large share of its misfortunes.

Mr. Stewart states, in "The Shepherd's Manual," that the sheep's hoofs do not grow from the coronet downwards, like a horse's hoofs, but from the whole inner secreting surface. The sheep's hoofs grow in length by a continual prolongation of the outer extremities, just like a horse's hoofs or the **human nails**; but if by any accident a hoof is destroyed, a new **one springs up** underneath, as Mr. Stewart correctly says, from the whole surface at once. The hoof proper is soft and thin, probably softer than that of the other domesticated ungulate animals; but, in certain cases, when there is an excessive development, it is very hard; **as** much harder than the horse's hoof **as** in its normal condition it is softer.

Mr. Stewart disposes of this subject rather too summarily; it is evident that his experience has been mostly with the English **breeds**, which are not so subject to the foot-rot as the Merino.

ORIGIN OF FOOT-ROT.—It is a prevalent Western opinion that the harmless fouls or scald-foot (sometimes called "ground-itch") will eventually terminate in the malignant, contagious **foot-rot**, if allowed to run its course. Some shepherds express themselves metaphorically, saying that the scald-foot is the "blossom" of foot-rot. There is probably a grain of truth **in** this belief; that is, there seems to be little doubt that fouls generally precedes foot-rot, if not always, **and** forms a predisposing condition thereto. I believe it is pretty generally conceded, by most students of animal pathology, that foot-rot is produced by a species of parasite. Now, why may it not be that the fetor of the scald is the source of attraction to this parasite; **and** that the maceration, **the soft** parboiled condition of the **cleft between the hoofs affords it** a congenial refuge? I might assert, without fear of successful contradiction, that the foot-rot never springs up absolutely *de novo*, from healthy tissues; that it never shows itself where scald has not been present before it. **And** I say that no one could hope to gainsay this, because no one can tell for the first few days whether rot or scald is present. **And does** not the hypothesis above broached serve to explain the latter fact? On some soils the scald will never, **after** any lapse of time, resolve itself into the rot, while in others it will; which is to say—if the above theory is correct—on these rot-proof soils there does not happen to be the parasite present to fasten upon the milder malady and convert it into the greater, while in the others **there is.** We are painfully aware that the pest of the human family known as the itch is fostered by filth and sores. Why may not this be the case with the ovine affliction under consideration?

There are certain animals which are so liable to the scald-foot that they have to be subjected to the shepherd's knife every month or so, else they will begin to go about gingerly. These ought to be weeded out of the flock, fattened and sold; they are a nuisance and an eye-sore.

WHAT FOOT-ROT IS.—There is so much misinformation on this subject that it is essential to define with accuracy and detail the symptoms of this disease. The farmer is often imposed upon by an unscrupulous dealer, with whom foot-rot is a sort of "trade capital," as scab is to the buyer of Territorial sheep, or alkali to the dealer in Territorial wools. I have known a simple-hearted old farmer submit to a dockage of twenty-seven cents a head on a flock of well-fattened Merino wethers, on account of alleged foot-rot, when there was no disease present whatever except the simple scald!

Perhaps it would be best to state briefly first, what foot-rot is not. It is not a disease of the interdigital canal. This is a duct which has its mouth about a finger's breadth above the cleft of the hoof, in front, and extends down and back toward the heel to a small sack which is doubled back, giving rise to the name employed by Youatt—"biflex canal." This duct secretes a whitish, viscid unguent for the lubrication of the inside surfaces of the segments of the hoof. When sheep run in a very damp pasture, this unguent, owing to the sluggishness with which the blood circulates, becomes thickened, and upon pressure of the foot being made it exudes in a ductile, vermicular string, which ignorant men pronounce "the seed of the foot-rot," or even "the rot itself!"

Neither does foot-rot in the Merino consist in "blisters" or boils between the segments of the hoof or about the heel. Boils sometimes occur on sheep's feet, it is true, and cause much soreness, fever and lameness; but if, when ripe, they are lanced and the pus thoroughly pressed out, they will give no further trouble, unless they are attacked by maggots.

Dr. Randall gives a very minute and, in the main correct, description of the true malady, which I will condense: "Foot-rot begins in the bridge or junction of the cleft, and its primary stage consists in a transformation of the skin from its normal smoothness, dryness and pink color to a whitish, parboiled and somewhat wrinkly condition, accompanied by the fetor common to this malady and the scald-foot. A thin, serous effusion sets in, which, as the disease advances in malignancy, assumes something more of a muco-purulent character. This corrosion of the tissue, accompanied with fever, proceeds downward until it reaches the line of junction of the skin and the horny walls of the inside of the hoofs, when it dives under the latter and attacks the body of the foot. Soon it completely invests the foot within the covering of the hoof, which it causes to cleave from the foot and hang only by the skin at the coronet, so that it can easily be wrenched off by the hand. The foot becomes a mass of hideous ulceration and is totally consumed, if, indeed, the sheep has not already perished miserably from the migration of the virus from the hoof to the side and its consequent invasion of the entire body, with its army of destroying maggots."

The most important error in Randall's unabridged account is the statement that "The offensive odor of the ulcerated feet, almost from the beginning of the disease, is so peculiar that it is strictly pathognomonic." The most experienced shepherd cannot, by the odor, detect any difference whatever between foot-rot and the common scald for the first week or ten days.

TREATMENT.—There is a large number of prescriptions for this malady, for it is so universally diffused east of the one hundredth meridian, that the ingenuity of many thousands of men has been brought to bear in combating it. Blue vitriol (sulphate of copper) is so easily obtainable, so cheap, and so efficacious, if rightly applied, that I do not deem it worth while to describe any other remedy. If any flock-master has ever applied blue vitriol, and afterward resorted to something else, that fact is strong presumptive evidence that he either did not sufficiently prepare the feet beforehand, or did not apply it with thoroughness.

First, if the disease has made such progress as to have passed under the horny shell of the hoof, it will be necessary to hunt it out thoroughly. The ulcerative matter may be so accumulated and hardened in the track of the malady as to prevent any remedy from reaching the real seat of the disease, where it is feeding on the fresh, healthy tissues under cover of the hoof. Hence, the knife must be employed to lay bare the virus where it is at work. All scraps and shells of horn, rendered useless by having been separated from the membranes which secrete them, should be cut away; also all remnants of the fleshy sole which the disease has killed. The only safe guide for the shepherd is to keep cutting off thin slices until there are very plain indications that the next stroke would draw blood; in other words, that healthy tissues are near at hand. If a little blood is drawn it should be stopped at once by an application of butter (chloride) of antimony; a flow of blood washes away the vitriol.

Second, the vitriol ought to be applied in such form that it will penetrate most readily to the seat of the contagion. Hence, it ought to be dissolved in water—a saturated solution, all it will dissolve—rather than in such viscid, gummy substances as red or white lead, tar, etc. Hence, too, the water when applied should be hot—as near the scalding-point as possible without taking off the hair or wool—say one hundred and thirty-five degrees Fahrenheit. A kettle ought to be kept boiling near by, from which hot liquor can be dipped into the bath when needed to raise the temperature.

For a foot-bath let a box be made six feet long, two feet wide, one foot deep, water-tight. Let it be placed about a foot from the wall, with a framework or fence at each end and one across it at intervals of fifteen inches. This will afford each sheep a standing-place two feet long and fifteen inches wide. Let a slat be nailed on top of the box, lengthways; this will pass under the flanks of each sheep, forward of its hind legs, as it stands in its place, and will prevent it from getting its hind-quarters down into the vitriol-water, which would stain the wool. The opposite side of the box will sustain the breast of each sheep and keep its fore-quarters out of the water. It requires one man to attend the sheep in the box and keep the solution hot. Some vitriol ought to be added occasionally to keep up the strength. Two men will be required to clean the sheep's feet with wet rags and pare away the diseased and dead matter.

This brings us to another point of the highest importance, which is, to make the vitriol solution stay where it is applied until it does its work. Hence, the hoofs should be as clean as possible from dung and dirt before the application is made, and be kept out of water for a day or two afterward. The knife must be applied thoroughly—yet not so as to cause a troublesome effusion of blood—to lay bare the disease in all its hiding-places, cutting away the hoof and the gristly integuments wherever any virus may possibly lurk beneath. To this end any measure which will fetch the sheep's feet much in the water for a day or two previous to the operation not only cleanses them, but softens the hoof, which is an important matter, since after some hours' soaking, the pocket-knife will readily pare away a hoof which when indurated by several days of dry weather will yield only to the chisel and mallet. As the operation generally has to be performed in summer, it is well to keep the flock on dry feed a day or two beforehand, so that the dung underfoot may not be so diffusive when the time comes for operating. If they can be kept standing on wet straw their hoofs will be soaking in the meantime. Then, if driven through high wet grass, the feet will be partly washed, and the cleansing can be completed with a swab in a tub of water. After the paring has been done, let the sheep stand fifteen or twenty minutes in a shallow vitriol foot-bath, say two inches in depth, strong and hot, as above described, and kept hot by the occasional removal of some of the liquid and the replacing of it with some freshly heated. After leaving the bath the sheep should be confined on a dry, hard floor for one or two days, where, if they have been previously kept on dry feed for a short space, the manure on the floor will not seriously abate the effects of the vitriol on the feet. It may be necessary to repeat the operation in two weeks. It is a great preventive and mollifier of foot-rot, to drive the sheep often over plowed ground or a dusty road; this serves as a disinfectant. Forty-eight hours after the sheep leave the foot-bath they may go into the dust with advantage.

FOR SMALL FLOCKS.—Where there are only a few sheep the vitriol may be dissolved by another formula and applied with a horse-hair brush, which the farmer can make for himself. This formula is as follows: One ounce of verdigris, two ounces of sulphate of potash, three ounces of blue vitriol, four ounces of nitric acid, four ounces of rain water. Mix in a glass bottle with a glass stopper. The horse hair for the brush should be tied on the handle with a woolen thread.

SCALD-FOOT (OR FOULS).—This is a slight galling or maceration in the cleft of the foot, generally produced by wet grass or dung. It is liable to trouble the Merino at all ages and all seasons of the year. In a wet spring, even sucking lambs will go around with the greatest apparent distress (it usually attacks the fore-feet first and most severely), stiff-legged, as if they were rheumatic, and spending nearly all their time lying down, rising only to suck and to follow painfully after the ewes. It will generally cure itself after a week or two, but if it lingers longer than this it ought to receive attention, especially if the sheep are fattening, as it undoubtedly pulls them down somewhat, and prevents them from feeding full. It is often caused by malformation of the hoofs—"bug-horned;" that is, touching only at the extreme points; turned under at the edges; thick and "clumped," etc.—which produces chafing, and needs only the knife to pare away or shorten the hoofs. A little finely powdered blue vitriol sprinkled in the cleft, well down into the bottom of it, is generally the only application needed.

As stated under the preceding head, this does not differ from foot-rot in its early stages, so far as the unassisted senses of the shepherd can discover. I never could detect the slightest difference for the first ten days at least. But on some soils, notably clay soils, there seems to be no doubt that the scald will eventually terminate in the foot-rot; while on others—as loamy, alluvial and limestone soils—I know, from many years' experience, that scald-foot will never, under the greatest neglect, become or lead to anything else. Foot-rot is contagious, and will soon run through the whole flock; but the simple scald will linger for months on such soils, confined to a single member of the flock, even though it may become so bad as to cause the miserable animal to graze on its knees. Sometimes maggots make their appearance on the scald-foot, and, if neglected, they will ultimately reduce the foot to a complete ruin (except that the hoofs do not come loose, as with foot-rot); and still there is no genuine rot, and never would be. But the popular ignorance and dread of the greater malady are so wide-spread that every lameness is at once pronounced "foot-rot;" and for appearance's sake, if not on the score of humanity, the scald-foot ought to be treated promptly, especially in little lambs and in fattening sheep.

INFLAMMATION OF THE INTERDIGITAL CANAL.—Mr. Stewart mentions that this is sometimes caused by sheep traveling on very sandy or dusty roads, as a result of which, dirt enters the canal and produces inflammation of the whole foot. I have seen it caused by sheep running in very wet, soft clay pastures; the feet were kept so constantly chilled that the circulation was retarded and the unguent secreted in the canal was not expelled. This retention brought on irritation. Probing with a small wire or the trimmed point of a feather, to remove the offending substance (if any is present); or fomentation of the foot with hot water and vinegar (if the trouble is a result of wet pastures) are all that is needed. The general treatment and the means of prevention are obvious.

"CANKER OF THE FOOT."—This, Mr. Stewart describes as ' a very obstinate disease," but I have never seen a case of it. He says: "It consists of inflammation of the sole of the foot, which gives way to a growth of spongy sprouts instead of the natural hoof, and a discharge of white curdy matter, which has a most offensive odor." It causes separation between the hoof and the fleshy or cartilaginous sole of the foot. From this and other statements made by Mr. Stewart, I am very strongly of the opinion that he has in mind a case of what in Ohio would be called the genuine foot-rot and treated accordingly. He recommends thorough cleansing, bathing in a solution of one dram of chloride of zinc in a pint of water, and a pledget of tow or lint dipped in a mixture of one part of common (not fuming) nitric acid with three parts of water and applied to the whole of the cankered surface. Repeat until cured.

FOOT-AND-MOUTH DISEASE.—This is rare in American flocks; it is very contagious, and may be communicated by cattle, hogs or sheep. Mr. Stewart thus describes it: "The first symptoms are a fit of shivering, succeeded by fever, cough, and an increased pulse. This is succeeded by a failing of the appetite, tenderness over the loins, flow of saliva from the mouth, and grinding of the jaws. Blisters, small and large, appear on the mouth and tongue, which break and become raw, causing great pain. The feet are swollen and also covered with blisters, which break and become sore, causing the animal to walk with difficulty and shake its feet or kick and lie down persistently." There is a simple form of it, which usually terminates favorably in ten or fifteen days, under a dose of two ounces of Epsom salts and a little ginger, once administered, a second dose being dangerous. The mouth should be washed with alum water (one ounce of alum in one quart of water), and the feet with soapsuds or a weak solution of sulphate of copper, then dressed with carbolic ointment and bound up in a cloth. If the malignant form is present, it is best to kill and bury the diseased animals and fumigate the quarters with sulphur fumes, removing the healthy sheep for a time.

CHAPTER XXXI.

DISEASES OF THE RESPIRATORY SYSTEM.

CATARRH.—This is known to most farmers as "snotty nose," and is called by some veterinarians "coryza." It is, perhaps, the most prevalent affection of the Merino in all parts of the United States—in the East, caused by dampness and cold, in the West, by alkali dust. It may appear paradoxical, but it is true that the very covering given to the sheep for protection, its fleece, which *is* a protection in a state of nature, becomes, under the capricious management of man and the artificial conditions of the sheep's life, the medium of its most common disease. When the wild Indians had no clothing to speak of, they had almost no consumption or catarrh; but when they donned the dress of civilization—on one day, off the next—they began to die of the galloping consumption. If sheep had no fleeces at all they would be less liable to catarrh; as it is, they are more liable to it than the other domestic animals which have no covering but short hair.

Dampness is the most pernicious enemy the sheep has; any degree of dry cold is not to be mentioned in comparison. Next to dampness, perhaps, is a warm and steamy sheep-house; it makes the sheep more tender and susceptible to cold when it goes out-doors. The best possible form of sheep-house is that which allows the wind freely at all times—unless in a hurricane or a driving rain—to blow through it, but high enough to be quite above the sheep's heads.

The fleece is a protection to the sheep when perfectly quiet, but let it become overheated from any cause, either by running or by crowding in a close stable, and it becomes a source of damage. A man can adapt the amount of his clothing to the weather; the sheep wears the same thick and heavy garment whether it is cold or warm. If this garment becomes wet to the skin it cannot change it, as the master would his overcoat; it clings close to the skin until it dries off. It is a cold, wet blanket; it is worse than nothing; worse than the shortest coat of hair would be, as a steady wear.

The Merino ought to be kept always dry in the winter, or else not housed at all. Let the air it breathes be dry and pure; above all, let the bedding and footing be dry; then there will be little trouble from the catarrh. It is the greatest folly to protect (?) sheep too much.

Catarrh is an inflammation of the mucous membrane lining the nostrils, throat, wind-pipe and nasal cavities. The inflamed condition causes an excessive secretion of mucus, which produces irritation and coughing.

TREATMENT.—In the first place, the causes must be removed; then the sheep should receive some warm mucilaginous drink, as slippery-elm or linseed tea, or a warm bran mash, with a little stimulant added, say a half-teaspoonful of ginger or gentian. A lump of pine tar, as large as a hazelnut, smeared on the root of the tongue is beneficial. Keep the nose clean by washing with warm water in which there are a few drops of *aqua ammonia*. If there is fever and the disease is likely to assume the more violent form described below, there may be given to the sheep the following: Epsom or Glauber salts, half an ounce; saltpeter, one dram; ground ginger, one dram; mix with molasses and place on the root of the tongue with a paddle. Hold the sheep's head up and jaws closed until the dose is swallowed.

COUGH.—If the catarrh is neglected, or additional cold is contracted, the cough will become worse—so bad as to be the most marked symptom of the case, showing that bronchitis has set in. In other words, the inflammation is going down toward the lungs, and there is danger. The sheep's appetite begins to fail, there is perceptible quickening of the pulse, denoting fever.

For this there may be administered: Linseed oil, one ounce; saltpeter, one dram; powdered gentian, one dram. Give in the same dose as above, in the same way, gradually reducing the amount of saltpeter one half.

OZŒNA.—Sometimes the catarrh assumes the form of a chronic ulcer in the nose, discharging constantly a whitish, fetid matter for months, without cough or fever.

Take sulphate of iron, four ounces; sulphur, two ounces; catechu, one ounce; mix in four pounds of oat-meal, with two tablespoonfuls of common salt. Give a heaping teaspoonful, mixed with molasses, to each sheep twice a day. Keep the sheep in the stable until cured.

INFLUENZA.—Sometimes sheep have a persistent running at the nose and eyes for a long time, accompanied generally with a cough. Once in a while, there is a case where the watery discharge from the eyes seems to be poisonous, making an inflamed streak down each cheek. There is a copious discharge of mucus, which is occasionally stringy and colored with blood. The deaths which occur are among the best-conditioned members of the flock rather than the poorest.

Where these conditions obtain, the following is the treatment: Take thymo-cresol, one ounce and a half; mix with one gallon of soft water; give each sheep a wineglassful twice a day, and sprinkle them with the same. Take the affected sheep into a tight room, and fumigate them by heating a gallon of the above mixture. Or the following may be given: Powdered rhubarb, three ounces; chlorate of potash, four ounces; nitrate of potash, three ounces; bicarbonate of soda, six ounces; cream of tartar, four ounces; sulphur, four ounces; mix; give a teaspoonful two or three times a day in enough molasses to make a paste.

PNEUMONIA (SPORADIC PLEURO-PNEUMONIA). — *Pleuro-pneumonia contagiosa*, is essentially a bovine disease and not infectious to other animals. It can be communicated to sheep only by inoculation. Sporadic pleuro-pneumonia may occur in all the domestic animals; but the veterinarian who asserts that the contagious pleuro-pneumonia ever attacks sheep is in error.

This is essentially the same as inflammation of the lungs in the human subject, and is consequently a very dangerous disease—usually the result of culpable carelessness on the part of the master in exposing the sheep to cold storms, especially soon after shearing. It may be induced by any sudden and violent change from hot to cold, even when the sheep are still protected by their fleeces. The symptoms are, a very high pulse and hot, quick breath, with the nostrils expanded, thin and tense; a short, hacking cough: grinding of the teeth; great restlessness. Generally considerable urine is voided.

The *post-mortem* examination reveals substantially the same pathological condition as that of a man who has died of pneumonia; the lungs in the stage of exudation or red hepatization, being like the liver; the windpipe full of false membranes, the air vesicles almost plugged up with them; the lobules of the lungs perfectly consolidated and separated from each other by lighter streaks of reddish-yellow lymph, occupying the interlobular areolar tissue. Lungs almost rotten, and so heavy that they will sink in water; the surface of them specked with dark clots. About twenty-four hours before death (which in this disease comes speedily), the sheep sometimes scour, voiding a greenish substance.

TREATMENT.—If the sheep is small and the wool short, it will be greatly relieved by poulticing on the breast over the heart with Indian meal mush as hot as the hand can bear, renewed every ten minutes for an hour. The poultice should be a foot square and an inch or more thick. It will be troublesome to apply, but is one of the very best remedies, and sometimes this is the only thing that will save the sheep. Bleeding from the jugular vein may also be employed. Directly after the bleeding, give two ounces of Epsom salts, in warm water, from a long-necked bottle or a horn. Injections of warm water are also beneficial. As soon as purging has taken place freely, let the following be given: Arsenic, five grains; powdered muriate of iron, two drams; gentian, one dram; in oat-meal or linseed gruel. Or the following: Powdered digitalis, one scruple; nitrate of potash, one dram; tartar emetic, one scruple; twice a day in gruel.

As soon as the sheep is out of danger the process of restoring the waste caused by the loss of blood and the fever must be begun. A pint of well-cooked gruel (it should be strained if not made free from lumps), to which is added half a dram of powdered gentian or ginger, may be given four or five times a day. As with all diseases of the lungs and throat, cold water should be freely supplied, and the mucus collecting in the nostrils should be sponged away.

PLEURISY.—This is an inflammation of the pleura or membrane covering the lungs and lining the inside of the thorax or lung cavity. It is produced by the same causes as pneumonia, and frequently accompanies that disease.

The symptoms are about the same as for pneumonia, but there is greater pain, the sheep sometimes moaning in agony. Pleurisy may be distinguished from pneumonia, however, by the rattling or gurgling of the lungs when the breath is thrown out; this is caused by the serous effusion in the lung cavity or thorax. After death this cavity is found full of water, the lungs are covered with livid patches like flakes of bran, but their substance is not impaired unless pneumonia was also present.

The prescription of digitalis and potash (one scruple of digitalis, one dram of nitrate of potash) may be given, with two drams of spirits of nitre substituted for the tartar emetic, in linseed gruel, twice a day for four or five days. Or this: Give five drops of tincture of aconite in a tablespoonful of cold water three times a day.

After a sheep has had an attack of pleurisy and recovered, there is apt to be an adhesion of the lungs to the sides of the chest, and this will forever after prevent it from thriving. If possible, it ought to be fattened and sold to the butcher.

PREVENTION.—See general remarks preceding the paragraph on pneumonia. All sudden changes ought to be avoided with sheep; even a change from poor to rich feed ought to be made gradually. Sheep that are shorn very early in the season for market, if it is cold, should be crowded close together at night in a small room to keep them warm. Yet all sheep will endure the loss of their fleeces better before they have become debilitated by the heat of the advancing season than they would later —that is, they will withstand a greater relative change in temperature.

Lambs frequently contract some disease of the throat or lungs, when following the ewes on windy days in April, after they have been turned on grass. Not being occupied in grazing and having no exercise, they stand doubled up, or curl down on the ground in the most protected situations they can find, and thus contract violent colds. If the ewes cannot be driven to some hillside, protected from the wind, they had better stay in the sheep-house through the day, fed on bran and clover, even if they go a little hungry, rather than expose the lambs to the danger of protracted naps on the cold, damp ground.

CHAPTER XXXII.

DISEASES OF THE ALIMENTARY SYSTEM.

CONSTIPATION OR "STRETCHES."—Early spring or winter lambs often suffer from costiveness if kept wholly on dry feed. The excrement becomes very dry and hard, it is voided with difficulty, the lamb sometimes bleats with pain, fever of the bowels comes on, and death will ensue if relief is not afforded. A half-ounce dose of Epsom salts in water ought to be given to the lamb every six hours, until the natural consistency of the fæces is established; then give the ewe and lamb wheat bran. Dry bran is always better than a bran mash, because the sheep being compelled to eat it more slowly, it is better mingled with saliva, and will be more readily and more completely digested. Apples, roots, refuse potatoes, corn-stalks or leaves, clover hay, or similar laxative food will prevent the trouble; if the ewe is kept healthy the lamb will not suffer.

When sheep are brought up in the fall and confined in winter quarters on dry feed, they are liable to colic or "stretches" unless great pains are taken to make the transition gradually.

The shepherd seldom perceives that the sheep is ailing until the trouble (which is simple costiveness at first) has sufficiently advanced to assume the form above designated. The sheep is observed to stand still and neglect its feed; it looks round at its sides; lies down and rises; then stretches itself very much as if it had just risen from a sleep, with the back rounded down. Occasionally it has violent colic, lies on the ground, struggles, and frets away the earth with its feet in a semicircle.

If nothing else is convenient, a handful of common salt will often suffice to relieve the sheep. Let it be turned upon its back and a tablespoonful at a time dropped well down upon the root of the tongue, the jaws being held until the salt is swallowed. Salt and water would be better.

Two ounces of Epsom salts or a half-teacupful of raw linseed oil will serve the same purpose; give the salts in water, from a long wine-bottle or a cow's-horn. Or give a teaspoonful of sulphur mixed with molasses or lard and placed on the root of the tongue. If the sheep is valuable and seems to be in great distress, it is well to reduce the risks as much as possible by giving an injection of warm water. It may be necessary to repeat it several times before the engorged bowels are fully relieved. The hind-quarters should be held up almost perpendicular to allow the enema to work down and dissolve the hardened fæces.

When sheep are brought up in the fall and turned on dry feed, if they have free access to salt and sulphur (one part sulphur to four of salt), it will assist in preventing constipation. But the better course is, if practicable, to give the flock a few hours' run every day, for a week, on rowen, rye, turnip-tops, or some other green feed, reserved for this occasion, to break the suddenness of the change. Two ounces of linseed meal, or a pint of wheat bran per day, will answer the same purpose very well. Corn-fodder is better than hay of any kind, except clover, on which to make the transition from grass to dry feed.

DIARRHEA OR "SCOURS."—Sheep which take the weather as it comes through the winter, running out every day, and are well fed, seldom scour when green grass comes, or at any time. But those which have been housed more or less, unless they have received roots or apples all winter, or at least for several weeks preceding the change, will show a greater or less percentage of fouled posteriors soon after they are let loose in the spring. Lambs are more liable to scours than grown sheep; the liability to this trouble steadily diminishes with age, until the teeth are broken, when it begins to increase. Very rank grass in a wet season, the pasture on low, sour lands, frozen clover in the fall, frozen turnip-tops, weeds which have grown watery in the shade of corn, are among the causes that produce diarrhea.

I have seldom failed to arrest diarrhea with dry wheat bran; indeed, bran is the sheet-anchor of successful American sheep-husbandry. It is not desirable to stop the scours too suddenly; it is nature's method of expelling from the intestines something which is offensive to them. An animal in poor condition is more subject to the scours than one which is robust—the bowels will brook less strain. If the discharge continues beyond a day or two, it ought to be checked, for it will then begin to interfere with nutrition and may terminate in the much worse disease, dysentery, which is a species of blood-disorder.

The sheep affected with diarrhea should be separated from the flock and kept in a lot with little green feed in it, and fed on bran until the looseness of the bowels is corrected. If the disease is persistent and mucus is voided, give a tablespoonful of castor-oil (two to a grown sheep), to remove any matter which may be irritating to the bowels; then follow this up in three or four hours with two teaspoonfuls of a strong decoction of oak bark or blackberry root, with half a teaspoonful of prepared chalk or baking soda, morning and evening.

The above remedies are simple and easily prepared, and are generally all that is required. It is well for the shepherd to keep in stock the following mixture or cordial, prescribed by Mr. Stewart: Prepared chalk, one ounce; catechu, four drams; ginger, two drams; opium, one dram and a half; to be mixed with half a pint of peppermint water and bottled for use. When needed, shake well, and give a lamb a tablespoonful twice a day; a grown sheep twice as much.

Diarrhea is generally a sign of weakness and poverty; in-bred lambs are subject to it. It is one of the numerous indications of faulty management. In some wet seasons it is so prevalent as to become almost enzoötic. Exposure to storms, by weakening the vitality of the animal, assists in bringing it on. In such seasons, lambs ought to be kept in condition by grain-feeding and judicious housing. The shepherd's good sense will tell him what kinds of vegetation are to be avoided; if not, he will discover by observation.

This being a disease of weakness, there is special need of stimulation. Hence, to the above purgative, as always when this class of medicine is given, there should be added a teaspoonful of ginger or gentian.

TYMPANITIS OR HOVEN.—This is an unnatural distension of the paunch or rumen with gas, caused by a diseased condition of that organ or by a too rapid swallowing of very green, succulent feed—as clover, rape, turnips, etc. It is not so common with sheep as with cattle, yet the shepherd will sometimes find a ewe (suckling ewes being especially liable to it, owing to their constant strong appetite and voracity), to all appearances healthy enough before, dead and much bloated, with a greenish fluid flowing from the mouth. I have lost a number of ewes from this disease (or, rather, this chemical process which usurped the place of digestion), on a field of red clover (which is worse than white clover), though they were on it constantly, and the soil was a very dry, sandy plain. They would rise at daylight with a voracious appetite, eat greedily of the clover while it was wet with dew, and be dead before noon. They had free access to salt.

After sheep have been confined to dry feed all winter, especially if they have been too closely housed or in a damp stable, there is frequently a diseased condition of the rumen, which is manifested by a lessened secretion of the alkaline gastric juices. Consequently there is a failure of digestion; the ingested grass, instead of being seized upon and submitted to the digestive processes, begins to undergo the simple chemical process of turning acid. This condition points at once to an alkaline remedy. A teaspoonful of water of ammonia in a pint of water may be given from the bottle or the horn.

The puncture of the paunch with a penknife or with a trocar and canula, is a desperate resort, and may be needed to save the life of the sheep, but should not be employed until other measures have been tried. A small stick placed in the mouth like a bridle-bit and the ends tied over the head will sometimes cause the animal to belch, especially if the paunch is kneaded. But hoven is more speedy with a sheep than a cow, suffocation comes sooner, and there must not be too much delay in resorting to the puncture.

The knife is to be inserted on the left side of the spine, close to it, and half way between the last rib and the hip-bone, because at that point the paunch is suspended by adhesion to the wall of the abdomen. If the sheep has previously been laid on its right side the discharge of gas will be unchecked. The puncture will generally heal of itself, but it is better to take a few stitches, as described in the paragraph on Worms.

Sheep fed on buckwheat and cotton-seed are more subject to hoven than those fed on hay, ground oats and bran. Give linseed oil, three ounces; turpentine, one dram. Then the following may be given twice a day: Linseed oil, six ounces; nitre, one ounce; mix, and add glycerine, four ounces; chloral hydrate, two drams; mix. Dose, one tablespoonful.

If nothing else is at hand, and the case does not yet call for the puncture, a tablespoonful of soft soap, diluted with half a teacup of water, may be given by means of a bottle; or a teaspoonful of carbonate of soda (commercial or washing soda), or a teaspoonful of chloride of lime in a half-pint of water. A tablespoonful of common salt, dissolved in a little water, will be beneficial.

Sometimes, when a puncture has been made, it closes before entire relief is given; it may be necessary to reopen it with a penknife or insert a quill. After the effects of the hoven have passed away the animal should receive stimulating food, with a little ginger or gentian added.

POISONING.—There are many plants or herbs in different localities of the United States which are popularly supposed to be poisonous to stock—Buckeye (the young shoots and the nuts), Laurel (narrow and broad-leaved), St. John's Wort, Indian Pea (*Phaca Nuttallii*), Tarweed, Red Baneberry (*Actæa rubra*) and others.

First, I will present an extract from the United States *Agricultural Report* respecting the so-called "Loco-weeds" or "crazy-weeds" of the Far West: "A considerable number of plants has been received. Those most frequently complained of have been *Oxytropis Lamberti*, *Astragalus mollissimus*, and *Sophora sericea*. In addition, there have also been mentioned, and some samples also have been obtained of, *Oxytropis multiflorus*, *Oxytropis deflexa*, *Malvastrum coccineum*, and *Corydalis aurea*, variety *occidentalis*.

"The reports from various correspondents and from widely-separated regions agree closely as to the injurious and frequently fatal effect upon animals of eating these 'loco-weeds.'

"The habit of eating these weeds seems to be formed because of the scarcity, at certain seasons, of nutritious grasses. All or nearly all of these plants, except *Oxytropis*, have a bitter, disagreeable taste, yet after the habit has once been formed the animals reject the sweetest grasses. Among the symptoms first noticed are loss of flesh, general lassitude, and impaired vision; later the animal's mind seems to be affected; it becomes often vicious and unmanageable, and flesh and strength are both raidly lost. When approaching some small object it will often leap into the air as though to clear a high fence. Frequently in these paroxysms horses have died from falling backward.

"The time required for these weeds to kill animals varies greatly, some dying within three or four days, others lingering for a year or longer. Some correspondents state that horses seem more susceptible to the influence of these plants than either cattle or sheep; others report that all are affected similarly.

"There is some difference of opinion as to the real cause of the diseases commonly attributed to 'loco.' Some think that the animals suffer not so much from direct poisoning as from lack of nutritive food and water. Mention is made of buttermilk as an antidote, but it seems not to have proved valuable."

The first thing to be done in case of poisoning is to remove the offending matter from the stomach. To effect this give two ounces of Epsom salts in a pint of warm water, and, if the animal is very valuable, injections of soap suds may be administered. After this it remains to counteract the depressing effect of the poison upon the nervous system by means of nervine stimulants. From my experience with common green tea, I would recommend a trial of it in any case where poisoning is suspected.

BUCKEYE AND LAUREL.—The symptoms of poisoning in general are reeling or staggering, frothing at the mouth, grinding of the teeth, rolling-up of the eyes, nervous twitching of the muscles of the neck and the legs. When a sheep has eaten Buckeyes there is a partial paralysis of the muscles of the legs —that is, the animal is wholly unable to stand, and lies helpless on its side, fretting away the earth in a semicircle with its legs; but all the other functions seem to remain; the sheep notices everything about it, bleats for its lamb, eats greedily when feed is offered, twitches its ears to drive off flies, makes an effort to leap up and run when a dog comes about, etc., etc. I have found common tea an almost unfailing remedy for both Buckeye and Laurel. It will afford more speedy relief from the latter than the former; with the Buckeye the doses may have to be continued two or three days. The infusion should be made strong—say three heaping teaspoonfuls of the best green tea in a pint of water, to be boiled (not merely steeped) twenty minutes. This will make one dose; it may be given in the morning, and another like it in the evening.

SORE MOUTH.—This is sometimes thought to be caused by St. John's Wort (*Hypericum perforatum*) in the hay, sometimes by any dry feed in the winter. Not only will the mouths be sore, but there will be heavy scabs at the corner of the mouth, on the lips and face, extending even up to the nose. Two or three applications of copperas water, or of iodine ointment or carbolic ointment will generally cure these sores. To remove any irritating cause which may exist in the bowels, give two ounces of Epsom salts, or a teaspoonful of cream tartar or of sulphur, mixed in molasses and laid on the root of the tongue.

INFLAMMATION OF THE BOWELS (*Enteritis*).—This is called "braxy" by the English shepherds. It is not common in the United States; I have seen it but once. I copy from Mr. Stewart's description: "The first symptoms are weeping and redness of the eyes, weakness and staggering, loss of appetite and rumination, inaction of the bowels, swelling of the flanks, high fever and difficult breathing; a puckered-up appearance of the mouth and nostrils, which gives a peculiar woe-begone and pained expression to the face; a tight skin and rapid emaciation. After death the stomach is found filled with putrid food and distended with gas; the bowels are gangrenous and in a state of decomposition; the liver is partly decomposed and filled with degenerated bile; the spleen is gorged with blood, softened, enlarged, and not unfrequently ruptured, ulcerated, and and exhibiting a seriously diseased condition."

In addition to this it may be mentioned that the bowels are sometimes marked with yellowish spots, as if stained with bile. Again, the intestines and lungs are full of blood; the gall bladder very large and full of liquid bile. This is a cause of sudden death to lambs sometimes, and is classed by shepherds loosely under the indefinite designation of "lamb cholera."

The cause must be sought in conditions of soil and water and climate. Low, sour lands, very hot sunshine and bad water seem to be the principal causes. Plenty of salt with wood ashes in it, at the rate of one part of ashes to four of salt, is often beneficial. Or the sheep may be dosed once a day with two tablespoonfuls of salt to which have been added a twentieth part each of copperas and powdered ginger; the whole dissolved in a half-pint of water. To try a change of pasture and water, with protection from the hot sun, would be the obvious dictate of prudence, when the disease is prevalent and quickly fatal.

English sheep brought from Canada to the sweltering, dry heat of an Ohio midsummer wilt down quickly and perish rapidly from a disease which appears to be enteritis, and which the shepherds call "black rot."

FOREIGN SUBSTANCES IN THE STOMACH.—Sheep sometimes exhibit a depraved appetite, as, for instance, a ewe will nibble at the end of her lamb's tail, biting off joint after joint until it is eaten nearly or quite to the root. At other times they will pull out their own or each other's wool in little locks, and swallow them; or they will swallow mouthfuls of earth, or gnaw rotten wood or other substances not called for by a healthy appetite.

It is seldom that these abnormal manifestations appear when the flocks are on pasture; they generally occur in the winter, from which it may be inferred that they are caused by something injurious taken into the stomach. I have found in the second stomach of a lamb clots of sand, curd, hay and other substances held together with wool. The lamb had swallowed the wool when sucking; it had never been properly shorn off the ewe's udder at the tagging season in spring. Grown sheep eat wool when nibbling in their fleeces to rid themselves of ticks, lice, etc.

These extraneous matters remain in the stomach because it is unable to digest them; they derange its action and produce irritation. The sheep seem to be led by instinct to swallow earth, sand, rotten wood, as a purge. They act strangely, lose their appetite, mope long periods in silence, turn up the upper lip, thrust out the nose, throw back the ears, etc.

Their instinct points in the right direction; they require a purge. Give the usual dose of salts heretofore mentioned as a purgative, in thin corn-meal or oat-meal gruel.

On this subject Mr. Stewart says: "In eating hay or other dry fodder, foreign matters, such as nails, pieces of wire or glass will sometimes find their way into the stomachs." I cannot conceive of a Merino taking its feed in such coarse, ravenous fashion as to be able to accomplish this feat!

CHOKING.—It is risky to allow sheep to run in an orchard where they are liable to swallow small, hard apples or clingstone peaches; or to feed on turnips or other roots cut into pieces. They will also sometimes swallow dry shelled corn so fast and with so little mastication as to become choked, though in this case there is little danger of a fatal result. But with fruit or roots there is liable to be a permanent stoppage which must be removed. When a sheep is first choked it dashes violently about, shaking its head and striking into the air with its forefeet. After awhile it stands with its head down, breathing hard and with saliva running from the mouth; the stomach is apt to become swollen with gas or with air swallowed in the effort to free its throat.

If the object cannot be worked up to the top of the gullet by the thumb and fingers pressing on the outside of the neck, it will have to be pushed on down into the stomach. For this purpose employ a small, flexible rod, like a ramrod, well oiled and with a little ball of tow wrapped smooth and tight around the end to prevent it from lacerating the gullet. Draw the sheep's head and neck out as nearly straight as possible, introduce the rod and carefully feel for the obstruction. When found, if it does not give way readily, let a few gentle taps be given to the upper end of the rod with a hammer.

CONGESTION OF THE LIVER.—The causes which are said by Mr. Stewart to produce this disease have never developed a case within my personal knowledge; but they have, instead, led to a congestion of the brain—in other words, apoplexy. The indications of a congested liver are obvious, though not always plainly manifest: Constipation, moping, a yellow tinge in the eyes. An ounce of Epsom salts and three grains of calomel may be given, mixed in molasses and laid on the tongue, every morning until the tone of the liver is restored. The sheep must not be allowed to drink much while taking calomel.

INFLAMMATION OF THE LIVER.—This, too, I have never seen; I do not think it occurs often among American Merinos. The symptoms are thus described by Stewart: "The system becomes fevered; the nose and mouth hot and dry; the breath fetid; the ears cold; the eyes pale and glassy; the pulse is irregular; breathing is slow, and the expirations short and sudden; the dung is dry, hard, black, and glazed with a greasy, yellowish-green mucus; the urine is highly colored, scanty, hot, and smells disagreeably."

Give, twice a day, in a warm, linseed infusion, the following: Sulphate of potash, two drams; calomel, five grains; powdered opium, one grain. Warm injections may also be given. When the animal begins to mend, give light and easily-digested feed, as wheat bran, crushed oats, bright clover hay, and in the drinking-water drop a few teaspoonfuls of vinegar or ten drops of aromatic sulphuric acid.

DYSENTERY.—I am not certain that I have ever seen a case of genuine dysentery, as there is some confusion respecting this and diarrhea. Randall says: "The stools are as thin or even thinner than in diarrhea, but much more slimy and sticky;" while Stewart and Dadd state that in dysentery the stools are generally hard, scanty and mixed with mucus and blood. Dadd gives several rules by which they may be distinguished, of which I quote a part:

"1st. Diarrhea most frequently attacks weak animals; whereas dysentery ofttimes attacks animals in good condition.

"2d. Dysentery generally attacks sheep in the hot months; on the other hand, diarrhea terminates at the commencement of the hot season.

"3d. In diarrhea there are scarcely any feverish symptoms and no straining before evacuation, as in dysentery.

*   *   *   *   *

"6th. In dysentery the appetite is totally gone; in diarrhea it is generally better than usual.

"7th. Diarrhea is not contagious; dysentery is supposed to be highly so."

It should be stated that the contagion can be imparted only by the dung, but it is better always to remove an affected sheep from the flock to enable better treatment to be given. The whole flock ought to be looked to, and probably some change made as to pasture or water, and some better provision of shade be made. The treatment required is, first, a dose of oil (two tablespoonsful of castor or linseed, with thirty drops of laudanum added), in thin, warm, strained gruel made of Graham flour. The gruel may be continued for several days as a nourishment and emollient to the bowels, with a teaspoonful of laudanum added once a day, the gruel to be given in half-pint or pint doses, according to the size of the sheep, three or four times a day. The quantity of laudanum may be diminished daily as improvement takes place.

HEMORRHOIDS.—A considerable part of the rectum sometimes

becomes everted or turned inside out, protruding as a red and inflamed tube to the length of several inches. Occasionally it will be in the form of a lump as large as a hen's egg or larger. The parts are to be carefully cleansed in warm water and treated daily with the following ointment: Powdered nutgalls, two ounces; oxide of zinc ointment, six ounces; mix. It may be necessary to put the sheep in a dark place by day to protect it from flies. After a few applications of the above, an effort should be made to return the everted intestine to its proper position, with the hind parts raised above the head. Mark the sheep to be drafted from the flock, and get rid of it.

CHAPTER XXXIII.

BLOOD DISEASES.

CONGENITAL RHEUMATISM.—Acute rheumatism, in the form generally seen in lambs, has been sufficiently treated already. But constitutional rheumatism is a far more serious and complicated ailment, and is one of that list of diseases which speak strongly of mismanagement, of neglect or penuriousness on the part of the owner. The shepherd is sometimes puzzled at the appearance in his lambs of a disease commonly called "joint-ail." The joints of the legs, especially the knees, are hot and swollen, and the lamb goes about gingerly or stiffly; there is a chalky secretion about the joints; they are sore to the touch. This is one of the manifestations of congenital rheumatism, and in all probability the lambs are indebted for it to the ram, perhaps, sometimes, to the mothers.

This kind of rheumatism lurks in the serous membrane; hence, besides the coverings of the joints, we may find it in the tendons and ligaments, and in the membranes which cover the heart, lungs, spinal marrow, bones, muscles and brain. It often passes rapidly from one of these parts to another, from one joint to another. The animal affected with this disease is uneasy and yet unable to walk naturally; if a ewe, she neglects her lamb; the appetite is irregular, sometimes accompanied by "loss of the cud;" the dung is hard, the urine hot, high-colored and scanty.

This disease is sometimes a result of the weakened vitality caused by in-breeding; it is in general a disease of poverty and penury. A rheumatic ram or ewe is likely to transmit to the lamb a susceptibility to rheumatism; it is a blood disease. An overworked ram is liable to it. A half-starved flock of lambs may show symptoms of it in the spring, especially if they have been exposed to cold storms. If a ram is turned into a large flock of ewes and allowed to run with them unchecked, his lambs the following spring may be expected to be rheumatic.

Tonic and stimulating treatment is indicated, with protection from storms. Give linseed gruel, bran or crushed oats. Begin with a mild purge—say two ounces of Epsom salts and one dram of ginger. Then follow, twice a day, with: Sulphate of potash, two drams; sulphuric acid, twenty drops; to be given in a half-pint of warm water.

A pint of strong tea of pennyroyal and sassafras is often beneficial in this disease.

ANTHRAX FEVER OR "MURRAIN."—This terrible malady has appeared sporadically in the Western States, where the rich, rank herbage and the sweltering heat of the sun, offer favoring conditions; it is sometimes complicated with inflammation of the bowels, and is known locally as "Black Rot." It is related to the Texas fever or Splenic Apoplexy, to the "*Loodiana*" of India, the "Horse Sickness" of South Africa, malignant sore throat, etc. The herbivora and birds are more especially its subjects. It mostly affects young stock in the Western States, and appears either in the spring or in the early fall, particularly when there has been a drought and the recurrence of rains produce rank vegetation. It may appear as a result of a too sudden change from poor to rich pastures, or *vice versa;* or from dry highlands to low ground.

The sheep gorge themselves on the watery grass in unaccustomed plenty; the stomach and bowels are overtaxed, and unable to carry forward the work of digestion; they become overloaded with a mass of half-rotten matter, which, instead of affording nutrition to the blood, loads it with poison; and the flaming heat of the sun completes the mischief by fevering the foul, black blood. The eyes are red, the mouth and tongue inflamed and blistered; the flanks and quarters are swollen, and the skin is disorganized, so that the wool comes out at a slight pull. It is chiefly the thriving stock that is attacked; the old and the poor escape,

After death the internal organs rapidly decay; the body, when opened, is found full of black blood, and there are large, black patches just beneath the skin. If left a few hours untouched the body becomes enormously swollen, so that the legs stand out horizontally, the tongue protrudes, and a quantity of dark mucus or slime is slowly discharged from the nostrils. The flesh and fluids of the body are highly poisonous, hence flies may spread the disease.

The healthy sheep ought to be removed at once from the diseased, and receive different feed, say crushed oats or bran, for a few days, with a limited allowance of short, sweet grass growing on a limestone soil, or on healthy, rolling lands, with shade and pure water. In the early stage of the disease the following may be found efficacious: Carbolic acid, one ounce; bicarbonate of soda, four ounces; water, two quarts; mix. Give a tablespoonful three times a day. Or the following may be tried: Sulphate of soda, two ounces; sulphur, one ounce; powdered myrrh, one scruple. To be given in oat-meal or any other gruel. In six hours follow this latter prescription with a teaspoonful of spirits of nitrous ether (sweet spirits of nitre) in a pint of water. Burn the dead or bury them deeply and securely.

"PELT-ROT."—There is a diseased condition of the system which manifests itself by a loss of wool, generally at first around the hind-legs, then on the sides, and so up to the backbone. The wool peels off clean; there are no scales or sores, but the skin looks reddish, probably from cold or sunburn. It occurs most frequently in ewes. It may be the result of puerperal fever, or of over-feeding with corn, or there may not be any assignable cause, as it will sometimes occur with sheep in good condition. Treatment: Bathe the sheep with a solution of saltpetre, one ounce to the quart of water.

PLETHORA.—This is not properly a disease; it is simply an excess of blood, though it may, all the same, result fatally. Sheep which have been fed to a very high condition and kept with a very limited amount of exercise, or none at all, are most subject to plethora. It is evinced by a flushed and fevered condition in general, distended nostrils, labored breathing, bloodshot eyes, etc. The remedies are too obvious to require anything more than simple mention, such as bleeding (from the facial vein is the best), purging with salts or castor-oil, reduced feed and more exercise, given gradually.

But the most serious effects of plethora generally manifest

themselves in the lambs of over-fed and over-housed ewes. In New York State in 1861-2, and again in Ohio, twenty years later, there was a fatality among lambs which amounted almost to an epizoötic, caused, it was very generally believed, by long confinement and dry-feeding of the ewes, and a consequent plethora in their systems. A great many lambs those two seasons had very sore mouths and eyes; they were sore when dropped, and at the age of a day or two they frequently became blind. Many of them died; some recovered under treatment, which consisted simply in an application of alum water to the eyes and some healing ointment to the sores. The plethora of the internal organs of the ewes prevented the lambs from expanding to their normal size and strength; and the fevered, impure blood of the dams, by a well-known law of embryology, was, so to speak, strained of its impurities while circulating through the unborn offspring.

ANÆMIA OR "PINING."—This is a bloodless condition of the system, like that induced by parasitism, but without the parasites. It is a disease which affects flocks in the Scotch Highlands; but it has never, so far as my observation and reading have extended, developed itself in the Western United States. Hogg, the "Ettrick Shepherd," says: "The hair of the animal's face becomes dry, the wool assumes a bluish cast, * * * and when dead, there is found but little blood in the carcass, and even the ventricles of the heart become as dry and pale as its skin." It occurs in dry seasons, he states, whereas parasitic anæmia is most prevalent in wet ones.

Of pining in the Eastern States, Mr. Stewart says: "When from continued wet weather the pasture becomes rank and watery, the flock appears at first in an excellent and thrifty state, but in a few days the animals are found lying listless, with drooping heads and ears, watery eyes, and the expression of the face miserable and painful. A few days afterwards the skin is tightly drawn, the wool becomes of a peculiar bluish cast, the skin beneath of a pearly white color, the eyes are also of a pearly, bloodless appearance, and death is busy in the flock." The cause of pining is thought to reside in the character of the soil itself; almost the only fact which is known with certainty concerning it is that it does not occur on limestone soils. The obvious remedies are a change of pasture to the short, sweet grasses of some hillside, together with high feeding, giving bran, cotton-seed meal, oil-cake meal, crushed oats, etc.

SCROFULA.—This is not only a noxious blood disease, but it is also generally hereditary, though there are reasons for believing that under certain circumstances, as, for instance, in the progeny of sheep too closely in-bred, it may be developed even where the ancestors were healthy. Its essential element consists in a taint in the blood, which, though it may for a long time lie dormant, will finally manifest itself in the form of tubercles in various parts of the body, chiefly in certain glands in the neck—the parotid and submaxillary; less frequently in the lymphatic glands, in the mesenteric glands, and some others. When these tubercles are deposited in the lungs, we have something very much like consumption in the human subject; and it is sometimes called by the shepherd "the thumps," from the loud and labored beating of the heart. Indeed, the scrofulous condition is very similar to that of the same name in man, in whom it is interchangeable with consumption, rickets, etc.

This is a disease of poverty, overcrowding, underfeeding, neglected stables, etc., which, happily, is very rare—at least, in its malignant forms—among the flocks of the United States. Under the form in which it is most frequently found it is generally denominated by the veterinarians "tuberculosis." Sheep are often slaughtered—oftener of the English breeds than of the Merino—which have the mesentery (the membrane which connects the bowels together) flecked with small tubercles, even to the lower end of the rectum. These are sometimes confounded with parasitic cysts. Probably the flesh of a sheep in which scrofula has no further extent than this is not seriously impaired for consumption as food—at any rate, not more unwholesome than the flesh of "rotted" sheep freely sold in England; but the farmer who witnesses or performs the disembowelment, and is aware of the nature of the phenomenon, would not care to bring the flesh to his own table. And what the farmer would not offer to his own family, he has no moral right to offer to the community.

When the disease assumes the more active and aggressive form of tumors in the glands of the neck, the sheep has passed out of the blind or obscure stage of tuberculosis, and other very obvious symptoms appear. A cough is heard, the appetite is feeble and capricious, fever sets in, the eyes and nose begin to discharge, and emaciation comes on. The sheep is in a decline it has consumption. The skin is drawn and pallid, the body almost bloodless. It is now too late to save the sheep; earlier in the progress of the disease the following may be given: Powdered iodide of iron, ten grains; mix with molasses and place on the root of the tongue once a day. Tincture of iodine or iodine ointment should be applied to the external ulcers.

From what has already been stated as to the nature of this malady, it is very obvious that a scrofulous animal should never be used as a breeder.

GOITRE.—This is an enlargement of the thyroid gland, which is situated on the front part of the neck, beneath the skin and immediately over the windpipe. This gland is a little larger in females than in males; it is ductless, having no excretory function. Goitre in lambs differs from that in the human family, inasmuch as in the latter it only appears some time after birth, even in goitrous districts, while in lambs it is congenital. This seems to show that it is due to some condition of the ewe which affects the fœtus. That it is not due to lime in the water is shown by the fact that it occurs in localities with every kind of soil from which limestone is wholly absent and where water is supplied to the flock altogether from cisterns. It is probably due to a variety of causes, among which may be named sameness of feed, overcrowding, bad ventilation, dampness of the sheep-house, lack of exercise.

PREVENTION.—The breeding ewes ought to be subjected to the best possible hygienic influences, the most important of which are, good ventilation, dry quarters and abundant exercise. Some noted breeders think ewes should not receive grain-feed during gestation. There can be no reasonable objection made to wheat bran and oats, except the trivial one that bran may cause the shedding of wool on the wrinkles. Moderate feeds of shelled corn will not injure pregnant ewes if they are otherwise well managed, though, of course, heavy feeding with it may prove highly deleterious. I have for many seasons given from a half-bushel to a bushel of shelled corn to one hundred and twenty-five ewes, up to the time of lambing, without producing any ill effects. Beets and turnips will assist in preventing goitre and the accompanying big bellies and weak constitutions. There is no objection to clover hay, provided the sheep take sufficient exercise to keep in health. When a ewe fills up on clover hay she is apt to be lazy and inclined to take too little exercise unless forced to it. The water should be close at hand and temperate, to induce frequent drinking. But, to repeat, the matters of preëminent importance are dry quarters and sufficient exercise.

The treatment is to give some form of iodine, both internally and externally. A new-born lamb can take about a grain of iodide of potassium three times a day; it should be given in a little warm water. The common iodine ointment of the drug-stores may be rubbed on the tumor.

HYDROCEPHALUS.—This is simply a form of scrofula, in which the disorganized blood, instead of secreting its serous or watery portion in the glands of the body or neck, as in other forms of scrofula, deposits it in the head (brain). It is present in the lamb at birth, the head being sometimes enormously enlarged, as if the lamb had been nearly strangled in birth. Indeed, it is necessary to wait a few hours before a decision is made, for when a ewe has had a very difficult and protracted labor, the lamb is likely to have a head so enlarged as to pass readily for a case of "water on the brain." If, after six or eight hours have elapsed, the head does not assume normal proportions, the lamb ought to be killed. If there are many cases of hydrocephalus, either the ram or the ewes or both ought to be changed.

DROPSY OR "RED WATER."—In the Far West this disease is brought on by the sheep feeding on frosted or snow-covered grass, in eating which they swallow a good deal of snow. They become "water-bellied," as it is sometimes termed, or have the "murrain;" they are dull and stupid, and stagger, carrying the head to one side; the eyes are staring, sometimes blind, and there is obstinate constipation. Death is speedy, and the autopsy reveals a quantity of red water (not blood, as is erroneously supposed) in the abdomen, secreted from the peritoneum or lining of the belly, which is inflamed and red in consequence.

Prevention is all-important. A free use of salt is recommended, also tar. In the case of large flocks the latter could not be given readily unless it was smeared in the salt-troughs. Isolated cases might be treated with the usual dose of Epsom salts and ginger. If there are symptoms of inflammation of the bowels, they may be treated as recommended heretofore (see that heading).

CHAPTER XXXIV.

DISEASES OF THE NERVOUS SYSTEM.

PARASITES IN THE HEAD.—Grubs and hydatids have been already treated under "Parasitic Diseases." But hydatids are sometimes the cause of obscure nervous affections which may be denominated :

PARASITIC PARALYSIS.—When lodged in any part of the spinal cord, working their way to the brain, they sometimes cause erratic movements, staggering, partial paralysis, which the shepherd is at a loss to explain. When the hydatid is in the left hemisphere of the brain (and, as far as my observation extends, the same is true of a grub in the nasal sinuses), the sheep will have a tendency to go in a circle toward the left; when it is in the right hemisphere, toward the right; when on the medium line, the sheep will move somewhere near in a straight line, but with the head held high. In other words, there is a certain "method in its madness," which distinguishes this species of cerebral or nervous disturbance from the diseases of the brain which are produced by congestion or undue pressure of the blood.

APOPLEXY.—The owner of Merinos is sometimes surprised, on going to his flock which has lately been confined to winter quarters, to find perhaps one of his best, fattest wethers lying on its side in an unconscious condition or already dead. When it is skinned the blood will be found settled thick and dark just beneath the skin, more especially on the side which was underneath. In all probability the sheep died of apoplexy. The sudden cessation of the exercise which it had freely taken before it was confined (and more freely, perhaps, than at any other time of the year, owing to the growing scantiness of the grass) caused an undue increase of blood, which, not being called into play in the legs, was determined to the head or congested about the body, as above mentioned. A sheep laboring under a stroke of apoplexy sometimes seems to be almost, if not quite, blind; the eyes are dilated and staring, but the pressure of blood on the optic nerves suspends the sense of sight. It reels and staggers, and finally falls helpless on its side.

Apoplexy principally attacks sheep in the opposite extremes of condition—plethora and poverty; the latter less frequently. In a very poor sheep certain disturbances of the digestive functions sometimes result in a determination of the blood to the head. Naturally, the treatment of the two cases will vary; the fat sheep will be bled and purged; the poor one ought to be nourished and stimulated, although gradually and carefully, until the stomach is able to bear the greater burdens.

When neglected, this form of disease may terminate in another still more violent and fatal, viz.:

INFLAMMATION ON THE BRAIN.—When the congestion is long continued, finally the brain itself becomes inflamed, and the animal becomes frenzied. Apoplexy is intoxication, but inflammation of the brain is delirium tremens. The mad and violent antics of a sheep in this condition are without any more system than those of a decapitated chicken. Immediate bleeding is called for, and in considerable quantity; the blood had better be drawn from the neck. Active purging ought also to be resorted to, by means of the usual (or even one a half larger) dose of salts, administered in solution with the bottle or horn to insure speedy action. Warmth and perfect quiet in a dark place would also commend themselves to the judgment of the shepherd.

PARALYSIS.—In apoplexy and inflammation of the brain, the affection is limited to the brain itself; in paralysis a part or the whole of the spinal cord is involved. Thus, in the first two maladies there is irrational action, or a total and sudden suspension of action; in paralysis there is a gradual suspension of action in a special function or part, or of the whole system. A sheep may have a paralysis of one side of the head, in consequence of which one eye will be partly or wholly closed, one ear will be lopped down, the head will be carried inclined to one side, and the cud will remain unmasticated in the side of the mouth. This would be classed as facial paralysis.

Then there is another species of paralysis which may be denominated paraplegia, or paralysis of the two hind legs. Old and poor ewes are subject to this when in an advanced stage of pregnancy, especially with twins; it is the result of an injury to or imperfect nutrition of so much of the spinal cord as supplies stimulus or life to the posterior portions of the body.

Special or local paralysis may exist in any other part, resulting from injury to or atrophy of the nerve leading to that part. Thus, injury to the pneumo-gastric nerve-trunk will cause difficulty in breathing and in swallowing; the breath will be labored and stertorous.

It is not always easy for the shepherd to determine whether it is the brain or the nervous system that is affected. If it is the brain (apoplexy or inflammation), there will generally be extreme violence in the actions, followed by collapse or coma; if it is the spinal cord or nervous system, there are not generally such exhibitions of insanity, but a gradual loss of some function or of nearly all the functions. Lambs which are the result of in-breeding, or which have been exposed to cold and wet, or their mothers starved or kept in damp, cold stables, become crippled, change from one leg to another, become helpless. Sometimes grown sheep in high condition, but which have been kept in a damp stable, will fall on their sides, helpless; if raised to their feet they will take a few steps, stiff-legged, then fall down in a tremble, grind their teeth, and froth at the mouth. Ewes which have come to parturition poor and have had a prolonged labor; sheep exposed to cold winds soon after being washed or shorn, are likely to suffer from paralysis.

In short, poverty and exposure are the prolific causes of paralysis; and this simple statement of itself suggests the general remedial measures of warmth, nourishment, stimulation. A half teaspoonful of powdered ginger or gentian, given in warm milk (if for a lamb), or a teaspoonful in warm gruel (for a grown sheep), or the same amount of aromatic spirits of ammonia and water of ammonia, rubbed on the spine, are recommended.

In mild cases this may be given: Spirits of nitrous ether (sweet spirits of nitre), two drams; powdered ginger, one dram; in warm gruel of some kind.

PALSY.—Same as paralysis.

EPILEPSY.—In this the behavior of the sheep is very nearly the same as in apoplexy, only the attacks are more frequently recurrent. The sheep stares about, staggers, falls in convulsions; then after a time it may rise to its feet and stand for a while in a stupor. It is thought to be generally caused by a large amount of very cold feed, such as frosty grass, taken into the stomach suddenly. The only thing to be done is to ascertain the causes and remove them.

LOCKJAW OR TETANUS.—This is related to the foregoing, but the causes are different, and are confined almost to a single one in practice—namely, exposure after castration. I have castrated many hundreds of young lambs, and find that it matters little how the operation is performed or to what amount of dry cold (not wind), they are exposed to afterward; but with mature rams much more care must be exercised, though the idea entertained by many shepherds, that the spermatic cords can safely be severed in no other way than by scraping, is superstitious. When a mature ram is to be castrated, the two points of the scrotum ought to be cut off, to allow accumulating pus to escape while the wound is healing; then a long slit should be made down the front side of each testicle. If necessary, let the knife be drawn down a second time and cut into the body of the testicle itself; this will insure the slitting of the membrane which envelops it, and which ought to be allowed to remain.

Lockjaw has few external symptoms except the immovable closing of the jaws, with now and then a bent neck and a rigidity of one or more of the limbs. The first thing to do is to relax, if possible, the spasmodic contraction of the muscles, so that the jaws can be opened and medicine administered. A hot bath of some duration will assist in doing this, but great care must be taken lest the animal should contract additional cold after it. If the jaws can be opened, give the usual purge of salts, dissolved in a half pint of warm whisky; then follow it, after two or three hours, with two drams of laudanum. A small teaspoonful of ginger, mixed in warm slippery-elm or linseed tea, or oat-meal gruel, given two or three times daily, will be beneficial.

CHAPTER XXXV.

DISEASES OF THE URINARY AND REPRODUCTIVE ORGANS.

MAMMITIS OR GARGET.—Inflammation of the udder and the loss of one or both of the teats, are a somewhat common trouble with Merino ewes. It is oftener the free milkers than the scanty ones which are thus affected, as a result partly of this full supply of milk and partly of neglect. High feeding on corn during pregnancy is a not infrequent cause of garget. If, when the lambs are weaned, the ewes are running on flush pasture and the free milkers do not receive some attention, their udders are apt to become painfully distended and swollen, and thus a foundation is laid for mammitis next spring. For, from whatever cause arising, one attack of this affection predisposes the ewe to a second. The Merino ewe which has borne two or three lambs, is also apt to have a baggy enlargement of one teat, which causes the lamb to neglect it, and unless much care is used an inflammation will be allowed to set in which will make matters worse. Garget is caused, too, by ewes sleeping in cold, damp places too soon after parturition.

A flock of ewes once gargeted will frequently show an obstinate recurrence to it at the next lambing-time, even when the management in the meantime has been faultless. It is not worth while to repeat the experiment; a ewe subject to this trouble had better be drafted, fattened and sold.

The inflamed udder should be treated with hot fomentations of spirits of camphor, and if the ewe is very valuable, a solution of carbonate of soda may be injected into the teats with a small syringe and then milked out. Give the following: Epsom salts, two ounces; nitrate of potash (saltpetre), two drams; ginger, one dram; may be given in water once a day.

RETENTION OF THE FŒTUS.—Dr. Edward Moore states that there is a case on record where a fœtus was retained three years, and yet the ewe survived! But generally, if a ewe goes beyond one hundred and fifty-eight days, the fœtus is dead and must be removed, or the ewe will die. When the lamb dies before birth Nature generally corrects the mistake by an abortion; that is, the fœtus is expelled as soon as life is extinct. But if the ewe is weak or unhealthy the expulsion may not take place, and relief must be afforded at once. The ewe will give notice of a calamity having occurred by moping, refusing feed, by a twitching of the ears and hind legs, and, more than all, by an exceedingly offensive, dark discharge from the vagina. Mr. Stewart recommends the dilatation of the mouth of the womb with the extract of belladonna, to facilitate the expulsion, but I have not found this necessary. I have had uniform success by proceeding as described on page 89, in an ordinary case of false presentation.

The treatment is as follows: Take calomel, eight grains; extract of hyoscyamus, one dram; linseed tea, half a pint; mix, and give two tablespoonfuls twice a day. With this alternate the following: Epsom salts, eight ounces; nitrate of potash, half an ounce; carbonate of soda, two ounces; water, one pint; give a quarter of a pint twice a day. Shake up before using. As soon as the bowels have been moved, omit the above and give this: Nitrate of potash, half an ounce; carbonate of soda, one ounce; camphor, one dram; gum water, eight ounces; dose, an eighth of a pint twice a day. The ewe should be fed on warm, thin oat-meal gruel, and if the discharge is very offensive, a dram of chloride of lime in a pint of warm water may be injected into the vagina.

EVERSION OF THE UTERUS.—A ewe which has a false presentation, or a stricture of the uterus, will sometimes continue her efforts at expulsion until the uterus is completely everted or turned inside out, so that it protrudes from the vagina as a red, inflamed sack as large as a child's head. It augurs very ill for the vigilance and keen-sightedness of the shepherd that he should allow labor to proceed so long and so violently without giving help, perhaps not even perceiving that there was a case of distress. An inexperienced shepherd is excusable for not seeing or understanding what is about to happen; but it is safe to say that such a deplorable accident need never occur in a flock which is carefully watched. A uterus can hardly be everted in less than ten or twelve hours of severe labor.

In the first place, the ewe must be delivered, carefully and gently; then the uterus should be washed with warm water or alum water and returned to its place by some one having a small hand, and with the finger nails well pared off. It may be everted a second time, and the shepherd must determine whether the ewe is sufficiently valuable to justify more thorough measures. With a curved spaying-needle let two stitches be taken across the vagina with a strong linen thread, each stitch being independent, with the thread cut off and tied for each stitch. Let the ewe be laid on her side on a decently soft bed, with the head considerably lower than the hind parts (not more than a foot lower), and the hind legs be padded and tied with a rope to a beam or something, to keep her from sliding down. Let her be turned over every few hours, for comfort. When she takes feed, which should be some nutritious gruel, given in a bottle or horn, care must be taken that she does not struggle or get into a position to endanger her. After two or three days she may be tried on her feet—the stitches still being in—but should be kept very quiet for some time longer.

Where straining is continued, a truss or pad is sometimes applied, and tied forward to a collar around the neck to keep it from slipping back.

RETENTION OF URINE OR DIURESIS.—This is not a common disease among Merinos, being confined mostly to the root-fed

English breeds. It is caused primarily by improper feed or water, or by sleeping in damp places. In other words, these causes produce cystitis or inflammation of the bladder, and this in turn brings about a retention of the urine. And when the urine is retained in the bladder until it becomes full, the further secretion of it is measurably or wholly stopped, and excretion of urine (diuresis) sets in—that is, a part of the urine is removed from the system through the skin. Moldy hay, wheat straw in too great quantities, second-growth clover containing ragweed, lobelia, and other irritating weeds, are among the causes of inflammation of the bladder. Excessive feeding on corn or meal and the drinking of hard water also sometimes cause it.

The ram or wether is more subject to urinary troubles than the ewe, on account of the greater length of the urinary canal and the peculiar vermiform or worm-like appendage at the end of the penis, through which the passage is so small that it easily becomes obstructed. The symptoms are uneasiness, stepping or stamping, striking at the belly with the hind legs, looking around at the sides, a bending of the back downward, a constant dropping of urine from the pizzle, a spreading of the legs apart and straining to urinate. As a remedy, take of creosote, half an ounce; acetic acid, two ounces; water, one pint; mix and keep well corked, away from frost. Give one teaspoonful in the water the sheep drinks, twice a day; if necessary, give it by means of the bottle. If the sheep is vigorous it may be bled in the neck, taking a half-pint. During treatment it should be kept dry and warm and fed on crushed oats or bran, with sweet hay and plenty of salt. If there is much fever, give the following to a strong ram: Linseed oil, two ounces; laudanum, two drams. If the fever continues, give the dose a second time, one-half reduced. For a small sheep diminish the dose proportionately.

GRAVEL OR STONE.—A ram which habitually drinks limestone water may have a chalky deposit in the bladder which effectually stops the passage of urine, and this, of course, is fatal. Rams running at large are less apt to be thus affected. On a limestone farm a housed ram ought to receive cistern-water.

STOPPAGE OF URETHRA.—Rams or wethers on a short allowance of water, summer or winter, or drinking water that is strongly charged with earthy matters, may have a sediment deposited in the urinary canal. The symptoms will be the same as those described for "retention of urine," and will therefore probably be treated the same way at first. If no improvement occurs, the shepherd may take it for granted that a stoppage exists. The ram will have to be set on his rump, the penis withdrawn from the sheath, and the "worm" or vermiform appendage cut off. This may seem unnecessarily summary and severe, but it is an operation not nearly so painful to the ram as castration, and it does not impair his usefulness in a majority of cases. The operation of slitting the penis or urethra lengthwise is one which should not be undertaken except upon the advice and with the assistance of an experienced veterinary surgeon. After the vermiform appendage has been cut off, a gentle pressure with the thumb and finger will generally remove the sediment by causing the urine to wash it out. If after this operation the animal appears to have still some retention of urine, the following may be given to act on the neck of the bladder: Linseed oil, three ounces; extract of belladonna, ten grains.

CLAP.—The ordinary symptoms of this disease are the same as in the foregoing urinary troubles, but a white acrid discharge finally sets in from the penis or vagina, which is pathognomonic. It may occur in either sex, by contagion, or arise *de novo* from excessive work, as when a ram is turned into a flock of ewes to serve them promiscuously. If allowed to run its course unchecked, this white acrid discharge would ultimately cause ulceration and destruction of the parts. The remedy is as follows: Spirits of camphor, four ounces; sugar of lead, one ounce; sulphate of zinc, two ounces; water, one quart; mix and put into a bottle. Then place the ram on his rump, draw the penis carefully from the sheath, holding it in a soft old piece of linen wet with the above liquid, until the ulceration can be traced to its upper limit. Bathe thoroughly with the lotion once a day. If a ewe has contracted the disease from a ram, a small sponge tied securely to the end of a smooth stick and saturated with this lotion may be introduced several times into the vagina; and for greater thoroughness some of it may be injected into the uterus with a syringe.

BLOODY URINE OR ALBUMINURIA.—This is often erroneously called "red water," a designation which correctly belongs to the abdominal affection already noticed. It is a disease of the kidneys, caused by improper feed or water, exposure, damp stables, etc. Sheep thus troubled are generally also weak, feverish, and seem to lose control of their legs; the urine is actually stained with blood. Sometimes this disease is very prevalent in a flock of ewes—generally in the winter. A change of feed should be adopted; clean, bright hay, or cornstalks with crushed oats, bran, linseed meal, and plenty of salt (but no sulphur) should be given. Chronic albuminuria is incurable: ewes afflicted with it are unfit for breeders, and should be fattened and sold to the butcher.

CHAPTER XXXVI.

MISCELLANEOUS.

FRACTURES.—On the Pacific coast sheep sometimes break their legs in the deep cracks formed in summer in the black "adobe" soil. In driving a flock through "a pair of bars," which are a nuisance on a sheep-farm, the careless shepherd often lets down two or three bars at one end only, thus compelling part of the sheep to jump or tumble over, at the risk of fracturing a bone or two. Unless the sheep is very valuable, and the owner has a special faculty for surgery, it had better be let alone, mostly, for no broken bones heal so rapidly as those of the sheep. Splints or a ligature are apt to do more harm than good, unless carefully watched. About all that is necessary is to put the sheep into a separate pen or yard, to prevent it from being jostled. I never knew a broken bone, even in an old sheep, to fail to knit without any further attention. When the break is in the fore-leg, below the knee, three splints may be applied, with padding between them and the leg; then the latter may be bent up and the lower end of it securely tied to a lock of wool on the shoulder. If the break is above the knee, no splints are needed, but the leg may be tied to the wool, as above, to prevent dangling.

WOUNDS.—A wound in a sheep will heal more readily than one in any other domestic animal, if it can be secured from the attacks of maggots; but this is the great difficulty. The sheep's flesh seems to be more attractive to these vermin than that of any other stock. If a wound occurs in the winter, it is a simple and easy matter to heal it; but if it is in the summer the shepherd will be obliged to expend a great deal of labor and care to save the sheep, and it is for him to decide whether it is valuable enough to justify the trouble. If a wound penetrates the flesh even very slightly, it must suppurate in healing, and suppuration will infallibly attract maggots. Hence, thorough precautions must be adopted against these. The best thing to do is to put a wide cloth bandage around the body, neck or limb, reaching beyond the wound two or three inches on each side; sew it securely and keep it saturated with oil of sassafras and whiskey, kerosene, benzine, or some other substance offensive to flies. Otherwise they will infallibly "blow" the wound or the pus just below it, and then the sheep is gone.

If the wound is a simple cut, a clean stitch about every inch —each stitch being independent, made straight across, and the thread tied and cut off for each stitch—may be taken in it just tight enough to bring the lips of the wound together, but not tight enough to pucker the flesh. If the wound is so shaped that it has a pocket anywhere, a slit ought to be made with a sharp knife as far down as the bottom of the pocket, so that all the pus can escape. The lower stitch may also be left a little slack to secure the same object. A puncture or stab-wound may be treated the same way, especial care being taken to provide an exit for all the pus quite down to the bottom of the wound. A torn and ragged wound—for instance one made by a dog—will cost more to cure than the sheep is worth, unless it is in the winter or the animal is exceptionally valuable. I have known a dog-bite in a perfectly healthy yearling, to suppurate for more than three months. Even if the sheep can be saved, a dog-bite anywhere about the legs or quarters generally cripples it for life, it is so poisonsus. All small fragments of skin and flesh should be trimmed off smooth, and the wound stitched up as above described. If much skin is gone, the stitches may be put closer together and drawn considerably, so as to somewhat pucker the skin, as the latter will sooner knit over the wound this way than if a good deal of new skin has to be formed. A pursed or puckered scar is not a deformity in a sheep, as it would be in a horse, and the wound will heal sooner if everything except the very largest fragments of flesh (and all the loose skin) are cut away, than it would if an attempt were made to save and stitch together the hanging shreds of flesh. The wound should be bathed at least once a day with carbolic acid greatly diluted with water or glycerine (one part of acid to twenty-five of glycerine), or with compound tincture of benzoin. Careful provision must be made for the escape of the pus, as above directed.

SORE EYES.—Sheep are sometimes seen with red eyes, matter formed at the corners, tears flowing from them. These indications are proof that they have taken cold or have irritated their eyes by thrusting down their heads into stubble or among briers; a ram sometimes has sore eyes, caused by a too tight-fitting cap placed over his face to keep him from fighting. The remedy is simple; it consists in the application, once or twice a day, of a wash made as follows: Sulphate of zinc, four grains; warm water, one ounce. Pure comb-honey is also good.

INSECT PLAGUES.—In some parts of the Southern States, especially in Louisiana, the Buffalo gnat is exceedingly troublesome to sheep, as well as all other stock. It makes its appearance usually about the first of August and continues about two months, though in very mild winters it lingers until in December. It is troublesome for only a short time. It tortures horses, cattle and sheep fearfully. Great numbers of horses and mules are destroyed by them on the "swamp plantations." On the uplands they are not so troublesome. Whiskey is the remedy. Grease, mixed with a little tar, pennyroyal, or other stinking stuff, and applied about the flank, throat, etc., the preventive. The ammonia of stables is repellent to the gnat and in these animals are safe from its attacks.

In the Far West, in Southern Oregon, Northern California, Washington, Idaho, Nevada, there are certain regions in the mountains where the Ear-fly is a great pest to stock. The preventive against this, too, is grease, which is rubbed thoroughly on the inside of the ears.

The prairie-dogs east of the Rocky Mountains, and the ground-squirrels of the Pacific coast, are a great annoyance to the shepherd; they destroy thousands of acres of valuable grass, gnawing the ground as bare as the public highway. Smoking with sulphur, poisoning with strychnine, drowning out with water and other expedients have been tried, but they all avail little against their countless numbers. Probably they will never be wholly subdued until the stock ranges give place to dense settlements and the constant plowing of the ground by farmers.

VEGETABLE FOES.—In the Far West there is a kind of grass, botanically called *Stipa spartea*, and somewhat resembling oats, and popularly known as "Weather grass," or "Needle grass." Caught in the wool of the sheep, the beard is propelled by the alternations of wet and dry, so as to cause the needle-like point at the lower end of the portion, which encloses the seed, to penetrate the hide of the animal; the beard breaks off, and the needle, continuing on its course, penetrates the vitals of the animal, causing painful death. The harmless silky growth tending backward from the needle acts as a barb to prevent any retrograde movement of the penetrating needle. These points also stick into the nostrils, nose and lips, and are also eaten, and, going into the stomach, cause death. The shepherds protect their flocks against it by the following method: Selecting a tract on which they wish to destroy the "Needle grass," they make a "fire-break" in the spring by plowing a number of furrows around the tract. This preserves the dead grass from the ordinary prairie fires of spring; then in June, when the needle grass is well started, they fire this reserved tract, and this destroys the needle grass for that season.

The minute prickles of some species of cacti trouble sheep in much the same way; they enter the skin wherever they touch it and penetrate until they reach some obstruction—for instance, a bone, against which they work their way in until they lie flat alongside of it. An ingenious citizen of Texas has invented an apparatus for burning off these prickles, which, it is claimed, leaves the cactus unharmed, thus affording a large amount of excellent feed for sheep on the desert.

SHEEP-PELTS—MODE OF TANNING.—Sheep which die in the winter, unless affected with some such disease as the rot, can be skinned; but in summer this is not so practicable, unless the carcass is found immediately after death. It is always best to skin the sheep if possible; then the wool can be pulled afterward or sold with the hide, if near a good market. If the market for pelts is not good, the wool can be loosened by a thick sprinkle of sharp, fresh lime on the flesh side, moistened with water and left to soak for about twenty-four hours. But this is objectionable; the lime corrodes the wool. To tan a sheep-skin, after the fat and flesh are well cleaned off, it should be put to soak in a bath composed of one pound of alum and one quart of salt (this is enough for a medium-sized skin), dissolved in sufficient milk-warm water to cover the skin. Set it away in some warm place, turn the skin every day for a week; then take out and wash in warm water, and hang up to dry. When partly dry rub and stretch it until dried; the more rubbing and stretching the whiter and better the leather will be. This will tan a skin with the wool on or off.

RAVAGES OF DOGS.—I have no statistics at hand but those of

Ohio, as the Compendium of the Tenth Census gives no information on this subject. In 1882 the number of dogs in Ohio was returned at one hundred and seventy thousand nine hundred and eleven; the value of the sheep killed and injured by them, was one hundred and seventy-three thousand nine hundred and seventy-six dollars. Therefore, a tax of one dollar on each dog, if fully collected, would not have reimbursed this loss, though it would probably have sufficed to pay all the actual claims for damages under the existing law.

Dogs always have been and always will be kept by mankind. The only practical question for legislators is, How to assess upon and effectively collect from their owners, the damage caused by them? A law requiring the personal presence of the sheep-owner and one or more of his neighbors as witnesses, at the meeting of the commissioners at the county-seat, is burdensome and unjust. These claims ought to be relegated to the local officers, and the latter ought to be authorized to make allowance for the following items: (a) value of animals killed; (b) value of those injured beyond recovery; (c) damage to animals bitten; (d) full amount of general injury to the flock from worrying and fright, including the estimated results in checking growth and fattening, and injury to ewes in lamb; (e) value of missing sheep, where the evidence that the loss is caused by dogs is as strong as in most circumstantial cases; and (f) full compensation for time lost to the owner because of the attack, this to include time used in getting his scattered flock together, finding missing and dead animals, presenting his evidence, working up his case and collecting his claim. Therefore, let owners insist upon a full and liberal appraisal by their neighbors.

A DOG-TRAP.—When a sheep has been killed a square enclosure of rails may be built around it, twelve feet high, ten feet square at the bottom, and the sides sloping inward until an opening is left at the top about five feet square. Any dog can easily climb and enter such a sloping pen, but not even a greyhound can get out. Poisoning often makes bitter enmities between neighbors, but when a dog is caught alive in a trap and his owner confronted with him, the shepherd has a great advantage over him.

DOG-GUARD.—Mr. James Wood gives the following experience: "Dogs rarely do serious injury to sheep in the day time. The flock requires protection at night. This can be conveniently and securely afforded by enclosures made of wire netting fastened to posts. Netting made of No. 15 wire is so strong that no dog will tear it with his teeth, and, if six feet high, no dog will jump over it. Some dogs might jump a fence of that height, but they will not try the netting. The mesh may be two or three inches. Such netting is now sold at a very low price, and the whole enclosure will cost but little. The door should be made of the netting stretched upon a frame. It is well to nail a board on the inside of the posts, a foot from the ground, to prevent the sheep pressing against the netting, and bulging it so far out as possibly to lift it from the ground.

"For summer folds, the highest and driest available grounds should be selected as affording free currents of air, because such sites are in every way better for the flock. For other seasons, these enclosures should be near the barns or feeding sheds. Indeed, there should be sheds affording shelter from storms and cold winds within them, or forming the most exposed two sides of the enclosures.

"Of all the domestic animals, sheep are the most easily folded at night. They naturally seek some high and dry resting place. Upon the Scotch moor lands they invariably leave the pastures upon which they have spent the day, to spend the night at their accustomed sleeping places. It is the same everywhere. If they have free access to the fold, it will be found only necessary to close the door at night, for all the flock will be there, quietly ruminating. It is well to tempt them with a little grain at first, and they can be regularly salted there. Unless their pastures are very rich and abundant, a little grain through the summer will give a profitable return. If the fold is not in the range, it is no more trouble to bring in fifty or a hundred sheep than to drive one cow home to the milking. A boy eight years old has folded a valuable flock for two years for the writer."

TRAINING A SHEPHERD DOG.—The winter is a good time for attending to this matter, for sheep are at this season more quiet and tractable than when full of the fire of summer on a lush feed of grass, and they will lend themselves more complacently to the work of breaking the pup. For it should be well understood at the outset, that it is a very uphill piece of work to attempt to break a green dog on a greener flock. An educated band of sheep is almost as necessary to the proper training of a pup, as a master of natural tact and liking for this business.

This last remark leads to the further one, that only a small minority of flock-masters are fitted by nature to handle well a dog already broken, and a still smaller minority who can bring up a pup from the start, and shape him into an instrument of good use. A man who has not this natural gift for training animals had better let the shepherd dog alone; it will prove a nuisance to him. Neither is it worth while for any one to bother with a shepherd dog, unless he has one or more flocks of at least a hundred head each. A few sheep are harder to handle with a dog than many.

My experience has been wholly with the English shepherd, or with a cross between the English and the New Mexican or Spanish; I never handled a Scotch Collie. The English dog, as imported into Southern Ohio, is slender in build—almost as slender as a hound—with long, silken, black hair, white belly and white tip to the tail. He is an animal of remarkable sagacity and energy, and his native force must be guided with great discretion, or it will develop into a scourge. The pup is perhaps the most restless of all animals, and one of the first things he is likely to learn is, to suck eggs. He must be broken of this habit with the utmost rigor, or he will have to be shot. Let him never see an egg until he is two months old; then take one boiled hard and hot, put it into his mouth, and hold his jaws tightly closed over it until he yells with pain. Every few days try him with another; if he shows the least disposition to tackle it, repeat the above dose, or crush between his jaws one filled with pepper.

The pup should have one master, and only one; all the other members of the family should be strictly forbidden to give him orders or cultivate his affections. This applies especially to the children; they will eventually make a fool of any dog. He should also be restrained at all hazards from chasing rabbits; not only preserved from the temptation as much as possible, but punished for the offense whenever he perpetrates it. A dog that will break away from his charge, or perhaps dash headlong through it, in pursuit of the first cotton-tail that jumps up, is of no account and might better be killed. A dog of the purest blood can be perverted by the boys to this wretched business, to the utter neglect of his own proper calling.

He should be kept from the sheep until he is a year old or thereabout. If he is of any value, he will be so frisky at a younger age as to be unmanageable. At first the master may attach a long line to him and teach him to come to heel at the word. He must be taught absolute and unhesitating obedience at all cost: yet great care must be exercised not to whip him too much; this would break his spirit—make him discouraged.

He must be made to come to heel at command, without being pulled or struck; and this must be done on many days. Every operation which it is necessary for him to go through at all must be repeated hundreds of times; nothing will impress a command on a dumb brute like continued repetition. The most successful trainer I ever saw would never go up or down the steps without compelling his young dog to follow him, even if he had to drag him every one of his fifty or a hundred trips.

There will be occasional times when a shepherd dog, even when well trained, will require a little punishment; he must be taught to come to heel and receive it. But he must be made to come up a hundred times, to be petted and rewarded where he comes up once to be whipped, and this leads to another remark. The master should always give his orders in an even, calm voice, devoid of passion; then the dog cannot tell from his tone whether he is to be scutched or not. The best of dogs is greatly tempted to run away when he knows from the angry bellowing of the master that he is to be chastised. One thing more: No dog should ever be allowed to go off after a flogging until the master and he have "made up." Some dogs are of such a sullen, unforgiving disposition that they will not make up. On such it is not worth while to waste time any further; they will never do any good.

At the second stage of the pup's education, he may be taken into the barnyard with a flock, the rope still around his neck, and made to go around them. The master must go around himself; if the pup does not follow he must be dragged with the line. This must be done scores of times. Then he must be made to lie down at some place where it is desired to have him stay to watch a gap, or the master's coat, or something, while the master goes off to the end of the rope. The sheep may then be crowded toward him a little, and if he flinches he must be made to return to his post, and this over and over again. Then let him get up and come to heel.

The greatest desideratum, perhaps, is to have the dog trained to bring sheep to you; and to do this he must be taught to "get out wide." He must never, on any account, be allowed to go straight toward the sheep—this is one of the most difficult things to prevent—but, if he does it, he must be called back and compelled to circle out wide. Sheep accustomed to a dog will run straight to the master if the dog will give them half a chance to do so.

As soon as he is thought to be sufficiently advanced to go

without the rope, the master may take him into a small field where there is a flock of gentle sheep, and manœuvre to get the sheep between the dog and himself. Then he can call and toll them around after him, compelling the dog to follow up; he will try to get around the sheep and come to his master, but the latter must so manage all the while as to keep the flock interposed between himself and the pup. After a while he will learn to follow quietly along, bringing up the stragglers.

In crowding sheep into close quarters, a dog that barks is far better than one that bites; and, indeed, a dog disposed to the latter course must be restrained and punished. It is sometimes a very difficult matter to teach a young dog to let the sheep's heels alone, and confine himself to barking. If he is held back with a rope, and a great noise and hurrah created, he will get to barking in his excitement; and, once the ice is broken, the way will be easier thenceforth.

The necessary words of command are few and simple, and they should never be varied: "Head away!" (head the flock); "Get out wide!" (go around 'em); "Hold!" (stop); "Fetch 'em up!" "Get over!" (mount the fence), will suffice.

WOOL WASTE AND SCOURINGS AS A FERTILIZER.—According to experiments made at the New Jersey Experiment Station, in 1884, wool waste (rags, shoddy, flocks), contained about eight per cent. of organic nitrogen; other analyses give rather less. Its value will of course depend considerably on the character of the soil. Experiments under charge of that Station show that wool waste alone did not produce the same effect as when used in combination with stable manure, but this result might vary on other soils. In another experiment, three loads of barn manure, one ton of wool waste, and two hundred pounds of bone black superphosphate, proved more profitable than fifteen loads of barn manure.

An analysis at Rothamsted, England, gave the following proportion in one thousand pounds of wool, unwashed: Nitrogen, seventy-three pounds; phosphoric acid, one pound; potash, forty pounds; lime, one pound, and magnesia, seven-tenths of a pound. Pure, washed wool has about sixteen per cent. of nitrogen, or more than double the unwashed. Washing removes most of the potash which is mostly in the "yolk." One hundred pounds of wool waste, therefore, containing sixteen or seventeen per cent. of nitrogen, should be over thirty times as strong as one hundred pounds of fresh cow dung.

As to "wool sweat" or "suint," or wool scourings, it is an

imperfect soap, consisting chiefly of potash, lime and magnesia united to a peculiar animal oil. It is remarkable that this soap of lime, insoluble in all other cases, is here soluble in water. Professor S. L. Dana states that the washings from wool annually consumed in France are equal to the manuring of three hundred and seventy thousand acres of land. Yet this substance must be applied judiciously; if scattered freely on the ground in a liquid form it has been known to sheet over the soil with a crust impervious to air or water, and so remain for years except where comminuted by the plow. In the United States *Agricultural Report* for 1870 there is a suggestion that these scourings should be reduced by heat to ashes before they are applied to the soil.

It is stated that in France a process has been discovered whereby a large amount of potash and valuable lubricating oil can be extracted from wool sweat, or yolk; and that this promises to give a very considerable value to a by-product which has heretofore been regarded in the United States as an almost total loss.

SUNDRY UTENSILS.—Stock registers, ear-tabs, toe-shears, etc., can be had of the makers who advertise them, as may shepherds' crooks. Any blacksmith can make a branding-iron (when this is wanted, though paint is better), of flat iron, five-eighths of an inch wide and one-quarter of an inch thick, bent into the required letter, with an iron handle eighteen inches long. Iron is better than wood, as it will permit the accumulating wool, grease and flesh to be burned off.

INDEX.

Barns, Houses and Appurtenances. 165
  House for Breeding Ewes, Smith's......165
  Cisterns for Houses......174
  Cisterns, How to Build......175
  Doors and Gates in......171
  Dust Bath in......123
  Mr. Frink's......195
  Feed Racks......172
  Fodder Ricks......186
  Grouping of......169
  Group of Three......170
Barn, a General Purpose......167
  For a Small Flock......169
  Double Racks in......173
  Single Racks for Lambs......173
  Table for Shearers......177
  Watering Troughs in......175
  Watering Troughs, Covered...176
  Wool Room in......176
Burrs in Pastures......127
Constitution in Sheep......48
Crossing and Cross-Breeding......52

DISEASES OF SHEEP AND REMEDIES.

  Acarus scabiei......301
  Actæa rubra......332
  Albuminuria......352
  Alimentary System, Diseases of. 328
  Anæmia......341
  Anthrax Fever......339
  Apoplexy......345
  "Bane"......287
  Baneberry......332
  Bleeding Sheep......279
  Blind Staggers......294
  Blood Diseases......338
  Bloody Urine......338
  Brain, Inflammation of......346
  "Braxy"......334
  Buckeye......332
  Buckeye and Laurel, Symptoms of Poisoning by......332
  "Buck-fly Grub"......236
  Buffalo Gnat......355
  Canker of the Foot......322
  Catarrh......324
    Treatment of......325
  Cercaria......288
  Choking......335
  Clap......352
  Constipation......328
  Cough......325
    General Remarks on......277
  Diuresis......300
  Dropsy......344
  Dysentery......337
  Enteritis......334

Diseases of Sheep.
  Epilepsy......343
  Eyes, Sore......355
  Feet, Diseases of......316
  Fluke......287
  Fluke, Eggs of......287
  Fluke, Its Changes......289
  Foot and Mouth Disease......323
  Foot Rot, Origin of......317
    Treatment of......319
    Foot-bath for......320
    Treating Small Flocks......321
    What is it?......318
  Foreign Substances in the Stomach......335
  "Fouls" in Sheep's Feet......321
  Fractures......353
  Gad-fly, Grub or Larva of......298
  Gad fly of Sheep......298
  Garget......349
  Gid......294
  Gid Hydatid, History of......294
  Goitre......343
    Prevention of......343
  Gravel......351
  Grub in the Head......298
    Preventive of......300
    Symptoms......299
    Treatment......299
  Head, Parasites in......345
  Hæmorrhoids......337
  Hoove......201
  Hoven......331
  Hydrocephalus......344
  Indian Pea......332
  Inflammation of Bowels......334
  Influenza......325
  Insect Plagues......355
  Interdigital Canal, Inflammation of......322
  Laurel......332
  Liver, Congestion of......336
    Inflammation of......336
    Rot......287
    Rot, Prevention of......291
  Linnæus truncatulus......288
  Lock-jaw......347
  "Loco Weeds"......332
  Louse, Sheep......314
  "*Lumbriz*"......297
  Lung Parasites......280
  Maggots......314
  Mammitis......349
  Miscellaneous......353
  "Murrain"......339
  "Needle-grass"......355
  Nervous System of......345
  Œstrus ovis......298

364   INDEX.

Diseases of Sheep.
 Ozæna 325
 Palsy 347
 Paperskin 277
  Copperas for 284
  Chlorine for 285
  High Feeding for 285
  Pumpkin Seeds for 285
  Symptoms of 282
 Paralysis 346
 Paraplegia 346
 Parasites 287
 Parasites in the Head 345
 Parasitic Diseases 287
 Parasitic Paralysis 341
 "Pelt Rot" 340
 "Pining" 341
 Plethora 340
 Pleurisy 427
  Prevention of 333
 Pneumonia 326
 Poisoning 332
 Purgatives 279
 Redia 288
 "Red Water" 344
 Respiratory System, Diseases of ... 324
 Retention of Fœtus 349
 Retention of Urine 350
 Rheumatism, Congenital 338
 Rot, Symptoms of 290
 St. Johnswort 332
 Scab, Destroying Infection 311
  Dipping Appliances 304
  Dips for 303
  Fences to Prevent 310
  "Handling" for 302
  Insect 301
  In the East 312
  Patent Dips for 302
  Prevention of 311
  Symptoms of 301
 Scald-foot 321
 Screw-worms 315
 Scrofula 342
 Snail for Fluke 288
 Snake Bites 316
 Staggers, Operation for 296
  Prevention of 297
 Stipa spartea 355
 Strongylus filaria 280
 Stone in the Bladder 351
 "Stretches" 328
 Sore Eyes 353
 Sore Mouth 334
 Sporadic Pleuro-pneumonia 326
  Treatment 327
 Sporocyst 288
 Tape-Worms 291
  Symptoms and Cure 292
 Tarweed 332
 Tetanus 347
 Thomas, Prof. A. P., on Fluke 287
 Ticks and Lice 312
  Eradicating 313
  Prevention of 314
 Tympanitis 331
 Urethra, Stoppage of 351
 Urinary and Reproductive Organs of 348

Diseases of Sheep.
 Uterus, Eversion of 350
 Vegetable Foes 322
 "Weather Grass" 355
 Worms, Other Intestinal 293
 Wounds 353
Dipping, Boilers and Vats for 309
 Draining Yards 309
 How Often 310
 Pens for the Flock 305
 "Spotting" 309
 Swimming Vats and Pens 307
 The Process 306
 Thoroughness Necessary 306
 The Swimming Method 307
 The Vat 309
 When to Dip 310
Dogs, Ravages of 356
 Trap for 357
 A Guard Against 357
 Training Shepherds' 358

Ewes.
 Acorns, Injurious to 161
 Age as Breeders 47-163
 Breeding, Selecting 155
 Breeding, Points in which the
  Ewe Prevails 156
 Condition at Coupling 157
 Condition, Maintaining in 159
 Defective Teats in 160
 Discarding, Causes for 157
 Drafting, Best Time for 156
 Ergot, Effects of 99
 Exercise of, Necessity for 159
 Spurred Rye, Effects of 99
 Feed for 95
 Feeding for Milk 159
 Feeding Suckling 160
 Gestation, Period of 157
 Getting Cast 164
 Green Rye for 99
 Lambing, Time of 158
 Missing 162
 Recurrence of 167
 Running Farrow 157
 "Teasers" for 160

FEEDING AND FODDERS.
 Alfalfa 66
 Analytical Tables, Value of 62
 Feeds, Correlation of 63
 Corn, Sheep in 138
 Corn-fodder for Sheep 204
 Feed 59
 Feeding, Experiments in 64
  Crout's, W. D. 73
  Kirby's, W. G. 73
  Sanborn's, J. W. 73
  Watkin's, O. M. 73
 Feeding, Merino Taste from 75
 "Sheepy Flavor" from 75
 Fodders for Sheep 203-286
 Fodder-corn for Sheep 205
 Feed, a Perfect 60
 Feeds, Mixed 63
 Feeds, Grain 61
 Mineral Matters Needed in 66
 Prairie Hay 66

INDEX. 365

Feeding and Fodders.
  Red Clover for Sheep........203
  Roots for Sheep... 62
  Sheep, Timothy for.......206
  Sheep, a Variety of Feed for...199
Feeding for Mutton ... .189

FLEECES.

Burs, Thistles, etc., in......... 39
Clouded 38-183
How to Fold......:...118
"Fribs" in............... 39
Mold in................. 38
"Old Sue's"............156
Sorts in 30
Strings in 38
Stuffing of... 38
Unevenness in............... 39

FORAGE PLANTS, NOT GRASSES.

"Alfileria",................ 71
"Bur-clover "............... 71
Erodium cicutarium 71
Eurotia lanata.............. 71
"Greasewood"............. 71
Medicago denticulata........... 71
Abione canescens... 71
Phaca Nuttalli..................332
"Pin-clover "............... 71
Prosopis juliflora. 71
Purshia tridentata.......... 71
"Sage "................. 71
Sarcobatus vermiculatus....... 71
Trifolium Andersonii.. 71
"White Sage".............. 71
"Winter Fat"............... 71
Frank, J. H., Methods of Feeding.194
Full Blood and Thoroughbred..... 44
Grades.................. 45

GRASSES, BOTANICAL NAMES IN *Italics*.

Agrostis vulgaris............70-208
Aira cæspitosa................ 70
Atropis tenuifolia.............. 70
Bermuda grass................221
"Blue Joint"................. 70
Bouteloua hirsuta............ 69
Bouteloua oligostachya........ 69
Buchlo dactyloides...... 69
"Buffalo Grass".............. 69
"Bunch-grass"............. 69
"Broom-grass" 69
Calamagrostis Canadensis...... 70
Dactylis glomerata 207
Eriocoma cuspidata........... 70
"False Red-top" 69
Festuca occidentalis........... 70
Festuca ovina.............. 70
Festuca scabrella... 70
"Foxtail-grass ".............. 70
"Gallotte"................. 70
"Galleta "................. 70
"Grama-grass "............. 69
"June "................68-208
Hilaria rigida 70
Hordeum murinum 70
"Hungarian Grass "..........207

Grasses.
  Grass, Meadow 208
  Manisuris granularis...........220
  Mesquite grass. 71
  Muhlenbergia gracillima 71
  Munroa squarrosa........... 69
  Panicum Germanicum........207
  Poa alpina 69
  Poa annua 69
  Poa compressa................ 69
  Poa pratensis............... 69
  Poa serotina............... 69
  Poa tenuifolia 69
  Red-top for Sheep..........70-208
  "Sage-grass"............. 69
  "Sand-grass "............. 70
  Sheep's Fescue....... 70
  "Smut-grass ".............220
  Sporobolus airoides........... 70
  "Squirrel-grass"............. 70
  Stipa comata............... 69
  Stipa occidentalis............ 69
  "Vining Mesquite"......... 69
  "Wire Grass" 69
Hoofs, Clipping 105
Hook, Making a............... 57

LAMBING.

Artificial Nipples. 92
Assisted Labor in... 88
Chilled Lambs. 90
Corps', Geo S., System....... 91
Creeps for Lambs............. 92
In the Field 97
Fixtures and Preparations for . 57
Foster Mothers.. 89
General Management in........ 88
Milk, Excess of... 95
Panels for Pens at 87
Os Uteri, Schirrous 88
Sheep Hook, its Uses in.... ... 85

LAMBS.

Acorns for 140
Castration of ....:.......107
Cholera in.. 93
Cholera Preventives........... 94
Cossets...... 91
Cows' Milk, Feeding with...... 93
Docking................107
Full Feed for............140
Fall and Winter 163
Fouling of... 93
Goitre in................. 98
In May 159
Pumpkins for.. 141
Re-docking of... 103
Stiff Neck in..... 101
Tagging 104-135
Tail, Burning off of...........103
Turnips for...... 141
Twins.................101
Weaning................134
Winter Care of...........179

Letter of Presentation............. 8
Letter of Request............. 7
Maggots....................131

MERINOS.

Adams', Seth, Importation... 11
Aguirre.................. 14
American Merinos........... 18
Atwood's, Stephen, Purchase.. 15
"Basrom," Fleece of.......... 20
Beall's, Victor, Delaine Merinos 21
Black-top Merinos............ 26
"Buckeye," Heaviest Fleece... 20
Correlation of Carcass and Fleece 22
Dana's, George, Flock........ 20
Delaine Merinos............. 25
Embargo, Effect of........... 15
English and Merino Cross..... 53
"Escurials"................ 14
Etymology of Merino......... 11
Fearing, Paul, and B. F. Gilman's Flock............. 20
Fleece, Weight of............ 19
Feeders, Merinos as.......... 159
French Merinos.............. 279
Guadaloupe Merinos.......... 14
Davis', Col. Humphrey, Importation................. 11
Hammond, Edwin, His Flock. 17
Hammond's Merinos.......... 18
"Infantados"............... 14
Jarvis', Wm., Importation.... 15
Kelly, Daniel, His Flock...... 21
"Little Wrinkly"............ 17
"Long Wool"............... 18
Mather, Increase, His Flock... 20
Merino Mutton.............. 189
National Improved Saxony ... 36
Negrettis................... 14
"Old Black"................ 17
"Old Greasy"............... 17
"Old Wrinkly".............. 17
"Patrick Henry," Fleece of... 20
Panlars.................... 14
Putnam's, Israel, Flock...... 20
Quinn, J. B., His Merinos.... 17
Race Type.................. 24
"Rambouillet" Merinos...... 30
Rich's, Charles, Flock........ 16
Southdown and Merino Cross.. 52
Stone's, Col. John, Flocks.... 20
Sweepstakes................ 17
Washington Co., Pa., Flocks... 21
Wrinkles................... 56
"Wooster"................. 17
"Young Matchless".......... 17

Mutton, A Leg of............ 129
Mutton, Daniel Webster on... 130
Mutton, Braxy flavor in 130
Mutton Merino.............. 72
Orchards, Sheep in.......... 138
Prepotency................. 50

RAMS.

Blinders for 150
"Capt. Jack "........ .149
Constitution of .342
Feeding of. 151
Fighting, to Prevent in.. 148
Management of Cross 152
Management of Service.. ..151

Rams.
"The Ram More than Half the Flock"................. 45
One or More................ 153
"Patrick Henry"............ 146
Points of a Good 144
Selection and Care of 142
"Silver Horn"............... 142
Summer Management of...... 147
Tarring.................... 148
Tying a.................... 150
Winter Treatment of........ 153
Shearing Cards............. 176

SHEEP.

Blacking the Fleece of....... 154
Blanketing................. 155
"Breeder's Fancy" in........ 49
Breeding Flock............. 155
Choose with Fewest Defects... 49
Cleanliness, Importance of ... 99
Condition, Maintaining an Even 139
Condition, Good 123
Corn for................... 138
Corn Fodder for 184
Depasturing Wheat with..... 201
Dust-bath for 133
Feeder, Methods of a Noted... 194
Feeding, Manner and Material, 192
Feed Troughs for........... 180
From Grass to Hay.......... 138
From Hay to Grass.... 200
Gad-fly on 133
Good Growth............... 123
Grain Feed at Night......... 180
Grain, Necessity for......... 168
Housing, the Gain of 182
In-Breeding... 55
Losing Wool................ 157
Maggots In................. 131
Marking................... 117
Manure, Making and Saving. 185
Mutton Feeding for......... 189
Mutton, When to Feed for.... 191
Opposites to be Mated ..: ... 144
Orchards, In............... 138
Over-feeding of 137
Pampering and over-feeding.. 136
Pasture in the West......... 67
Pedigree 44
Pelts, Tanning of........... 356
Quiet, Importance of........ 199
Sales, Mutton, of at Chicago... 24
Salting of 133
Scavengers, As............. 125
Season, At the End of.... .141
Shearing................... 115
Shearing, Sorting at........ 117
Shearing, Speed in........ 122
Shelters, Temporary........ 181
Snow Eaters................ 188
Soiling 139
Sorting for Winter... 181
Stables, Cleaning out....... 185
"Stubbling," Blacking, etc... 154
Summer Housing and Feeding. 185
Summer Management of.. ...125
Ticks on................... 131

INDEX. 367

Sheep.
    To an Acre, Number 127
    Water for, Necessity of 128
    Washing, A. F. Breckenridge's
      Record.. 108
    Washing, Loss from 109
    Washing, Modes of 113
    Washing, Policy of 105
    Washing, Shearing without ... 112
    Winter Management of 177
    Working off the Culls 128
    Yarding in Winter 177

SHEEP HUSBANDRY, SYSTEMS OF:

Arizona, In........................ 233
    General Management.......... 233
Atlantic Slope, On................ 209
    Allis' J. F. C., Management ... 210
    Early Lambs................. 209
    Manure, Value of 210
    Roots, Feeding of............ 211

California, In..................... 234
    Alfalfa, Sheep on............ 243
    Breeding Flocks and Ewes.... 238
    Characteristics of California
      Sheep...................... 231
    "Dodge Gate," The, in....... 245
    Flock, A Sample............. 247
    General Management.......... 237
    Hay in California............ 248
    History of, in............... 234
    Herder, His Life in.......... 248
    Losses of Sheep in........... 247
    Mutton Sheep in California... 240
    Pastures, Effect of Sheep on.. 241
    Vineyards, Sheep in.......... 245
    Shearing in.................. 238
    Sierra Nevada, Sheep in...... 247
    Wheat Farms,Effect of Sheep on 241
    Wool, Grades of 239
    Preparing for Shipping....... 246
    Produce of................... 234

Colorado, In....................... 271
    "Alfalfa Mutton" in......... 272
    Bands, Division of Sheep into 272
    Corrals in................... 273
    Grasses in................... 271
    Increase from Irrigation..... 272
    Snow Storms in............... 273
    Summer Management............ 272

Dakota, In......................... 263
    Alfalfa in................... 263
    Alkali in.................... 263
    General Management in....... 263
    Milk-weed, Poisonous......... 264
    Ranges, Fenced and Open...... 264

Idaho, In.......................... 258
    Flocks, Size of.............. 258
    Grasses and Herbage.......... 258
    Hay Cut from "Claims"....... 258
    Lambing...................... 258
    Losses, Annual............... 258
    Merino Blood Predominant..... 258
    Mutton, Small Demand......... 259
    Snows, Early and Late........ 258
    Wheat in Dough, Cut for Winter 258

Kansas, In, **See Prairie Regions**... 214
Montana, In........................ 259
    Bunch Grasses................ 260
    "Chinook," The, in.......... 259
    Diseases in.................. 260
    Grasses in................... 259
    Winter, Hay for.............. 262
    Winter, Shelter in........... 262
    Wool, Clip of 1885 259
    Wool, Preparing for Market .. 260
Nebraska, In....................... 264
    "Blizzards" in.............. 267
    Elements, Dangers from the... 267
    Feeding in Winter............ 268
    Scab in...................... 269
    Water in Winter.............. 269
    Wind-Breaks.................. 267
Nevada, In......................... 249
    Drives of Sheep, Their Effects 249
    Lambing in................... 250
    Mutton in.................... 250
    Scab in...................... 251
    Shearing in.................. 250
    Systems of Management........ 249
    Wool in...................... 250
New Mexico, In.....................
    Breeding for Hardiness in.... 231
    Losses of Sheep in........... 231
    Pasturage and Forage......... 231
    Sheep Drives................. 232
    Taken on Shares.............. 233
Oregon, In......................... 251
    Alkali, Effects of, on Fibre.. 253
    Beasts of Prey in............ 255
    Conditions and Modes in...... 252
    "Chinook," Sheep in the..... 257
    "Dead Tip" in............... 254
    Long-wooled Sheep in......... 252
    Losses, Average Annual....... 250
    Merinos, Introduction of..... 251
Prairie Regions, In the............ 214
    Bad Management in............ 214
    "Buck Rake," Use of......... 217
    Foot Rot in.................. 216
    General Management in........ 217
    "Go Devil," Use of.......... 217
    Grasses, Beard or Broom...... 216
    Grasses, Natural............. 216
    "Mutton Corn"............... 216
    Prairie Hay.................. 216
    Sorghum Fodder............... 216
    "Stalk Pastures"............ 215
    Wool in...................... 215
    Wool. **Prices of**................ 215
Southern States, In................ 218
    Bermuda Grass in............. 221
    Cotton Seed for **Sheep**........ 220
    Department of **Agriculture**, Its
      Queries.................... 221
    Department of Agriculture, Re-
      turns Tabulated............ 223
    "Guinea Grass" in........... 226
    "Japan Clover" in........... 219
    "Japan Clover," Dr.Phares on 221
    Lambing in................... 226
    Lespedeza striata............ 219
    Liatris odoratissima......... 220

INDEX

Southern States, In.
  Manisuris granularis............221
  "Smut-grass"....................220
  Sorghum Halepense..............220
  Peters', Richard, Experience....219
  "Vanilla Plant" for Sheep......220
  Watts', Col. J. W., Experience..218
  Wool, Washing of................220
Submontane Region, In the.........212
  Extent of.......................212
  Flocks, The size of.............213
  Feed in.........................214
  Wool Rather than Mutton.........207
Texas, In.........................222
  "Chourros," The.................223
  "Corral," The...................224
  Fencing in......................224
  Flock, A Sample.................230
  General Management in...........224
  Herding in......................224
  Kendall, G. W., His Farm........222
  Lambing in......................225
  "Muttons," Feeding of...........229
  Pasturage in....................226
  Range in........................223
  Shearing........................223
  Shearing, Semi-annual...........235
  Sheep, Texan....................227
  Winter Ranges...................225
Utah, In..........................274
  Clip in 1879....................274
  Losses, Sources of..............275
  Losses in Winter of 1879-80.....274
  Merinos, Introduction of........274
  Poisonous Plants in.............275
Washington Territory, In..........257
  "Chinook," Sheep in the.........257
  Cattlemen vs. Sheep.............257
  Grasses in......................257
  Lambing, Time for...............257
  Scab in.........................257
  Shearing in.....................257
  Winter Management in............257
Wyoming, In.......................269
  Government Report on............270
  Laramie City a Wool Center......270
    Dipping Tanks at.............270
    Shearing Pens at.............270
Teaser............................163
Teeks.............................131

Teeth Indicating Age..............129
  Variation.......................51
Wethers, Tagging..................105
Wheat, Affects of Depasturing.....202

Wool.
  Analysis of.....................361
  And Yolk, Correlation of........146
  Australian......................83
  Author's Experience in..........84
  Black-Top and Clots in..........36
  Brashy..........................123
  California Grades...............239
  Cotting of......................35
  "The Crimp".....................28
  Dead Tip in.....................22
  Fiber, Effect of Climate on.....33
    Effect of High Feeding on....33
    Length and Density of........35
    Structure of.................27
  Grades in....................31-123
    Whence the Grades Come.......32
  Gray Shoulder Clot in...........36
  How Planted.....................29
  "Jar" in........................37
  Jointed.........................37
  "Kemp" in.......................37
  Length and Diameter of..........28
  Manufactures....................82
  Prairie.........................215
  Montana, Packing in.............261
  Preparation for Shipping........246
  Press...........................119
  Prices in Boston for Seventeen
    Years........................42
  Price Compared with Cotton......43
  Production of...................41
  Product of the United States....80
  Quarter Blood...................124
  Room in Sheep House.............176
  Round and Flat..................28
  Sacking and Transportation......124
  Sectional Prices of.............40
  Scourings, As Fertilizers.......361
  Shearing, Cards for.............176
  Strength of Dry and Yolky.......85
  Sunman's, T. W. W. Flock........83
  Waste...........................361
  Where to Sell...................122
Wool, Storing.....................121
Yolk and Wool, Correlation of.....146

Alphabetical Catalogue

— of —

O. Judd Co., David W. Judd, Pres't,

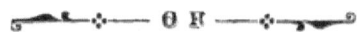 PUBLISHERS AND IMPORTERS OF

All Works pertaining to Rural Life.

751 Broadway, New York.

Agriculture, Horticulture, Etc.

FARM AND GARDEN.

Allen, R. L. and L. F. New American Farm Book	$ 2.50
American Farmer's Hand Book	2.50
Asparagus Culture. Flex. Cloth	.50
Bamford, C. E. Silk Culture. Paper	.30
Barry, P. The Fruit Garden. New and Revised Edition	2.00
Bommer. Method of Making Manures	.25
Brackett. Farm Talk. Paper 50c. Cloth	.75
Brill. Farm-Gardening and Seed-Growing	1.00
—— Cauliflowers	.20
Broom-Corn and Brooms. Paper	.50
Curtis on Wheat Culture. Paper	.50
Emerson and Flint. Manual of Agriculture	1.50
Farm Conveniences	1.50
Farming for Boys	1.25
Farming for Profit	3.75
Fitz. Sweet Potato Culture. New and Enlarged Edition. Cloth	.60
Flax Culture. Paper	.30
French. Farm Drainage	1.50
Fuller, A. S. Practical Forestry	1.50
Gregory. On Cabbages	.30
—— On Carrots, Mangold Wurtzels, etc.	.30
—— On Fertilizers	.40
—— On Onion Raising	.30
—— On Squashes	.30

Harlan. Farming with Green Manures	1.00
Harris. Insects Injurious to Vegetation. Plain $4. Col'd Engravings.	6.50
Harris, Joseph. Gardening for Young and Old	1.25
——— Talks on Manures. New and Revised Edition	1.75
Henderson, Peter. Gardening for Pleasure	1.50
——— Gardening for Profit. New and **Enlarged** Edition.	2.00
——— Garden and Farm Topics	1.50
Henderson & Crozier. How the Farm Pays	2.50
Hop Culture. New and Revised Edition. Paper	.30
Illustrated Dictionary of Gardening. Vols. I & II, each	5.00
Johnston. Agricultural Chemistry	1.75
Johnson, M. W. How to Plant. Paper	.50
Johnson, Prof. S. W. How Crops Feed	2.00
——— How Crops Grow	2.00
Jones, B. W. The Peanut Plant. Paper	.50
Lawn Planting. Paper	.25
Leland. Farm Homes, In-Doors, and Out-Doors. New Edition	1.50
Long, Elias A. Ornamental Gardening for Americans	2.00
Morton. Farmer's Calendar	5.00
Nichols. Chemistry of Farm and Sea	1.25
Norton. Elements of Scientific Agriculture	.75
Oemler. Truck-Farming at the South	1.50
Onions. How to Raise them Profitably	.20
Our Farm of Four Acres. Paper	.30
Pabor, E. Colorado as an Agricultural State	1.50
Pedder. Land Measurer for Farmers. Cloth	.60
Plant Life on the Farm	1.00
Quinn. Money in the Garden	1.50
Riley. Potato Pests. Paper	.50
Robinson. Facts for Farmers	5.00
Roe. Play and Profit in my Garden	2.50
Roosevelt. Five Acres Too Much	1.50
Silos and Ensilage. New and Enlarged Edition	.50
Starr. Farm Echoes	1.00
Stewart. Irrigation for the Farm, Garden and Orchard	1.50
Ten Acres Enough	1.00
The Soil of the Farm	1.00
Thomas. Farm Implements and Machinery	1.50
Tim Bunker Papers; or, Yankee Farming	1.50
Tobacco Culture. Paper	.25
Treat. Injurious Insects of the Farm and Garden	2.00
Villes. School of Chemical Manures	1.25
——— High Farming without Manures	.25
——— Artificial Manures	6.00
Waring. Book of the Farm	2.00
——— Draining for Profit and Health	1.50
——— Elements of Agriculture	1.00
——— Farmers' Vacation	3.00
——— Sanitary Drainage of Houses and Towns	2.00
——— Sanitary Condition in City and Country Dwellings	.50
Warington. Chemistry of the Farm	1.00
White. Gardening for the South	2.00

FRUITS, FLOWERS, ETC.

American Rose Culturist	.30
American Weeds and Useful Plants	1.75
Bailey. Field Notes on Apple Culture.	.75
Boussingault. Rural Economy	1.60
Chorlton. Grape-Grower's Guide	.75
Collier, Peter. Sorghum, its Culture and Manufacture	3.00
Common Sea Weeds. Boards	.50
Downing. Fruits and Fruit Trees of America. New Edition	5.00
——— Rural Essays	3.00
Elliott. Hand Book for Fruit-Growers. Paper 60c. Cloth	1.00
Every Woman her own Flower Gardener	1.00
Fern Book for Everybody	.50
Fuller, A. S. Grape Culturist	1.50
——— Illustrated Strawberry Culturist	
——— Small Fruit Culturist. New Edition	1.50
Fulton. Peach Culture. New and Revised Edition	1.50
Heinrich. Window Flower Garden	
Henderson, Peter. Hand Book of Plants	3.00
——— Practical Floriculture	1.50
Hibberd, Shirley. The Amateur's Flower Garden	2.50
——— The Amateur's Greenhouse and Conservatory	2.50
——— The Amateur's Rose Book	2.50
Hoopes. Book of Evergreens	3.00
Husmann, Prof. Geo. American Grape growing and Wine Making	1.50
Johnson. Winter Greeneries at Home	1.00
Moore, Rev. J. W. Orange Culture	1.00
My Vineyard at Lakeview	1.25
Origin of Cultivated Plants	1.75
Parsons. On the Rose	1.50
Quinn. Pear Culture for Profit. New and Revised Edition	1.00
Rivers. Miniature Fruit Garden	1.00
Rixford. Wine Press and Cellar	1.50
Robinson. Ferns in their Homes and Ours	1.50
Roe. Success with Small Fruits	2.50
Saunders. Insects Injurious to Fruits	3.00
Sheehan, Jas. Your Plants. Paper	.40
Stewart. Sorghum and Its Products	1.50
Thomas. American Fruit Culturist	2.00
Vick. Flower and Vegetable Garden. Cloth	1.00
Warder. Hedges and Evergreens	1.50
Webb, Jas. Cape Cod Cranberries. Paper	.40
White. Cranberry Culture	1.25
Williams, B. S. Orchid Grower's Manual	6.50
Wood, Samuel. Modern Window Gardening	1.25

Cattle, Dogs, Horses, Sheep, Swine, Poultry, Etc.

CATTLE, SHEEP, AND SWINE.

Allen, L. F. American Cattle. New and Revised Edition............	2.50
Armatage, Prof. Geo. Every Man His Own Cattle Doctor. 8vo..	7.50
Armsby. Manual of Cattle Feeding................................	2.50
Cattle. The Varieties, Breeding, and Management...............	.75
Coburn, F. D. Swine Husbandry. New and Revised Edition......	1.75
Clok. Diseases of Sheep..	1.25
Dadd, Prof. Geo. H. American Cattle Doctor. 12mo............	1.50
—— American Cattle Doctor. 8vo Cloth......	2.50
Fleming. Veterinary Obstetrics...................................	6.00
Guenon. On Milch Cows..	1.00
Harris, Joseph. On the Pig..	1 50
Jennings. On Cattle and their Diseases........................	1.25
—— On Sheep, Swine, and Poultry,.......................	1.25
Jersey, Alderney, and Guernsey Cow......................	1.50
Keeping One Cow..	1.00
Macdonald. Food from the Far West............................	1.50
McClure. Diseases of the American Horse, Cattle, and Sheep.......	2.00
McCombie, Wm. Cattle and Cattle Breeders....................	1.50
Martin, R. B. Hog-Raising and Pork-Making....................	.40
Miles. Stock Breeding..	1.50
Powers, Stephen. The American Merino for Wool and Mutton. A practical and valuable work.....................................	1.75
Quincy, Hon. Josiah. On Soiling Cattle........................	1.25
Randall. Fine Wool Sheep Husbandry...........................	1.00
—— Practical Shepherd....................................	2.00
—— Sheep Husbandry.....................................	1.50

Reasor. On the Hog...	1.50
Sidney. On the Pig...	.50
Shepherd, Major W. Prairie Experience in Handling Cattle...	1.00
Stewart, Henry. Shepherd's Manual. New and Enlarged Edition..	1.50
Stewart, E. W. Feeding Animals...............................	2.00
The Sheep. Its Varieties and Management. Boards............	.75
Youatt and Martin. On the Hog...............................	1.00
Youatt. On Sheep...	1.00

DOGS, ETC.

Burgess. American Kennel and Sporting Field. 8vo.............	3.00
Dog—The Varieties and Management.........................	.50
Dogs of Great Britain, America, and Other Countries, Compiled from Stonehenge and other Standard Writers. The most Complete Work ever Published on the Dog. 12mo.........	2.00
Forester, F. The Dog, by Dinks, Mayhew, and Hutchinson. 8vo...	3.00
Floyd, Wm. Hints on Dog Breaking. 12mo.....................	.50
Hallock, C. Dog Fanciers' Directory and Medical Guide. 18mo.....	.25
Hammond, S. Dog Training. 12mo.............................	1.00
Hill, J. W. Management and Diseases of the Dog. 12mo.........	2.00
Hooper, J. J. Dog and Gun. Paper...........................	.30
Hutchinson, G. N. Dog Breaking. 8vo.........................	3.00
Idstone. The Dog. Illustrated. 12mo...........................	1.25
Laverack, E. The Setter. 4to.................................	3.00
Mayhew, E. Dogs; Their Management. 16mo...................	.75
Points for Judging Different Varieties of Dogs. Paper..	.50
Richardson. Dogs: Their Origin and Varieties. Paper 30c. Cloth	.60
Shaw, T. Vero. Illustrated Book of the Dog. 4to...............	8.00
Stables, Gordon. Our Friend the Dog. 8vo....................	3.00
—— Practical Kennel Guide.................................	1.50
—— Ladies' Dogs as Companions...........................	2.00
Stonehenge. The Dog in Health and Disease. 8vo.............	3.00
—— Dogs of the British Islands. 8vo.......................	6.00
—— The Greyhound..	5.50
Youatt. On the Dog. 8vo.....................................	2.50

HORSES, RIDING, ETC.

Anderson, E. L.	Modern Horsemanship. 8vo	7.00
——	The Gallop. 4to. Paper	1.00
Armatage, Geo.	Every Man His Own Horse Doctor, together with Blaine's Veterinary Art. 8vo. ½ morocco	7.50
Armatage, Geo.	Horse Owner and Stableman's Companion. 12mo	1.50
Battersby, Col. J. C.	The Bridle Bits. A valuable little work on horsemanship. Fully illustrated. 12mo	1.00
Baucher, F.	New Method of Horsemanship. 12mo	1.00
Bruce.	Stud-Book. 4 vols	35.00
Chawner, R.	Diseases of the Horse and How to Treat Them. 12mo	1.25
Chester.	Complete Trotting and Pacing Record	10.00
Dadd, C. H.	American Reformed Horse Book. 8vo	2.50
——	Modern Horse Doctor. 12mo	1.50
Day, W.	The Race Horse in Training. 8vo	6.25
Du Hays, C.	Percheron Horse. New and Revised Edition. 12mo	1.00
Durant.	Horseback Riding	1.25
Famous Horses of America.	Cloth. 4to	1.50
Gleason, O. R.	How to Handle and Educate Vicious Horses	1.00
Going, J. A.	Veterinary Dictionary. 12mo	2.00
Herbert, H. W.	Hints to Horse Keepers. 12mo	1.75
Helm, H. T.	American Roadsters and Trotting Horses. 8vo	5.00
Horse, The;	Its Varieties and Management. Boards	.75
Howden, P.	How to Buy and Sell the Horse. 12mo	1.00
Jennings, R.	Horse Training Made Easy. 16mo	1.25
——	The Horse and His Diseases. 12mo	1.25
Law, J.	Veterinary Adviser. 8vo	3.00
Liautard.	Chart of Age of Domestic Animals	.50
——	Animal Castration. 12mo	2.00
Manning.	The Illustrated Stock Doctor	5.00
Mayhew, E.	Illustrated Horse Management. 8vo	3.00
——	" Horse Doctor. 8vo	3.00
McClure, R.	Diseases of American Horses. 12mo	2.00
——	American Gentleman's Stable Guide. 12mo	1.00
Miles, W.	On the Horse's Foot. 12mo	
Rarey.	Horse Tamer and Farrier. 16mo	.50

Riding and Driving...	.20
Riley, H. On the Mule. 12mo..	1.50
Russell. Scientific Horse-Shoeing.................................	1.00
Saddle Horse, The. Complete Guide to Riding and Training....	1.00
Saunders. Horse Breeding. 12mo....................................	2.00
Stewart, R. American Farmer's Horse Book. 8vo..................	3.00
Stonehenge. Every Horse Owner's Cyclopædia, 8vo.............	3.75
——— On the Horse in the Stable and the Field. English Edition. 8vo..	3.50
——— On the Horse in the Stable and the Field. American Edition. 12mo...	2.00
Teller. Diseases of Live Stock. Cloth, 2.50; Sheep.............	3.00
Wallace. American Stud Book. Per vol............................	10.00
Williams. Veterinary Medicine.....................................	5.00
——— Veterinary Surgery..	7.50
Woodruff. The Trotting Horse in America. 12mo................	2.50
Woods, Rev. J. G. Horse and Man................................	2.50
Youatt & Skinner. The Horse. 8vo................................	1.75
Youatt & Spooner. " " 12mo..................................	1.50

POULTRY AND BEES.

Burnham. New Poultry Book..	1.50
Cook, Prof. A. J. Bee-Keeper's Guide or Manual of the Apiary....	1.25
Cooper, Dr. J. W. Game Fowls....................................	5.00
Corbett. Poultry Yard and Market. Paper........................	.50
Felch, I. K. Poultry Culture.......................................	1.50
Halsted. Artificial Incubation and Incubators. Paper...........	.75
Johnson, G. M. S. Practical Poultry Keeper. Paper............	.50
King. Bee-Keeper's Text Book......................................	1.00
Langstroth. On the Honey and Hive Bee........................	2.00
Poultry. Breeding, Rearing, Feeding etc. Boards..............	.50
Profits in Poultry and their Profitable Management. Most complete Work extant............................	1.00
Quinby. Mysteries of Bee-Keeping Explained (Edited by L. C. Root).	1.50
Renwick. Thermostatic Incubator. Paper 36c. Cloth..........	.56
Root, A. I. A. B. C. of Bee-Culture...............................	1.25
Standard Excellence in Poultry.................................	1.00
Stoddard. An Egg-Farm. Revised and Enlarged..............	.50
Wright. Illustrated Book of Poultry.............................	8.00
——— Practical Poultry-Keeper....................................	2.00
——— Practical Pigeon Keeper.....................................	1.50

Our Sportsman's Books

ANGLING, FISHING, ETC.

Burgess, J. T. Practical Guide to Bottom Fishing, Trolling, Spinning, Fly, and Sea Fishing. 8vo........................	.50
Fish Hatching and Fish Catching. By Roosevelt and Green. 12mo..	1.50
Forester, F. Fish and Fishing. New Edition. 8vo...............	2.50
——— Fishing with Hook and Line. Paper....................	.25
Fysshe and Fysshynge, from the Boke of St. Albans........	1.00
Hamilton, M. D. Fly Fishing. 12mo........................	1.75
Harris. The Scientific Angler—Foster........................	1.50
Henshall, J. A. A Book of the Black Bass. 8vo.............	3.00
Keene, J. H. Fly-Fishing and Fly-Making. 12mo. Just Published..	1.50
——— Practical Fisherman. 12mo..........................	4.00
King, J. L. Trouting on the Brule River. 12mo...............	1.50
Norris, T. American Fish Culture. 12mo....................	1.75
——— American Angler's Book. 8vo.......................	5.50
Orvis, Charles F. Fishing with the Fly. Crown 8vo...........	2.50
Pennell, H. C. Bottom; or, Float Fishing. Boards............	.50
——— Fly-Fishing and Worm-Fishing. Boards............	.50
——— Trolling for Pike, Salmon, and Trout. Boards.....	.50
Prime. I go a Fishing....................................	2.50
Random Casts from an Angler's Note Book................	.50
Roosevelt, R. B. Game Fish of the Northern States and British Provinces. 12mo................................	2.00
——— Superior Fishing: or, the Striped Bass, Trout, Black Bass, and Blue Fish of the Northern States. 12mo........................	2.00

Roosevelt & Green. Fish Hatching and Fish Catching.......... 1.50
Slack, J. H. Practical Trout Culture. 12mo.... .... 1.00
Scott, G. C. Fishing in American Waters. 8vo 2.50
Walton & Cotton. Complete Angler. 8vo.......... 5.00
— " " Bohn...... 2.00
— " " Chandos...... .... 1.50
— " " 12mo .... .80

BOATING, CANOEING SAILING, ETC.

Canoeing in Kanuckia. 12mo...................... .... .75
Fellows, H. P. Boating Trips on New England Rivers. 12mo..... 1.25
Frazar, D. Practical Boat Sailing. 16mo............ .... 1.00
Henshall, J. A. Camping and Cruising in Florida. 12mo.... ... 1.50
Kemp, Dixon. Manual of Yacht and Boat Sailing (the Standard Authority). Royal 8vo. Illustrated. 10.00
Kemp, Dixon. Yacht Designing. Folio..... .... 25.00
Kunhardt, D. T. Small Yachts. 4to, 14¾ x 12¾. 7.00
Prescott, C. E. The Sailing Boat. 16mo................. .50
Steele, T. S. Canoe and Camera. 12mo...................... 1.50
Swimming. Routledge............................... .20

FIELD SPORTS AND NATURAL HISTORY.

American Bird Fancier. Enlarged edition............... .... .50
Adams, H. G. Favorite Song Birds............................. 1.50
Archer, Modern. Paper..... .... .75
Bailey. Our Own Birds............................... 1.50
Bird-Keeping. Fully Illustrated.... .... 1.50
Brown. Taxidermy 1.00
Canary Birds. New and Revised Edition. Paper, 50c. Cloth..... .75
Coues. Key to North American Birds............ .. .. 10.00
Cocker. Manual................................ 1.50
Edwards. Rabbits 1.25
Goode and Atwater. Menhaden.............. .... 2.00
Holden. Book of Birds................................. .25
Lawn Tennis Hand Book................... .... .75
Packard. Guide to Study of Insects.... .... 5.00
— Half Hour Insects.......................... 2.50
— Common Insects..... .... 1.50
Practical Rabbit Keeper... .... 1.50
Swimming, Skating and Rinking................ .25
Van Doren. Fishes of the East Atlantic Coast................... 1.50
Warne. Angling. Boards.................... .... .50
Wilson. American Ornithology. 3 vols 18.00
Wilson and Bonaparte. American Ornithology. 1 vol........ 7.00

HUNTING, SHOOTING, FISHING, ETC.

Adirondacks Guide. Wallace	2.00
Amateur Trapper. Boards	.75
Batty, J. H. How to Hunt and Trap. 12mo	1.50
——— Practical Taxidermy. 12mo	1.50
Barber. Crack Shot—the Rifleman's Guide. 12mo	1.25
Bogardus, Capt. Field, Cover, and Trap Shooting. 12mo	2.00
Bumstead. On the Wing	1.50
Dead Shot. A Treatise on the Gun	1.25
Farrow. How to Become a Crack Shot. 12mo	1.00
Forester, F. Life and Writings—D. W. Judd. 2 volumes. 8vo	3.00
——— Field Sports. 2 volumes. 8vo	4.00
——— Complete Manual for Young Sportsmen. 8vo	2.00
——— American Game in its Season. 8vo	1.50
Gildersleeve, H. A. Rifles and Marksmanship. 12mo	1.50
Gloan. The Breech-loader	1.25
Gould, J. M. How to Camp Out. 16mo	.75
Greener, W. W. Choke Bore Guns. 8vo	3.00
——— The Gun and its Development	2.50
Gun, Rod, and Saddle. "Ubique"	1.00
Hallock. Sportsman's Gazeteer and General Guide—A Treatise on all Game and Fish of North America. Instruction in Shooting, Fishing, Taxidermy, and Woodcraft, with Directory of Principal Game Resorts and Maps. New and Revised Edition. 12mo	3.00
Henderson, H. Practical Hints on Camping. 12mo	1.25
Lewis, E. J. The American Sportsman. 8vo	2.50
Murray. Adventures in the Wilderness. 12mo	1.25
Murphy, J. M. American Game Bird Shooting. 12mo	2.00
Newhouse. Trapper's Guide. 8vo	1.50
Pistol, The—How to Use. 12mo	.50
Prescott, C. E. Practical Hints on Rifle Practice with Military Arms	.50
Roosevelt, R. B. Florida, and the Game Water Birds of the Atlantic Coast and Lakes of the United States. 12mo	2.00
Samuels. Birds of New England and Adjacent States	4.00
Shooting on the Wing. 16mo	.75
Smith, George Putnam. The Law of Field Sports	1.00
Stonehenge. Rural Sports—The Standard Encyclopædia of Field Sports. ½ morocco. 8vo	5.00
Thrasher, H. Hunter and Trapper. 12mo	.75
Wingate, C. W. Manual for Rifle Practice. 16mo	1.50
Woodcraft. "Nessmuck." 12mo	1.00

ARCHITECTURE, ETC.

Allen, L. F	Rural Architecture..	1.50
American Cottages...		5.00
Ames.	Alphabets..	1.50
Atwood.	Country and Suburban Houses...........................	1 50
Barn Plans and Out-Buildings........................		1.50
Bell.	Carpentry Made Easy..	5.00
Bicknell.	Cottage and Villa Architecture,........................	4 00
———	Detail Cottage and Constructive Architecture.............	6.00
———	Modern Architectural Designs and Details.................	10.00
———	Public Buildings New..	2.50
———	Street, Store, and Bank Fronts, New.....................	2.50
———	School-House and Church Architecture...................	2.50
———	Stables, Out-buildings, Fences, etc........................	2.50
Brown.	Building, Table and Estimate Book........................	1.50
Burn.	Drawing Books, Architectural, Illustrated and Ornamental. 3 Vols. Each..	1.00
Cameron.	Plasterer's Manual..	.75
Camp.	How Can I Learn Architecture..............................	.50
Copley.	Plain and Ornamental Alphabets..........................	3 00
Cottages.	Hints on Economical Building............................	1.00
Cummings.	Architectural Details.....................................	6.00
Elliott.	Hand Book of Practical Landscape-Gardening.............	1.50
Eveleth.	School-House Architecture................................	4.00
Fuller.	Artistic Homes..	3.50
Gilmore, Q. A.	Roads and Street Pavements........................	2.50
Could.	American Stair-Builder's Guide...........................	2.50
———	Carpenter's and Builder's Assistant.......................	2.50
Hodgson.	Steel Square...	1.00
Holly.	Art of Saw Filing..	.75
Harney.	Barns, Out-Buildings, and Fences........................	4.00
Hulme.	Mathematical Drawing Instruments......................	1.50

Hussey. Home Building..	2.50
—— National Cottage Architecture	4.00
Homes for Home Builders. Just Published. Fully Illustrated.	1.50
Interiors and Interior Details	7.50
Lakey. Village and Country Houses	5.00
Modern House Painting	5.00
Monckton. National Carpenter and Joiner	5.00
—— National Stair-Builder	5.00
Painter, Gilder, and Varnisher's Companion	1.50
Palliser. American Cottage Homes	3.00
—— Model Homes	1.00
—— Useful Details	2.00
Plummer. Carpenters' and Builders' Guide	.75
Powell. Foundations and Foundation Walls	2.00
Reed. Cottage Houses	1.25
—— House Plans for Everybody	1.50
—— Dwellings	3.00
Riddell. Carpenter and Joiner Modernized	7.50
—— New Elements of Hand Railing	7.00
—— Lessons on Hand Railing for Learners	5.00
Rural Church Architecture	4.00
Scott. Beautiful Homes	2.50
Tuthill. Practical Lessons in Architectural Drawing	2.50
Weidenmann. Beautifying Country Homes. A superb quarto Vol.	10.00
Woodward. Cottages and Farm Houses	1.00
—— Country Homes	1.00
—— National Architect. Volumes 1 and 2. Each	15.00
—— Suburban and Country Houses	1.00

MISCELLANEOUS.

Collection of Ornaments	2.00
Common Sea Weeds	.50
Common Shells of the Seashore	.50
Corson, Miss Juliet. Cooking School Text Book	1.25
——— Twenty-five Cent Dinners. New Edition	.25
De Voe. Market Assistant	2.50
Dussauce. On the Manufacture of Vinegar	5.00
Eassie. Wood and its Uses	1.50
Eggleston. Roxy	1.50
——— Circuit Rider	1.50
——— School Boy	1.00
——— Queer Stories	1.00
——— End of the World	1.50
——— Mystery of Metropolisville	1.50
——— Hoosier Schoolmaster	1.25
Elliott, Mrs. Housewife. New and Revised Edition	1.25
Ewing. Hand Book of Agriculture	.25
Ferns and Ferneries. Paper	.25
Fisher. Grain Tables	.40
Fowler. Twenty Years of Inside Life in Wall Street	1.50
Gardner. Carriage Painters' Manual	1.00
——— How to Paint	1.00
Hazard. Butter Making	.25
Household Conveniences	1.50
How to Detect the Adulterations of Food. Paper	.25
How to Make Candy	.50
Leary. Ready Reckoner	.25
Myers. Havana Cigars	.25
Our Farmers' Account Book	1.00
Parloa, Miss. Cook Book	1.50
Ropp. Commercial Calculator	.50
Scribner. Lumber and Log-Book	.35

Ware.	The Sugar Beet	4.00
Weston, J.	Fresh Water Aquarium. Paper	.25
Weir, Harrison.	Every Day in the Country	.75
Wingate, Gen. G. W.	Through the Yellowstone Park	1.50
Williams.	Ladies' Fancy Work	1.50
——	Evening Amusements	1.50
——	Beautiful Homes	1.50
——	Ladies' Needle Work	1.00
——	Artistic Embroidery	1.00
Willard.	Practical Butter Book	1.00
——	Practical Dairy Husbandry	3.00

Warne's Useful Books. Boards. With practical illustrations:

The Orchard and Fruit Garden. By Elizabeth Watts	.50
Vegetables and How to Grow Them. By Elizabeth Watts	.50
Cattle and their Varieties	.50
The Dog and its Varieties	.50
Flowers and Flower Garden. By Elizabeth Watts	.50
Hardy Plants for Little Front Gardens	.50
Poultry—An Original and Practical Guide to their Management	.50
The Modern Fencer. By Capt. T. Griffith	.50
The Modern Gymnast. By Charles Spencer	.50
Cattle and their Varieties and Management	.75
The Horse and its Varieties and Management	.75
Sheep and its Varieties and Management	.75

Our Very Latest Publications.

Through the Yellowstone Park on Horseback.' By Gen. G. W. Wingate	1.50
Fly-Fishing and Fly-Making. By Keene	1.50
How to Handle and Educate Vicious Horses. By O. R. Gleason	1.00
The Law of Field Sports. By Geo. P. Smith	1.00
Bridle Bits. A Treatise on Practical Horsemanship. By Col. J. C. Battersby	1.00
The Percheron Horse in America and France	1.00
Profits in Poultry. Useful and Ornamental Breeds	1.00
Cape Cod Cranberries. By James Webb. Paper	.50
How to Plant. By M. W. Johnson	.50
The American Merino for Wool and Mutton. By Stephen Powers	1.75

New and Revised Editions.

Hallock. Sportsman's Gazetteer	3.00
Stewart. Irrigation for the Farm, Garden and Orchard	1.50
Farm Implements and Machinery. By Thomas	1.50
Egg Farm. By Stoddard. Cloth	.50
Play and Profit in My Garden	1.50
Silos and Ensilage	.50

Send Postal for Complete Catalogue of our Publications regarding Horses and Horsemanship, Hunting, Fishing, and all other Out-Door Sports and Pastimes.

O. JUDD CO.   DAVID W. JUDD, Pres't.
751 BROADWAY NEW YORK.